GENERAL CYTOCHEMICAL METHODS

VOLUME 1

GENERAL CYTOCHEMICAL METHODS

Volume 1

Edited by

J. F. DANIELLI

Department of Zoology

King's College, London

ACADEMIC PRESS INC., NEW YORK, 1958

111 Fifth Avenue, New York 3, N. Y.

Library of Congress Catalog Card No. 58–10409

PRINTED IN THE UNITED STATES OF AMERICA

PREFACE

This is the first of a series of volumes which will appear at intervals of about two years. Subsequent volumes will each contain a number of descriptions of additional techniques and supplements to previous volumes which will bring the earlier volumes up to date. The endeavor is to include techniques which have been sufficiently studied to eliminate most of the uncertain points which in cytochemistry usually beset a new method.

Ten years ago enough had been done in cytochemistry to show that in this field lay remarkable possibilities of understanding the chemical structure of cells. But most of the methods then available were the subject of sharp controversy. To assist in clarifying the issues involved, the International Society for Cell Biology set up a Cytochemical Commission. The Commission has held a number of round-table discussions which have been of the greatest value. Insofar as agreed common standards of exactness have been achieved, much must be attributed to these meetings. It is hoped that these standards are reflected in this volume.

The first volume of a series such as this is necessarily experimental. The treatment accorded varies in degree of detail in the different articles. It is hoped that readers and users of this volume will let both the authors and the editor know their views on the value and deficiencies of the different manners of treatment. This will be helpful, not only in amending the present volume, but also in designing subsequent volumes.

J. F. DANIELLI

March, 1958
London, England

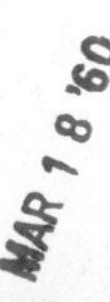

CONTRIBUTORS TO VOLUME 1

H. Stanley Bennett, *Department of Anatomy, University of Washington, Seattle, Washington*

Albert H. Coons, *Department of Bacteriology and Immunology, Harvard Medical School, Boston, Massachusetts*

J. F. Danielli, *King's College, London, England*

Howard G. Davies, *Medical Research Council, Biophysics Research Unit, King's College, London, England*

Arne Engström, *Department of Medical Physics, Karolinska Institutet, Stockholm, Sweden*

S. J. Holt, *Courtauld Institute of Biochemistry, Middlesex Hospital Medical School, London, England*

Cecilie Leuchtenberger, *Institute of Pathology, Western Reserve University, Cleveland, Ohio*

Bo Lindström, *Department of Medical Physics, Karolinska Institutet, Stockholm, Sweden*

S. R. Pelc, *Medical Research Council, Biophysics Research Unit, King's College, London, England*

P. M. B. Walker, *Medical Research Council, Biophysics Research Unit, King's College, London, England*

Ruth M. Watts, *Department of Anatomy, University of Washington, Seattle, Washington*

CONTENTS

Quantitative Determination of DNA in Cells by Feulgen Microspectrophotometry

Cecilie Leuchtenberger

Autoradiography as a Cytochemical Method, with Special Reference to C^{14} and S^{35}

S. R. Pelc

The Cytochemical Demonstration and Measurement of Sulfhydryl Groups by Azo-aryl Mercaptide Coupling, with Special Reference to Mercury Orange

H. Stanley Bennett and Ruth M. Watts

Indigogenic Staining Methods for Esterases

S. J. Holt

Fluorescent Antibody Methods

ALBERT H. COONS

The Calcium Phosphate Precipitation Method for Alkaline Phosphatase

J. F. DANIELLI

THE WEIGHING OF CELLULAR STRUCTURES BY ULTRASOFT X-RAYS

ARNE ENGSTRÖM AND BO LINDSTRÖM

Department of Medical Physics, Karolinska Institutet, Stockholm, Sweden

I. INTRODUCTION

The aim of modern developments in quantitative histo- and cytochemistry is to define the microscopic biological structures in terms of their chemical composition and enzymatic activity, and essentially two types of technical obstacles have to be mastered in order to analyze chemically as small volumes of biological tissues as one cubic micron (μ^3). The greatest of these difficulties is perhaps the development of analytical techniques sensitive enough to obtain a defined and measurable signal from the very small amounts of substance available for the analysis. Difficulty in obtaining the sample in such a form that it resembles the living tissue as much as possible is the other great troublemaker in quantitative cytochemistry. Naturally the ultimate goal of cytochemistry is to investigate the chemical characteristics of the living cell itself.

In quantitative cytochemistry the unit of weight is 10^{-12} gm. ($= 1\ \mu\mu$g. or pg.) which is the weight of 1 μ^3 of tissue with unit density. As most biological objects contain about three-quarters of water and only one-quarter of dry substance, the weight of the latter portion is about 2.5×10^{-13} gm. in 1 μ^3 of soft tissue. The major part of this dry weight consists of proteinaceous substances, and in many instances components such as fats, carbohydrates, and nucleic acids will *each* not amount to more than a few per cent of the total amount of dry substance. However, in special biological objects the concentrations of the latter compounds may reach high values. Therefore if compounds occurring in concentrations of a few per cent each are to be determined in a volume of tissue of 1 μ^3 with an accuracy of about 10%, the analytical methods must be as sensitive as 10^{-14}–10^{-15} gm.

A considerable number of the methods utilized in quantitative cytochemistry are based on the *absorption* of electromagnetic radiation of various wavelengths ranging from X-rays to ultraviolet, visible, and infrared light. The basis of the absorption methods is given by the fundamental relation between the thickness of the absorbing material present and the attenuation of monochromatic radiation according to the Lambert law,

$$I = I_0 \cdot e^{-\mu l} \tag{1}$$

where I_0 and I are the intensities of the incident and the transmitted radiation, respectively, μ the linear absorption coefficient, and l the thickness of the absorbing substance. Since in cytochemistry very small areas of the sample are to be analyzed, the measurements of the radiation intensities must be performed via an enlarging (microscopic) system.

In addition to the microabsorptiometric methods, techniques based on the *emission* of characteristic radiation (fluorescence) from the sample have found great use. In particular, X-ray microfluorescence procedures have been developed into precise methods for the localization and the quantitative estimation of minute amounts of trace elements in biological tissues although the most sensitive method for trace metal analysis may be activation analysis in a high intensity neutron beam.

Finally, the diffraction of radiation in biological samples is used to identify chemical components, and especially X-ray microdiffraction and electron diffraction techniques have proven most useful. These methods are of course only applicable to repetitive structures, extensive enough to fulfill the criteria of interference.

In the present communication certain aspects of X-ray microabsorption techniques as applied to cytochemical analysis will be discussed. In particular the determination of the total dry weight and the contents of water of cytological structures *in situ* will be described in detail.

The limit of resolution of present techniques of X-ray microscopy is about the same as that of light microscopy. The least complicated of the X-ray microscopic methods is the direct contact microradiography, generally abbreviated as C.M.R., where the sample, in direct contact with a fine-grained photographic emulsion, is imaged with soft X-rays. Extremely fine-grained photographic emulsions are now commercially available, and the resolution of the contact microradiographic technique is now only limited by the resolving power of the light optical microscope used to inspect the microradiogram. Projection X-ray microscopy and other techniques such as X-ray microscopy based on total reflection, and scanning X-ray microscopy, may, however, give a higher resolution when further developed. However, the very soft X-rays necessary for cytochemical analyses present many technical difficulties which must be overcome before these types of X-ray microscopes can be developed into routine instruments.

TABLE I

TERMINOLOGY OF X-RAYS

Wavelength	Terminology	Main use
< 0.1 Å	Ultrahard	Industrial and medical radiography
0.1–1 Å	Hard	
1–10 Å	Soft	Diffraction, X-ray microscopy
>10 Å	Ultrasoft	

Before discussing the technique of weighing cellular structures by means of X-rays, some definitions must be given regarding the properties of the radiation. The terminology useful for the classification of the wavelength regions utilized for X-ray microscopy is presented in Table I. Mainly, the soft and ultrasoft X-rays must be utilized for high resolution

X-ray microscopy of biological material as is evident from the following discussion.

In the original method for weighing cellular structures by X-rays (Engström and Lindström, 1950) a continuous X-ray spectrum, generated at 3000 volts and filtered through a 9 μ thick aluminum foil, was used. This X-ray spectrum contained wavelengths mainly in the region 8 to 12 Å. Recently, however, the technique has been improved to encompass also ultrasoft monochromatic X-rays, in which case the theory of the method can be mastered in an exact way. Moreover, it will be shown in the following section that the wavelengths of the X-rays utilized to determine the mass of biological material can be selected in such a way that a maximum of information can always be obtained from the absorption image of the specimen, whether the biological object has a high or low mass. This is one of the great advantages of the X-ray method in comparison with the microinterferometric mass technique in which the accuracy is determined by the smallest measurable phase shift, thus independent of mass.

II. THE ABSORPTION OF X-RAYS AND THE CONTINUOUS X-RAY SPECTRUM

A. The General Laws for Absorption

In the different wavelength regions of the electromagnetic spectrum the fundamental Lambert law (Eq. 1) is modified in different directions. For X-rays the absorption per unit of mass is usually given instead of the absorption per unit of length, which depends on the physical state of the absorber. If the density of the absorbing element is ρ gm. cm.$^{-3}$, the product $\rho \times l$ represents the mass m of absorbing material per unit of area, expressed in gm. cm.$^{-2}$. Equation 1 then takes the form,

$$I = I_0 \cdot e^{-\frac{\mu}{\rho} m} \tag{2}$$

where μ/ρ is the *mass absorption coefficient* having the dimension cm.^{2} gm.$^{-1}$.

The inner electron shells in the atoms of the absorber are of decisive importance for the absorption coefficient, which is almost completely independent of the physical and chemical state of the absorbing element. It is, however, a function of the frequency of the incident X-rays (wavelength dependence), and of the atomic number of the absorber. A close approximation to the composite mass absorption coefficient, $(\mu/\rho)_{\text{total}}$, of a chemical compound or a mixture, composed of n elements with percentages of weight a_i and individual mass absorption coefficients $(\mu/\rho)_i$, is given by

$$\left(\frac{\mu}{\rho}\right)_{total} = \sum_{i=1}^{n} \frac{a_i}{100}\left(\frac{\mu}{\rho}\right)_i \tag{3}$$

The validity of this expression depends on the specific characteristics of the mass absorption coefficients as stated above. Contrary to the case in other wavelength regions, the exact mass absorption coefficients for compounds or mixtures of known composition can therefore be calculated theoretically, once the mass absorption coefficients of the different elements have been determined and tabulated.

For X-rays with wavelengths longer than 0.1 Å passing through matter, the true photoelectric absorption and the scattering are the two processes causing the loss of energy, as expressed by the components of the mass absorption coefficient,

$$\frac{\mu}{\rho} = \frac{\tau}{\rho} + \frac{\sigma}{\rho} \tag{4}$$

where τ/ρ is the mass photoelectric absorption coefficient and σ/ρ is the mass scattering coefficient. Hence, τ/ρ is a measure of the fraction of incident X-ray quanta, the energy of which is consumed to eject photoelectrons from inner electron orbits of the atoms of the absorber, and σ/ρ is a measure of those X-ray quanta scattered by the electrons of the absorbing material. For X-rays with wavelengths longer than about 2.5 Å, i.e., in the greater part of the soft and in the whole ultrasoft X-ray region, σ/ρ can be neglected in comparison with τ/ρ, and the mass absorption coefficient μ/ρ can then be taken equal to the mass photoelectric absorption coefficient τ/ρ.

The mass photoelectric absorption coefficient is a function of the wavelength λ of the incident X-rays and of the atomic number Z of the absorbing element. An *approximate* empirical formula for this function is given by

$$\frac{\mu}{\rho} = k \cdot \lambda^u \cdot Z^v \tag{5}$$

where u has a value about three and v a value about four. With given constants the validity of this equation is, however, restricted to a limited range for λ and Z. For each particular element, μ/ρ is a continuously increasing function of the wavelength of the incident X-rays, except for characteristic discontinuities at certain wavelengths. At these critical absorption limits (edges) the total photoelectric absorption in the different orbits of the atoms of the absorber, as expressed by the mass photoelectric absorption coefficient τ/ρ, is greater on the short wavelength side of this limit than on the long wavelength side, as, on the latter side, no photoelectric absorption occurs in the orbit corresponding to the absorption limit.

B. The Numerical Values of the Mass Absorption Coefficients

When the intensities of incident and transmitted X-rays (I_0 and I in Eq. 2) have been determined experimentally, it is necessary to know

TABLE II

Mass Absorption Coefficients of Elements with Low Atomic Numbers [a]

Absorber	Al Kα 8.34 Å	Cu Lα 13.33 Å	Fe Lα 17.61 Å	Cr Lα 21.57 Å	Ti Lα 27.43 Å	C Kα 44.63 Å
1 H	7.5	30	70	130	260	1000
2 He	30	120	275	500	1000	4300
3 Li	78	280	640	1200	2300	9400
4 Be	152	581	1288	2292	4532	17,430
5 B	324	1233	2711	4784	9200	32,540
6 C	605	2290	4912	8440	15,760	—
7 N	1047	3795	7910	13,120	22,590	3647
8 O	1560	5430	10,740	16,610	1473	5470
9 F	1913	6340	11,600	1015	1949	7280
10 Ne	2763	8240	1079	1863	3575	13,180
11 Na	3129	661	1402	2429	4651	16,650
12 Mg	3797	981	2085	3601	6830	22,850
13 Al	323	1146	2441	4189	7840	24,910
14 Si	510	1813	3812	6420	11,510	33,840
15 P	640	2259	4661	7670	13,280	38,610
16 S	814	2839	5710	9160	15,520	45,230
17 Cl	990	3364	6530	10,210	17,330	50,100
18 A	1163	3795	7110	11,070	18,820	—
19 K	1429	4504	8310	12,960	22,030	—
20 Ca	1706	5150	9450	14,800	24,910	—
21 Sc	1819	5280	9750	15,210	—	—
22 Ti	2002	5680	10,480	16,300	—	—
23 V	2168	6100	11,230	—	—	—
24 Cr	2409	6740	12,360	—	—	—
25 Mn	2556	7160	—	—	—	—
26 Fe	2799	7850	—	—	—	—
27 Co	2956	—	—	—	—	—
28 Ni	3154	—	—	—	—	—
29 Cu	3346	—	—	—	—	—
30 Zn	3685	—	—	—	—	—

[a] From Henke *et al.* (1956).

the magnitude of the mass absorption coefficient μ/ρ in order to calculate the amount m of absorbing material. In the physics literature there are only scattered experimental determinations of the mass absorption coefficients (= the mass photoelectric absorption coefficients) in the soft and ultrasoft X-ray region. However, in a recent publication Henke *et al.* (1956) have devised a semiempirical method for the determination of mass absorption coefficients for X-rays with wavelengths between 5 and 50 Å. A universal function for the absorption by K shell electrons, and another for the absorption by L and extra L shell electrons are given, and these values permit the calculation of mass absorption coefficients for elements with atomic numbers <30. Available experimental and calculated absorption data were compared with those calculated from quantum theory by these authors, and the agreement was very good. Some of the new calculations of μ/ρ for certain important wavelengths in the soft and ultrasoft X-ray regions are presented in Table II.

C. Suitable Wavelength Ranges for Microradiography and the Continuous X-ray Spectrum

Assume a biological sample with the thickness l, where the water has been removed with no change of the dimensions of the specimen as compared with the fresh state. Furthermore, assume that in two adjacent regions there are slightly different total dry weights per unit area, and consequently different linear absorption coefficients μ_1 and μ_2 as the thickness is assumed to be constant. From Eq. 1 the difference between the transmitted intensities, I_1 and I_2 respectively, in the two areas is given by

$$I_2 - I_1 = I_0(e^{-\mu_2 l} - e^{-\mu_1 l}) \tag{6}$$

From this equation the value of l corresponding to the maximum difference between the transmitted intensities can be calculated, and considering a small difference between μ_1 and μ_2, the expression is obtained,

$$l_{\max} = \frac{\log_e \dfrac{\mu_1}{\mu_2}}{\mu_1 - \mu_2} \approx \frac{1}{\mu} \tag{7}$$

where μ is the mean linear absorption coefficient.

In average biological samples in the wavelength range between 5 and 20 Å the value of kZ^v in Eq. 5 is about 1.6. Eliminating the absorption coefficients in Eqs. 5 and 7 the following approximate expression is obtained,

$$\rho \cdot l \approx \frac{1}{1.6 \cdot \lambda^3} \tag{8}$$

where ρ is the mean of the density of the dehydrated biological sample.

The continuous X-ray spectrum has a sharp short wavelength limit λ_0 determined by the voltage V across the X-ray tube according to

$$\lambda_0 = \frac{12{,}350}{V} \tag{9}$$

The maximum intensity in the continuous X-ray spectrum is situated at about $3\lambda_0/2$, and then the intensity decreases asymptotically towards longer wavelengths.

Thus, putting $\lambda_0 = 2\lambda/3$, and eliminating λ between the Eqs. 8 and 9, the following approximate expression for V, the suitable voltage across the X-ray tube, is obtained,

$$V \approx 2.2 \cdot 10^4 \cdot \sqrt[3]{\rho l} \tag{10a}$$

when a continuous X-ray spectrum is utilized. For monochromatic X-rays the corresponding approximate voltage is given by

$$V = 1.5 \cdot 10^4 \cdot \sqrt[3]{\rho l} \tag{10b}$$

or the suitable wavelength can be obtained immediately from Eq. 8.

For biological specimens of commonly used thicknesses (from 1 to 10 μ) and consisting of between 20 and 35% of dry substance the continuous X-ray spectra must be generated at voltages between 500 and 2000 volts if small mass differences in the specimen shall be visible. Thus, by selecting suitable wavelength ranges (voltages), the X-ray absorption image of a specimen can always be registered under optimum conditions independent of the absolute value of the total dry weight per unit area in the specimen, as already pointed out in the introduction. Figure 1 presents suitable wavelength ranges and voltages for microradiography of soft tissues with varying sample thicknesses and percentages of dry substance.

III. CONDITIONS FOR OPTIMUM, VISUAL, AND PHOTOMETRIC CONTRAST IN X-RAY MICROSCOPIC IMAGES

A. The Definition of a Contrast Function

Before entering the detailed theory of mass determination of cytological structures by X-rays the optimum conditions for contrast in the X-ray microscopic absorption image will be discussed. The presentation will closely follow the contribution by Henke *et al.* (1957). From Eq. 2 it can be seen that the microradiogram "signal" may be described in terms of a variation in the mass m per unit area or in the mass absorption coefficient μ/ρ, or simply in terms of $(\mu/\rho \cdot m)$, which implies either a change of the mass per unit area or of the chemical composition or of both. The product $(\mu/\rho \cdot m)$ is therefore proposed as the correct number to use as an experimental variable in a microradiographic analysis. This variable may then be "opened up," when reducing the experimental data as allowed or suggested by the experiment itself or by other sources of information.

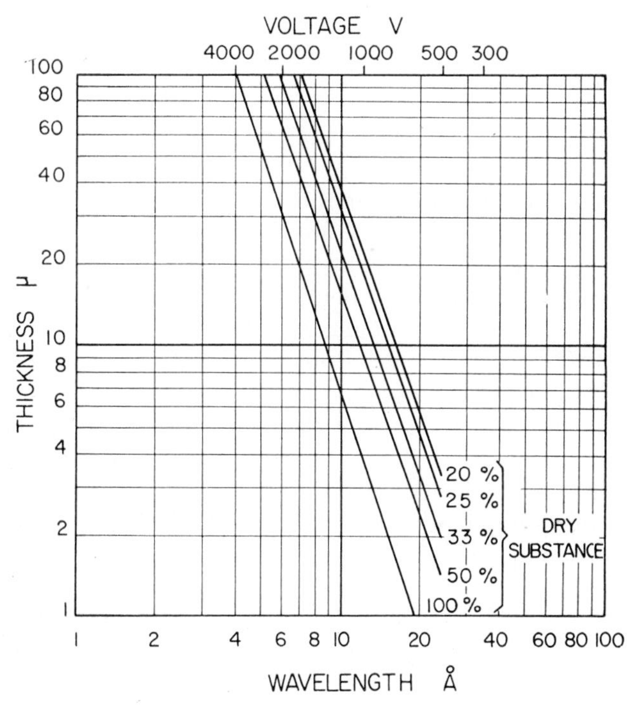

FIG. 1. Suitable X-ray energies for microradiography of soft tissues of varying thickness (Engström, 1955b).

The relative difference between the absorption indices

$$\Delta\left(\frac{\mu}{\rho}m\right)\Big/\left(\frac{\mu}{\rho}m\right),$$

of two specific areas in the specimen is usually the number of primary interest, as the absolute sample thickness is usually of little meaning because of the distortion of the biological structure during preparation, the unreliability of microtome calibrations, and inability to measure accurately the thickness of microscopic sections. The ratio gives relative variation of the mass or the mass absorption coefficient for the particular radiation—if either the chemistry or the mass per unit area can be assumed constant.

The "contrast" function C is therefore defined by the following relation,

$$C = \frac{-\Delta D}{\Delta\left(\frac{\mu}{\rho}m\right)\Big/\left(\frac{\mu}{\rho}m\right)} \tag{11}$$

where D is the density in a given region of the microradiogram, defined by the usual relation,

$$D = \log_{10}\frac{i_b}{i} \tag{12}$$

Here i_b and i are the light intensities at the photometric procedure transmitted through an unexposed area and the particular exposed area, respectively, in the microradiogram. Differentiation of Eq. 12 shows that the change in photographic density, ΔD, is proportional to the relative change in the transmitted light, $\Delta i/i$, in a given region of the microradiogram. Therefore ΔD is approximately proportional to the response of the eye to the microradiographic "signal" in this region, and then the "contrast" function is a measure of the visual response per unit relative change of the absorption index. Of course, the eye response is also dependent in a complicated manner on the area associated with the variation of photographic density and on the "distractions" in the neighboring areas of the microradiogram.

B. The Determination of the Contrast Function

The contrast, C, is a function of three variables, ultimately, considered as the wavelength of the X-rays used, the chosen photographic emulsion, and the exposure time. These must be expressed in terms of numbers which can be determined experimentally. The sample X-ray transmission, t, defined as $e^{-\mu/\rho \cdot m}$, can be used to measure the effect of wavelength. The average photographic density in the region of the microradiogram where a variation in density is being evaluated, measures the exposure time for a given sample transmission. Lastly, the density versus exposure data is used to calibrate the chosen photographic emulsion.

In the following discussion the density is expressed as a function of a variable x, proportional to the exposure and therefore also to the exposure time T, as the X-ray intensity is held constant and the reciprocity law is assumed to be valid. Hence,

$$D = f(x) \tag{13}$$

where x is defined by the relation

$$x = k \cdot I_0 \cdot e^{-\frac{\mu}{\rho}m} \cdot T \tag{14}$$

In Eq. 14 I_0 is the radiation flux per unit area at the photographic emulsion with the sample removed, and k is a constant of proportionality.

Differentiating Eq. 14 with respect to $(\mu/\rho \cdot m)$ and recalling that ΔD is equal to Δf according to Eq. 13, $\Delta(\mu/\rho \cdot m)$ and ΔD can be eliminated in Eq. 11. Hence the following expression for C is obtained:

$$C = \frac{\mu}{\rho} m \cdot x \cdot \frac{\Delta f}{\Delta x} \tag{15}$$

The factor $\Delta f/\Delta x$, i.e., the slope at x of the D-vs-x curve, can be determined by numerical differentiation from the D-vs-T table, T being considered numerically equal to x through the choice of the constant k in Eq. 14. This leaves the factor $x(\Delta f/\Delta x)$ in terms of the variable x. By replacing the variable x in Eq. 14 by the corresponding value of D (reading across the D-vs-x table), the table of values expressing the function $x(\Delta f/\Delta x)$ can be obtained as a function of D. Recalling Eqs. 15 and 11, the following relations are obtained:

$$x \frac{\Delta f}{\Delta x} = F(D) = \frac{C}{\frac{\mu}{\rho} m} = \frac{-\Delta D}{\Delta\left(\frac{\mu}{\rho} m\right)} \tag{16}$$

The function $F(D)$ is dependent only upon the emulsion and can be obtained from the D-vs-T calibration data. According to Eq. 16 this convenient function is equal to the contrast per unit absorption index and thus also equal to the contrast for a sample of transmission $1/e$ ($t \approx 37\%$). The contrast function for any other transmission is obtained by multiplying $F(D)$ by the absorption index which is given by $\log_e(1/t)$.

It is interesting to note that the value of $F(D)$, or the contrast per unit absorption index, is also given in terms of the slope of the curve for the density, D, plotted as a function of $log_{10}T$, (rather than simply T), by the relation,

$$\frac{C}{\frac{\mu}{\rho} m} = 0.434\gamma \tag{17}$$

where the slope number γ is used in analogy to the H and D convention of light photography.

C. The Optimum Sample Transmission and Exposure Time for a Fine-Grained Photographic Emulsion

For two different lots of Eastman Kodak No. 649 Spectroscopic Plates density versus exposure time calibrations have been made using 13.3 Å radiation (Cu $L\alpha_{1,2}$) generated in an aluminum cathode-copper anode gas discharge type X-ray tube with a voltage of 1.4 kv. and a copper foil as filter. This X-ray tube will be described in detail in Section VII. In Fig. 2, the density versus exposure time curve is shown along

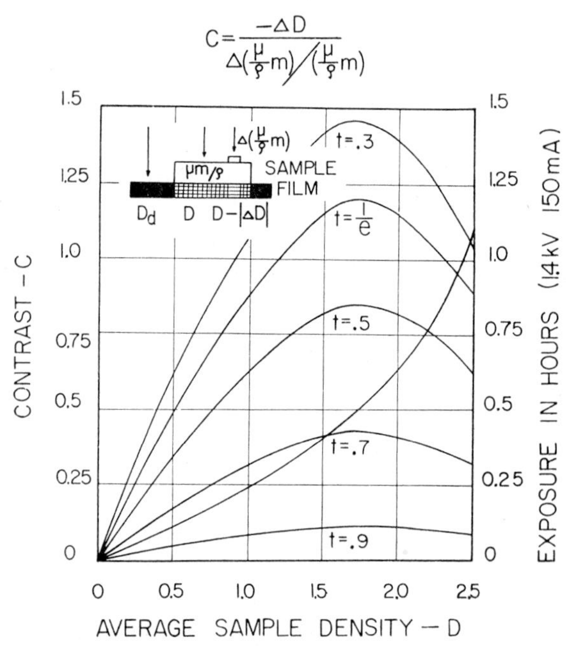

Fig. 2. Contrast functions for various sample transmissions (Henke *et al.* 1957).

with the derived contrast functions for several sample transmissions, using a common D axis. The optimum exposure time for this fine-grained emulsion and the radiation used is that which yields an average density of about 1.7 in the particular region of the microradiogram in which a variation in density is being analyzed.

As expressed in the contrast curves for the several sample transmissions t, there is no optimum transmission value, as such, for visual evaluation of microradiograms. This is also evident from Eq. 16. For a given average sample density D, the contrast increases with $\log_e (1/t)$ or, correspondingly, with the absorption index $\mu/\rho \cdot m$. However, if the absorption image of a structure in the specimen with a low transmission is to have the optimum density in the microradiogram, the exposure of the sample regions of relatively high transmissions can become very large, resulting in a low contrast for such regions. Therefore, a limit of low transmissions does exist, but the value of this limit may vary for different parts of the specimen. For such specimens either a series of exposures utilizing X-rays with different wavelengths may be necessary, or a single exposure with a "compromise" wavelength may be sufficient.

D. Optimum Conditions for the Microphotometric Evaluation of Microradiograms

In the preceding discussion the optimum conditions for the visual examination of small structures in the X-ray absorption images of different specimens have been analyzed. Here the wavelength of the X-rays and the exposure time ought to be chosen in order to create, in the microradiogram, a relatively great difference between the density of the particular structure and that of the surrounding region of the specimen. When the photographic densities approach either very high or very low values, however, the microphotometric determinations of the densities become more and more inaccurate. Therefore both an optimum sample transmission and an optimum exposure time must be determined for a maximum precision of the measurement of the absorption index $\mu/\rho \cdot m$. An analysis of these optimum conditions for microphotometric measurements of microradiograms has been made by Henke *et al.* (1957).

With the assumption that the density versus exposure curve is linear, an error function E has been derived. For the detailed derivation of the error function the reader should consult the paper quoted above. It was found convenient to define E as the ratio between the relative error of the absorption index

$$\Delta\left(\frac{\mu}{\rho}m\right)\Big/\left(\frac{\mu}{\rho}m\right)$$

and the relative error of the photometer reading for the unexposed photographic emulsion. In Fig. 3, E is plotted as a function of the density D_d for several values of the sample transmission. Here D_d is the density of an area in the microradiogram exposed to the incident X-ray intensity (schematically illustrated in Fig. 2). The error curves illustrate that the error function has a minimum for a sample transmission of approximately 12%. If the density D_d is within the range

0.5 to 1.5, however, the increase of the error function E is relatively small.

When a reference system is introduced, as discussed in the following section, the preceding analysis does not apply. In this case the microphotometric readings for the absorption images of the reference system and the specimen are compared, and the actual photographic densities are not calculated. This type of procedure increases the accuracy of the microphotometric evaluation.

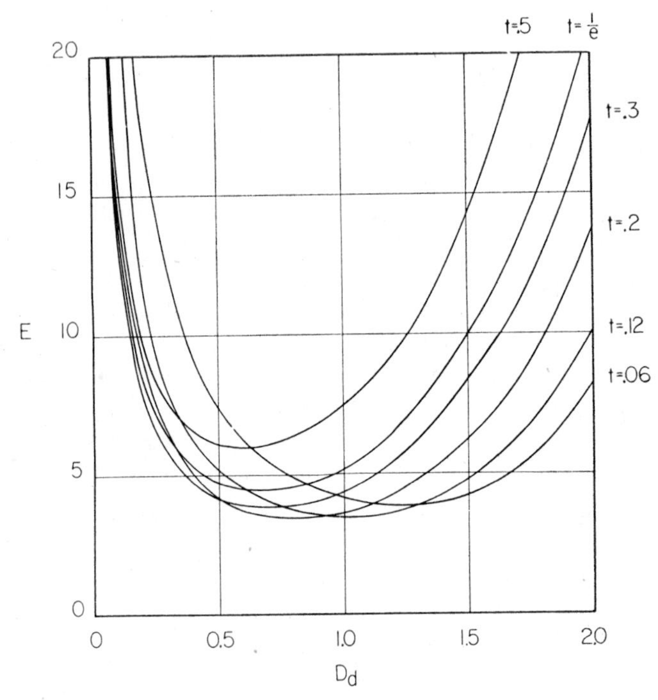

FIG. 3. The error function E for different sample transmissions (Henke *et al.* 1957).

IV. THE THEORY OF X-RAY WEIGHING OF CELLULAR STRUCTURES

A. GENERAL

The quantitative microradiographic procedure designed to determine the total dry weight of cytological structures is schematically illustrated in Fig. 4. The biological specimen, with a reference system consisting of a step wedge of nitrocellulose foils, is mounted over a slit on a metal disc covered with a supporting foil. In certain modifications of the technique the reference system is omitted. The biological specimen together with the reference system are exposed with a continuous soft X-ray spectrum on a fine-grained photographic emulsion in close contact with the specimen. By microphotometry of the small X-ray image the transmitted X-ray intensities, shown in the developed microradiogram, are obtained as a function of the weight per unit area of the different steps in the wedge. The photocurrents corresponding to the X-ray transmission of the steps in the wedge are plotted as a function

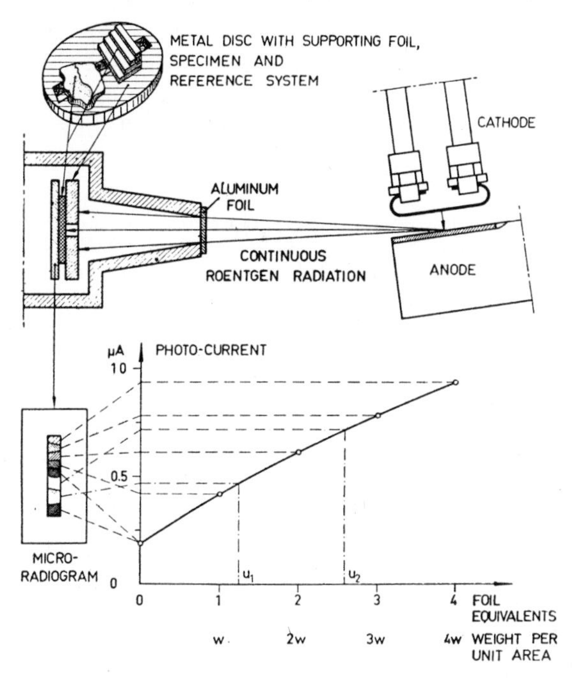

FIG. 4. Principle of quantitative microradiography (Lindström, 1955).

of the weight per unit area of the different steps. By means of this calibration curve the photometer deflections referred to biological structures to be analyzed are transformed to weights per unit area of the reference system (u_1w and u_2w, respectively). Then the weight per unit area of the biological object can be calculated.

The procedure is based on the fact that carbon, nitrogen, and oxygen in tissue proteins are the main X-ray–absorbing material in different biological structures. Other elements, such as iron, zinc, calcium, sodium, and potassium are present in relatively low concentrations in most biological objects, but as the values of the mass absorption coefficients of these other elements differ from those of protein, systematic errors are introduced. The complete theory of total dry weight analyses of biological objects will now be presented including the derivation of the systematic errors both for monochromatic and continuous X-rays. The discussion follows strictly the presentation by Lindström (1955).

As the absorption of X-rays is an atomic property, the composite mass absorption coefficients of different organic substances in biological objects can be calculated, provided the elementary composition of the compounds and the mass absorption coefficients of the different elements are known, as discussed in Section II. Proteins from different tissues and animals have a remarkably constant elementary composition though the amino acid composition varies considerably. In Table III, taken from

TABLE III
THE ELEMENTARY COMPOSITION OF SOME COMMON ORGANIC COMPOUNDS [a]

Compound	% C	% H	% N	% O	% S	% P
Animal proteins	52.5	7.0	16.5	22.5	1.5	—
Fibrin	52.68	6.83	16.91	22.48	1.10	—
Elastin	52.4	7.2	17.1	23.1	0.2	—
Collagen	50.2	6.7	18.2	24.3	0.6	—
Amyloid	49.6	6.8	14.0	26.8	2.8	—
Thymus histone	52.37	7.70	18.35	20.96	0.62	—
Plant proteins	52.0	6.9	17.6	22.5	0.9	—
Deoxyribonucleic acid	37.18	4.14	15.14	34.60	—	8.94
Ribonucleic acid	35.0	4.1	16.5	34.7	—	9.7
Glycogen	44.44	7.77	—	47.79	—	—
Tripalmitin	75.87	12.24	—	11.89	—	—
Tristearin	76.79	12.44	—	10.77	—	—
Triolein	77.32	11.84	—	10.84	—	—
Trilinolein	77.85	11.23	—	10.92	—	—
Cerebrosides	66.5	11.0	2.5	18.3	0.7	1.0
Parlodion	28.1	3.6	11.5	56.8	—	—
Zapon varnish	46.7	5.3	6.6	41.4	—	—

[a] From Lindström (1955).

Lindström (1955), the approximate elementary composition of several organic compounds in biological material has been compiled. The elementary composition of Zapon varnish and Parlodion, utilized as reference systems, is also included. Using the values of the mass absorption coefficients of low atomic elements for soft and ultrasoft X-rays (Table II), the mass absorption coefficients of the tabulated compounds have been calculated and are presented in Table IV.

TABLE IV
MASS ABSORPTION COEFFICIENTS OF SOME COMMON ORGANIC COMPOUNDS

Absorber	Al Kα 8.34 Å	Cu Lα 13.33 Å	Fe Lα 17.61 Å	Cr Lα 21.57 Å	Ti Lα 27.43 Å
Animal proteins	854	3095	6390	10,480	12,580
Fibrin	856	3102	6410	10,510	12,640
Elastin	859	3111	6420	10,530	12,510
Collagen	879	3179	6550	10,730	12,490
Amyloid	888	3204	6590	10,740	11,830
Thymus histone	842	3054	6320	10,380	12,820
Plant protein	858	3108	6420	10,530	12,660
DNA	981	3508	7160	11,560	10,990
RNA	988	3532	7210	11,630	11,050
Glycogen	1015	3615	7320	11,700	7730
Tripalmitin	645	2387	5010	8390	12,160
Tristearin	634	2347	4940	8290	12,290
Triolein	638	2363	4970	8340	12,380
Trilinolein	642	2379	5010	8400	12,460
Cerebrosides	727	2657	5520	9140	11,590
Parlodion	1177	4167	8390	13,320	7870
Zapon varnish	998	3571	7270	11,690	9470

B. THE SYSTEMATIC ERRORS UTILIZING MONOCHROMATIC X-RAYS

When monochromatic X-rays of wavelength λ_j traverse a homogeneous mixture of an average animal protein and n other elements or chemical compounds x_i, the following expression for the intensity of the transmitted X-rays is obtained:

$$I_{1j} = I_{0j} e^{-\left(\frac{\mu}{\rho}\right)_{\mathrm{prot}\,j} m_{\mathrm{prot}} - \sum_{i=1}^{n} \left(\frac{\mu}{\rho}\right)_{xij} m_{xi}} \tag{18}$$

I_{0j} = the intensity of the incident X-rays with the wavelength λ_j. I_{1j} = the intensity of the corresponding transmitted X-rays. $(\mu/\rho)_{\mathrm{prot}\,j}$ = the mass absorption coefficient of the average animal protein at the wavelength λ_j. $(\mu/\rho)_{xij}$ = the corresponding mass absorption coefficient of the element or compound x_i. m_{prot} = the weight per unit area of the average protein. m_x = the weight per unit area of the element or compound x_i. If the mixture of the average protein and the other elements or compounds x_i is assumed to have the same mass absorption coefficient as an average animal protein, a similar equation is obtained,

$$I_{1j} = I_{0j} e^{-\left(\frac{\mu}{\rho}\right)_{\mathrm{prot}\,j} m_{\alpha j}} \tag{19}$$

but here $m_{\alpha j}$ is the approximate weight per unit area of the mixture, when utilizing radiation with the wavelength λ_j. This weight must then have a value that equalizes the exponents in Eqs. 18 and 19. Thus

$$\left(\frac{\mu}{\rho}\right)_{\text{prot}\,j} m_{\text{prot}} + \sum_{i=1}^{n} \left(\frac{\mu}{\rho}\right)_{xij} m_{xi} = \left(\frac{\mu}{\rho}\right)_{\text{prot}\,j} m_{\alpha j} \tag{20}$$

From this assumption a systematic error, $\mathcal{E}_j$, is introduced, as $m_{\alpha j}$ generally differs from the correct weight per unit area. Then the following equation holds,

$$m_{\alpha j} = (1 + \mathcal{E}_j)\left(m_{\text{prot}} + \sum_{i=1}^{n} m_{xi}\right) = (1 + \mathcal{E}_j) m_x \tag{21}$$

where m_x is the correct weight per unit area. Eliminating $m_{\alpha j}$ in Eqs. 20 and 21, and regrouping, the following expression is obtained,

$$\sum_{i=1}^{n} \left(\frac{\mu}{\rho}\right)_{xij} m_{xi} - \sum_{i=1}^{n} \left(\frac{\mu}{\rho}\right)_{\text{prot}\,j} m_{xi} = \mathcal{E}_j \left(\frac{\mu}{\rho}\right)_{\text{prot}\,j} \left(m_{\text{prot}} + \sum_{i=1}^{n} m_{xi}\right) \tag{22}$$

Introducing m_x in Eq. 22 and solving for $\mathcal{E}_j$, gives

$$\mathcal{E}_j = \frac{\sum_{i=1}^{n} m_{xi} \left[\left(\frac{\mu}{\rho}\right)_{xij} - \left(\frac{\mu}{\rho}\right)_{\text{prot}\,j}\right]}{m_x \left(\frac{\mu}{\rho}\right)_{\text{prot}\,j}} \tag{23}$$

which expresses the resulting systematic error for a mixture of an average animal protein and n elements or chemical compounds x_i, exposed to monochromatic X-rays with the wavelength λ_j, when the mixture is assumed to have the same mass absorption coefficient as the protein. This expression is composed of the sum of n partial systematic errors $\mathcal{E}_{ij}$, given by:

$$\mathcal{E}_{ij} = \frac{m_{xi}}{m_x} \cdot \frac{\left(\frac{\mu}{\rho}\right)_{xij} - \left(\frac{\mu}{\rho}\right)_{\text{prot}\,j}}{\left(\frac{\mu}{\rho}\right)_{\text{prot}\,j}} \tag{24}$$

Thus the partial systematic error $\mathcal{E}_{ij}$ is proportional to the relative amount of the element or compound x_i, m_{x_i}/m_x, and to the relative difference between the mass absorption coefficients of x_i and of the average protein. As $m_{x_i}/m_x \leqslant 1$, the partial systematic error is restricted by the expression:

$$|\mathcal{E}_{ij}| \leqslant \left| \frac{\left(\frac{\mu}{\rho}\right)_{x_{ij}} - \left(\frac{\mu}{\rho}\right)_{\text{prot}\,j}}{\left(\frac{\mu}{\rho}\right)_{\text{prot}\,j}} \right| \tag{25}$$

In biological material practically all of the more common elements have atomic numbers lower than 30. For these elements Eq. 24 is presented graphically in two different ways.

In Fig. 5A, B the relative amount of different elements is plotted as a function of the atomic number Z when utilizing monochromatic X-rays with the wavelengths 8.34 Å (Al $K\alpha_{1,2}$) and 21.57 Å ($CrL\alpha_{1,2}$), respectively. The partial systematic errors $\mathcal{E}_{ij}$ are fixed at $\pm$ 5%, where the positive sign is used when

$$\left(\frac{\mu}{\rho}\right)_{x_{ij}} > \left(\frac{\mu}{\rho}\right)_{\text{prot}\,j}$$

and the negative sign when

$$\left(\frac{\mu}{\rho}\right)_{x_{ij}} < \left(\frac{\mu}{\rho}\right)_{\text{prot}\,j}$$

As $\mathcal{E}_{ij}$ is proportional to the relative amount of x_i, the shapes of the curves are independent of the absolute value of $\mathcal{E}_{ij}$. Where the slope of the curves is negative, the systematic error is positive, and vice versa. The two curves have a maximum between 6 C and 7 N, as the mass absorption coefficient of an average animal protein has a value between those of carbon and nitrogen. A second maximum is found for sulfur at 8.34 Å (Fig. 5A) and for chlorine at 21.57 Å (Fig. 5B). In fact, the partial systematic error for sulfur cannot exceed $\pm$ 5% when λ_j is equal to 8.34 Å, as the relative difference between $(\mu/\rho)_S$ and $(\mu/\rho)_{\text{prot}}$ is very small at this wavelength. The same result is found for chlorine at 21.57 Å. The discontinuity of the curve between the elements 12 Mg and 13 Al in Fig. 5A depends on the fact that elements with atomic number $Z \geqslant 13$ have their K absorption limits at wavelengths shorter than 8.34 Å. The discontinuity of the curve between the elements 8 O and 9 F in Fig. 5B has an analogous explanation.

In Fig. 6A, B the partial systematic errors $\mathcal{E}_{ij}$ of the more common elements in biological material are plotted as a function of the wavelength λ_j. The relative amount of each element is fixed at 10%. As, according to Eq. 24, the partial systematic error $\mathcal{E}_{ij}$ is proportional to

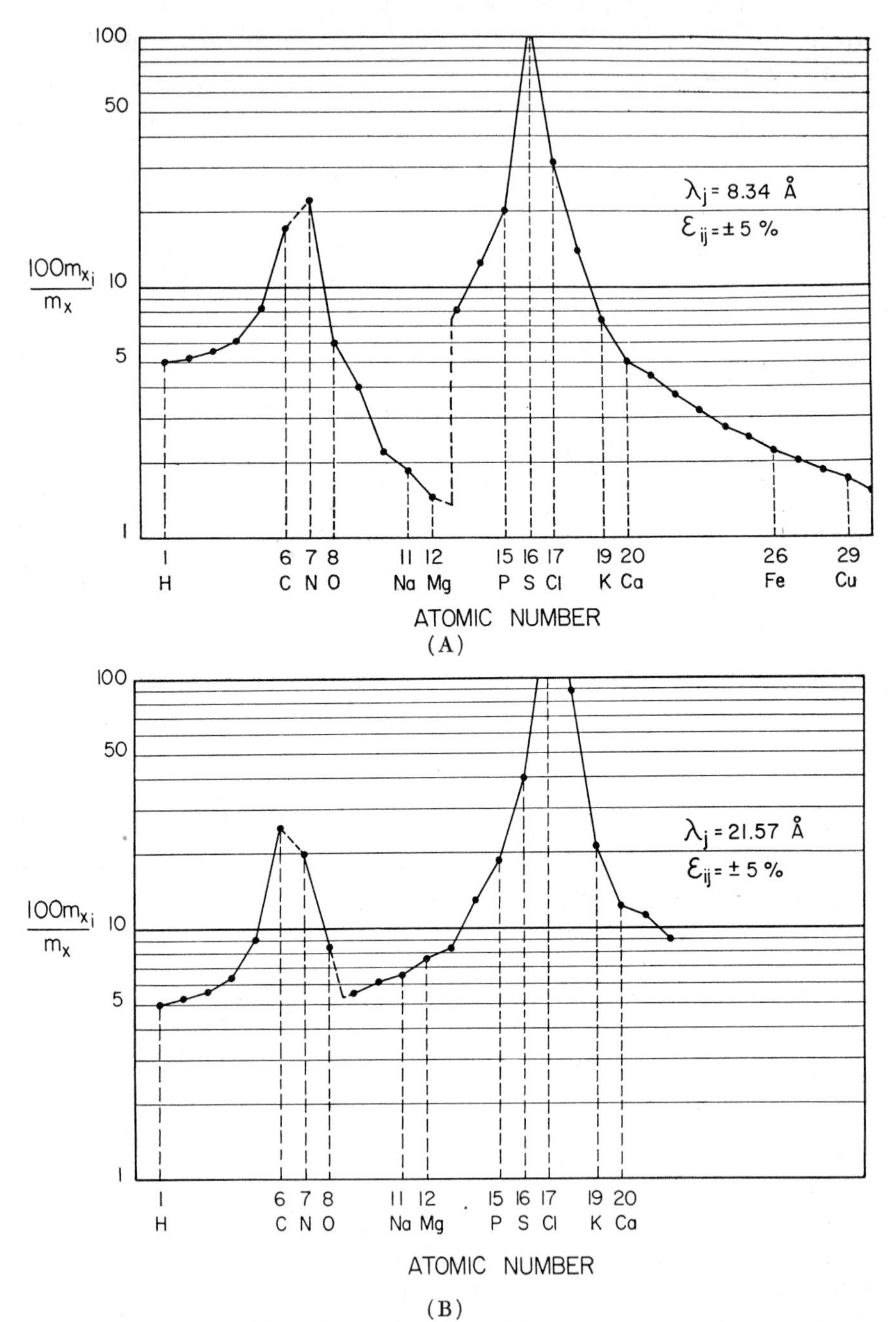

FIG. 5. A and B: The relative amount of different elements as function of the atomic number for two wavelengths, 8.34 and 21.57 Å.

the relative amount of the element x_i, the errors at other concentrations can also be obtained from the curves.

The curves in Fig. 6A, B show that, in general, the systematic error is smaller at longer wavelengths. All the curves have a discontinuity at the wavelength 5.02 Å, corresponding to the K absorption limit of sulfur.

With regard to the shape of the curves the elements can be divided into three classes. The first class includes hydrogen, carbon, nitrogen, and oxygen, having approximately constant systematic errors between 2.5 and 22 Å. The elements of the second class, sodium, magnesium, phosphorus, sulfur, chlorine, potassium, and calcium, have their K absorption limits in this wavelength range. A considerable decrease of the systematic error $\mathcal{E}_{ij}$ of each of these elements occurs at the wavelength corresponding to the respective K absorption limit. For phosphorus the systematic error is about 70% on the short wavelength side of the K absorption limit at 5.77 Å and about —2.5% on the long wavelength side. For sulfur the values are 82% and —0.5% at the K edge (5.02 Å). On the short wavelength side of each limit the systematic error $\mathcal{E}_{ij}$ of the element in question increases with decreasing wavelength, but on the long wavelength side of the edge the systematic error is small and approximately constant. Iron and copper, representing the third group, have L absorption limits in the wavelength region to be utilized for total dry weight determination of biological material. The influence of these discontinuities on the systematic errors is in principle the same as that of the K absorption limit of the elements in the second class.

For sodium, magnesium, phosphorus, sulfur, iron, and copper the sign of $\mathcal{E}_{ij}$ shifts at the discontinuity of the corresponding error curve. Hydrogen and carbon have negative systematic errors in the whole wavelength range, whereas nitrogen, oxygen, chlorine, potassium, and calcium have positive systematic errors. This indicates that in a mixed system the partial systematic errors of the different elements tend to compensate each other at wavelengths longer than 6 Å.

Equation 24 is also valid for different organic compounds x_i, mixed with the model protein. In Fig. 7 the partial systematic errors $\mathcal{E}_{ij}$ of some organic compounds are plotted as a function of the wavelength λ_j. The elementary composition of the compounds and the mass absorption coefficients have been presented in Tables II and IV. The relative amount of each compound is assumed to be 10%. These curves are similar to those in Fig. 6A, B. The error curve of carbon is also included in the diagram.

All the curves in Fig. 7 have a discontinuity at 5.02 Å, corresponding

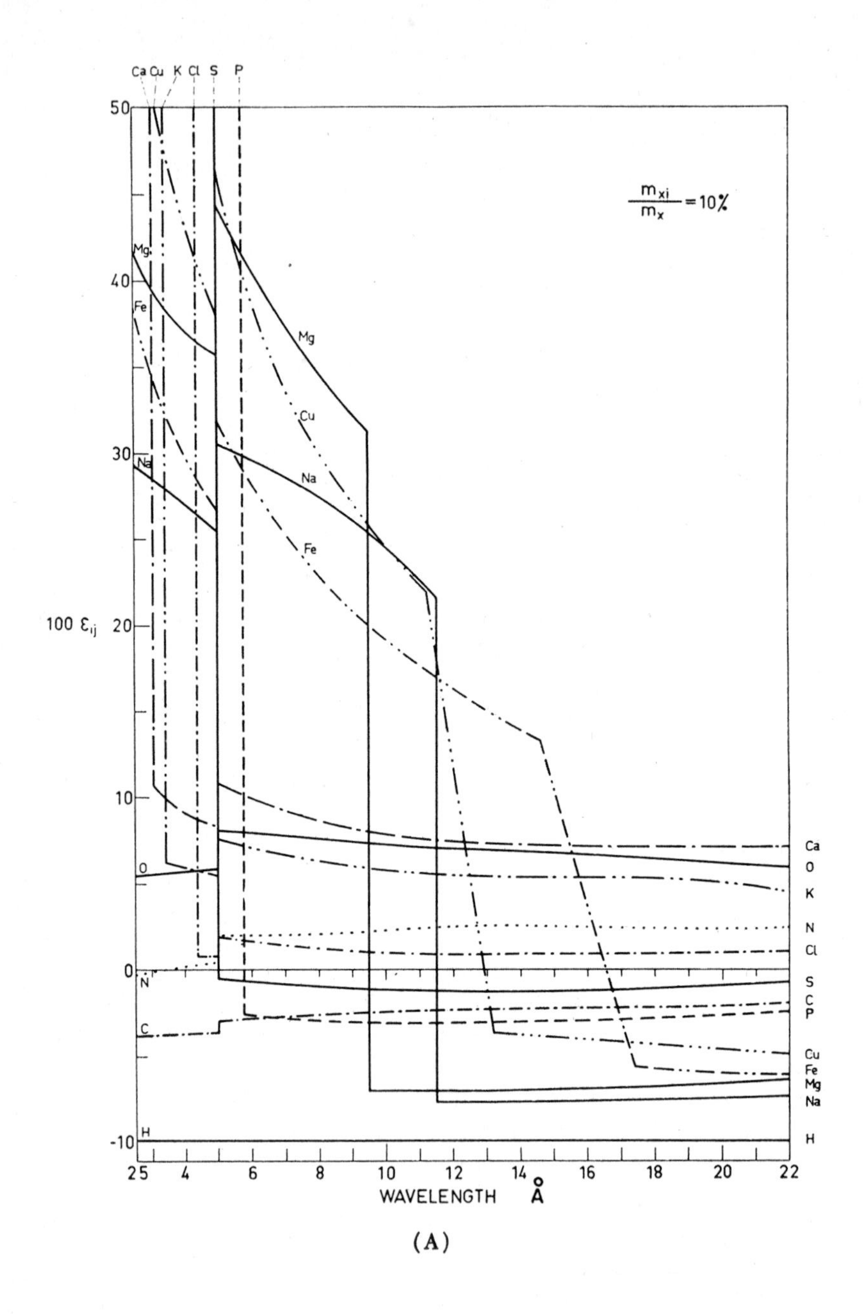

(A)

FIG. 6. A and B: The partial systematic errors ε_{ij} caused by common elements in biological material as function of the X-ray wavelength (Lindström, 1955).

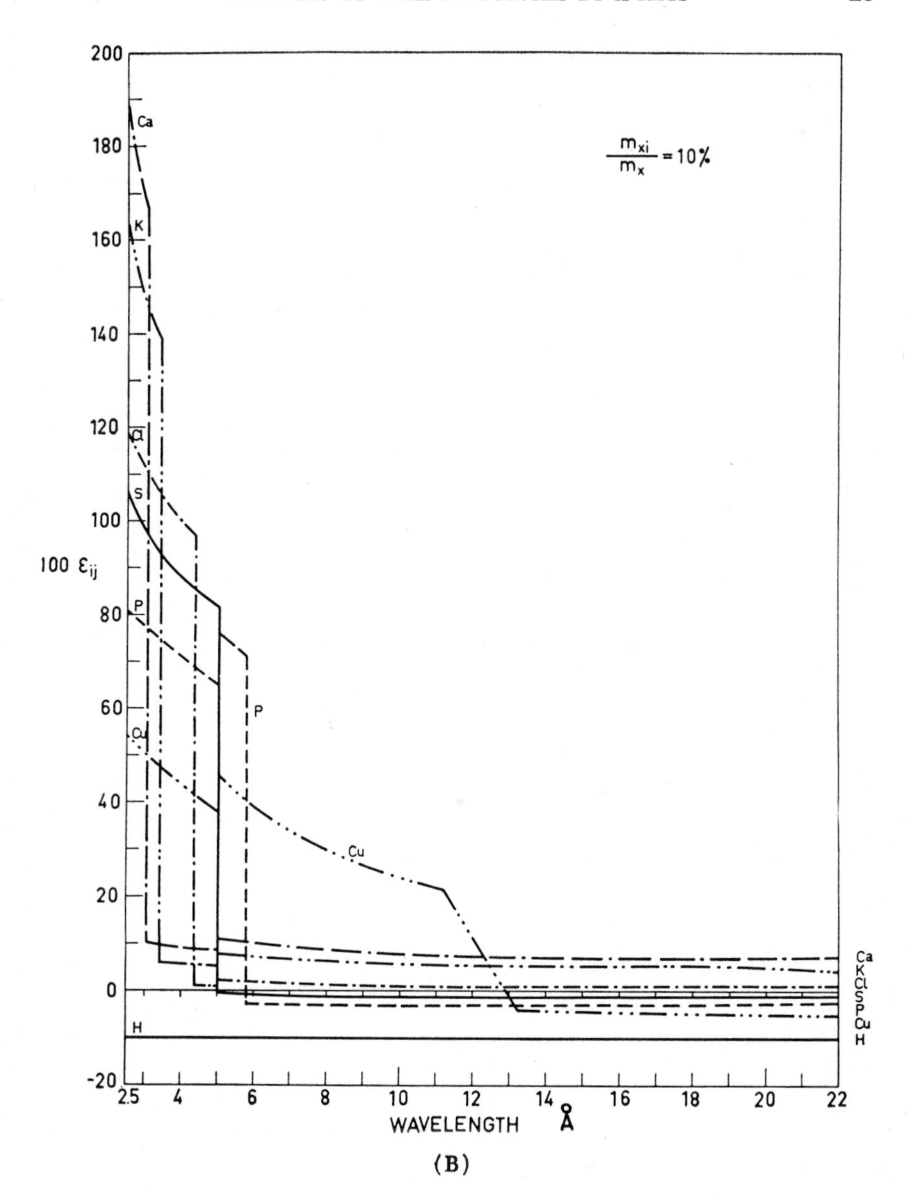

(B)

to the K absorption limit of sulfur. The curves of RNA, DNA, and cerebrosides have a second discontinuity at the K absorption limit of phosphorus (5.77 Å). In general the absolute values of ε_{ij} are smaller on the long wavelength side of these discontinuities. However, cerebrosides have an approximately constant systematic error in the whole

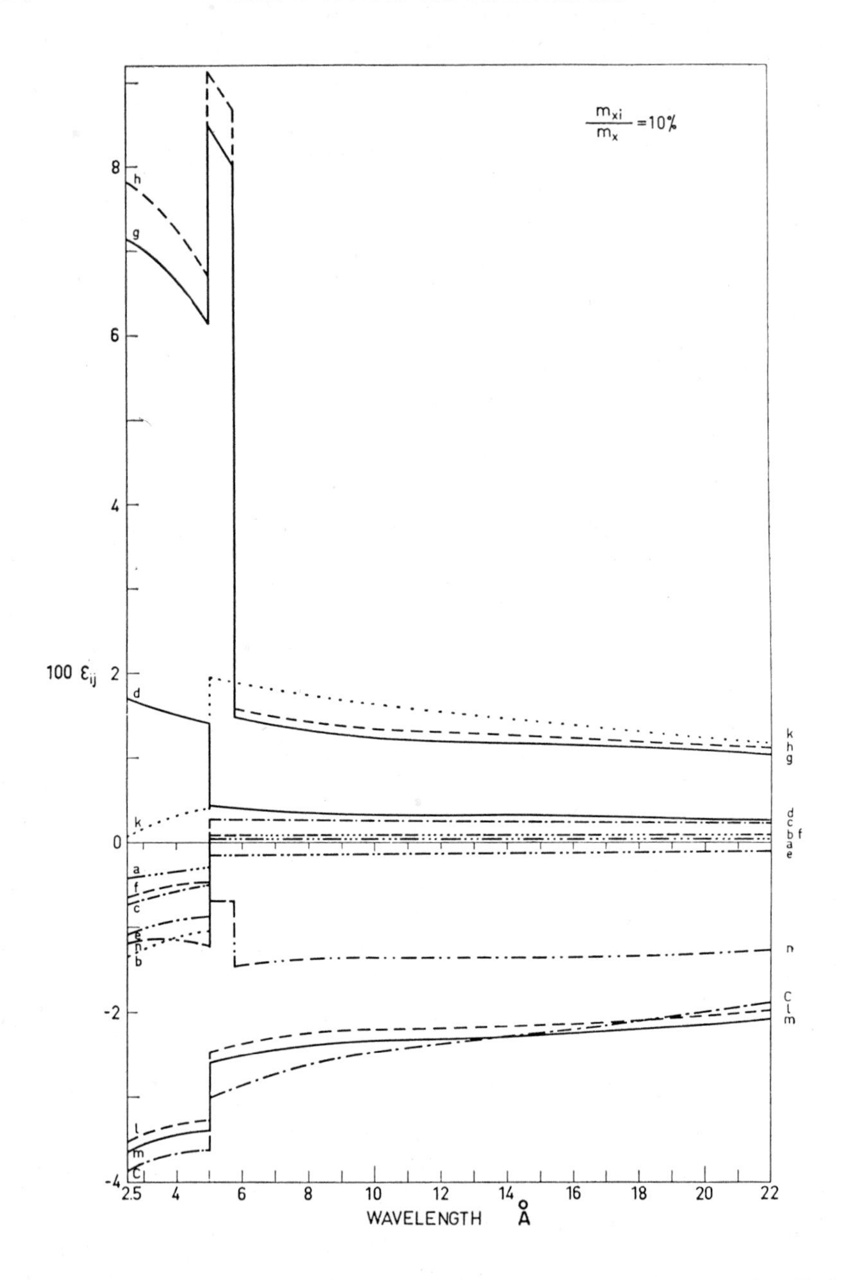

FIG. 7. The partial systematic errors ε_{ij} caused by organic compounds as function of the X-ray wavelength. a, fibrin; b, elastin; c, collagen; d, amyloid; e, thymus histone; f, plant proteins; g, deoxyribonucleic acid; h, ribonucleic acid; k, glycogen; l, tripalmitin; m, tristearin; n, cerebrosides (Lindström, 1955).

wavelength region, and the systematic error of glycogen is 0.4% on the short wavelength side and 1.9% on the long wavelength side of the sulfur discontinuity.

With regard to the magnitude of the systematic errors $\mathcal{E}_{ij}$ at wavelengths longer than those of the discontinuities, the curves can be divided into three groups. The first includes glycogen, RNA, and DNA which have positive systematic errors, varying between 1 and 2%. The second comprises amyloid, collagen, elastin, average plant protein, fibrin, and thymus histone. In this group the partial systematic errors are between —1.4 and 1.7%, depending upon the varying concentration of sulfur in the different conjugated proteins.

The third group includes the error curves of the lipids. The systematic errors of cerebrosides varies between —1.5 and —1.3% at wavelengths longer than 5.02 Å. In the same wavelength range the partial systematic errors of the most common triglycerides are between —2.6 and —2.0%. The sequence between the different triglycerides is tripalmitin, trilinolein, triolein, and tristearin, where tripalmitin has the smallest systematic error and tristearin the greatest one. Only the error curves of tripalmitin and tristearin are represented in Fig. 7, but the difference between the ordinates of these two curves is small and varies between 0.10 and 0.15%.

C. The Systematic Errors Utilizing Continuous X-rays

In the preceding calculations the partial systematic errors were calculated assuming monochromatic X-rays of wavelength λ_j. Due to the extremely low sensitivity of the fine-grained photographic emulsions, soft *continuous* X-rays have hitherto been used for most of the quantitative determinations of total dry weights of biological material. Suitable wavelength ranges for microradiography of thin sections of soft tissues can be taken from Fig. 1. Preferably, the X-rays should be generated at 1500 or at lower voltages. It is obvious from the diagrams in Figs. 6A, B and 7, that most of the partial systematic errors $\mathcal{E}_{ij}$ decrease at longer wavelengths. The resulting systematic errors have been calculated for continuous X-rays generated at 1500 volts.

The distribution of the X-ray intensity in the continuous spectrum is not accurately known at low voltages, and several expressions giving the intensity as a function of the wavelength have been proposed. According to the two extreme expressions the theoretical distribution of a continuous X-ray spectrum generated at 1500 volts has been calculated (Lindström, 1955). The filtration of the radiation in a thin aluminum foil, serving as a light filter, and in the supporting Zapon varnish foil on the metal disc (see Fig. 4) has also been included in the calculations in order to compensate for the influence of the experimental conditions. The distribution of the intensity in this filtered continuous X-ray spectrum is shown in Fig. 8, and the relative distribution of the intensity is given in Table V.

TABLE V
THE INTENSITY DISTRIBUTION IN THE FILTERED, CONTINUOUS X-RAY SPECTRUM, GENERATED AT 1.5 KV.[a]

Wavelength region Å	Distribution of intensity in per cent, according to	
	Fig. 8A	Fig. 8B
8.25– 9.5	11.5	23.8
9.5 –10.5	16.7	24.8
10.5 –11.5	17.0	19.2
11.5 –12.5	14.8	12.8
12.5 –13.5	11.9	8.1
13.5 –14.5	9.1	4.9
14.5 –15.5	6.6	2.9
15.5 –17.0	6.4	2.2
17.0 –19.0	4.3	1.0
19.0 –∞	1.7	0.3

[a] From Lindström (1955).

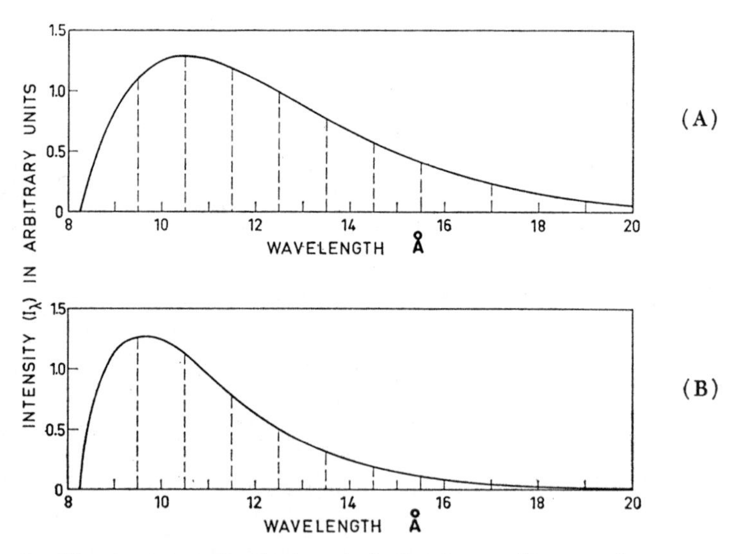

FIG. 8. The intensity distribution (calculated according to the two most differing formulae) in the continuous X-ray spectrum generated at 1.5 kv. after filtration in 0.33 mg. Al per cm.[2] (Lindström, 1955).

When utilizing the filtered, continuous X-ray spectrum, the resulting systematic error $\mathcal{E}$ of n elements or compounds x_i is determined in the following way. In Eqs. 19 and 21 $m_{\alpha j}$ is eliminated and then,

$$I_{1j} = I_{0j} e^{-\left(\frac{\mu}{\rho}\right)_{\text{prot}\,j}(1 + \mathcal{E}_j) m_x} \tag{26}$$

If the continuous X-ray spectrum is divided in n_1 regions, each with a mean wavelength λ_j, the X-ray intensities in the different regions can be added according to

$$\sum_{j=1}^{n_1} I_{1j} = \sum_{j=1}^{n_1} I_{0j} e^{-\left(\frac{\mu}{\rho}\right)_{\text{prot}\,j}(1 + \mathcal{E}_j) m_x} = \sum_{j=1}^{n_1} I_{0j} e^{-\left(\frac{\mu}{\rho}\right)_{\text{prot}\,j}(1 + \mathcal{E}) m_x} \tag{27}$$

where $\mathcal{E}$ is the resulting systematic error.

Expanding the exponential expressions, neglecting square and higher powers, the following is obtained,

$$\sum_{=1}^{n_1} I_{0j} \left[1 - \left(\frac{\mu}{\rho}\right)_{\text{prot}\,j}(1 + \mathcal{E}_j) m_x\right] = \sum_{j=1}^{n_1} I_{0j} \left[1 - \left(\frac{\mu}{\rho}\right)_{\text{prot}\,j}(1 + \mathcal{E}) m_x\right] \tag{28}$$

which can be transformed to,

$$m_x \sum_{=1}^{n_1} I_{0j} \left(\frac{\mu}{\rho}\right)_{\text{prot}\,j} \mathcal{E}_j = m_x \mathcal{E} \sum_{j=1}^{n_1} I_{0j} \left(\frac{\mu}{\rho}\right)_{\text{prot}\,j} \tag{29}$$

Eq. 29 solved for $\mathcal{E}$, gives,

$$\mathcal{E} = \frac{\sum_{j=1}^{n_1} I_{0j} \left(\frac{\mu}{\rho}\right)_{\text{prot}\,j} \mathcal{E}_j}{\sum_{j=1}^{n_1} I_{0j} \left(\frac{\mu}{\rho}\right)_{\text{prot}\,j}} \tag{30}$$

If $\mathcal{E}_j$ is constant, the equality $\mathcal{E} = \mathcal{E}_j$ is immediately obtained. For a single element x_i, i.e. for $n = 1$, Eq. 30 is also valid, if $\mathcal{E}_i$ and $\mathcal{E}_{ij}$ are substituted for $\mathcal{E}$ and $\mathcal{E}_j$, respectively. In this equation I_{0j} can be replaced by proper transmission value in Table V, and if the relative amount of the element or compound x_i is 10%, $\mathcal{E}_{ij}$ can be obtained from Fig 6A, B and Fig. 7, respectively.

The resulting systematic error $\mathcal{E}_i$ of individual elements and compounds, presented in Table VI, has been determined from Eq. 30. The two values for each element and compound correspond to the two different distributions of the intensity of the continuous X-rays in Fig. 8A, B. In Table VI the difference between the two values is small for the organic compounds and for most of the elements. Because of absorption discontinuities and a change of the sign of the partial systematic errors $\mathcal{E}_{ij}$ in the wavelength range from 8 to 20 Å, the corresponding values of sodium, magnesium, iron, and copper are more divergent. As only relatively low concentrations of these elements occur in biological material, the uncertainty of the resulting systematic errors $\mathcal{E}_i$ is without practical importance.

TABLE VI

THE RESULTING SYSTEMATIC ERROR $\mathcal{E}_i$ OF DIFFERENT ELEMENTS AND ORGANIC COMPOUNDS FOR THE FILTERED, CONTINUOUS X-RAY SPECTRUM, GENERATED AT 1.5 KV., THE RELATIVE AMOUNT OF EACH ELEMENT 10% [a]

Element	100 $\mathcal{E}_i$ Intensity distribution according to		Compound	100 $\mathcal{E}_i$ Intensity distribution according to	
	Fig. 8A	Fig. 8B		Fig. 8A	Fig. 8B
1 H	−9.9	−9.9	Fibrin	0.03	0.03
6 C	−2.3	−2.4	Elastin	0.06	0.06
7 N	2.4	2.4	Collagen	0.26	0.26
8 O	6.9	7.1	Amyloid	0.32	0.32
11 Na	0.2	7.7	Thymus histone	−0.12	−0.12
12 Mg	−5.1	−1.9	Plant protein	0.05	0.05
15 P	−2.9	−3.0	DNA	1.19	1.22
16 S	−1.1	−1.1	RNA	1.26	1.29
17 Cl	0.9	0.9	Glycogen	1.47	1.52
19 K	5.4	5.5	Tripalmitin	−2.17	−2.19
20 Ca	7.3	7.5	Tristearin	−2.29	−2.32
26 Fe	12.0	15.8	Triolein	−2.23	−2.26
29 Cu	8.8	16.0	Trilinolein	−2.18	−2.21
			Cerebrosides	−1.37	−1.37

[a] From Lindström (1955).

The common elements in biological material with positive and negative systematic errors, respectively, are about equal in number and therefore the systematic errors of the different elements will compensate each other to a considerable extent. Iron, copper, and other elements with atomic numbers above 19 are present in very small concentrations in biological soft tissues. The high amount of calcium in calcified tissues, however, involves a considerable error, and therefore special procedures have been developed for the analysis of such tissues.

The resulting systematic errors of the different proteins are negligible. Utilizing X-rays in this wavelength region, the positive systematic errors of RNA and DNA are caused by the high percentage of oxygen and not by the high relative amount of phosphorus in these compounds. However, 10% RNA or DNA gives a resulting systematic error of only about 1.25%. The same relative amount of glycogen involves a systematic error of 1.5%.

The lipids have negative systematic errors due to the high contents of carbon and hydrogen. For a mixture of 90% protein and 10% lipids the resulting systematic error is about —2.2%. It may therefore be necessary to introduce a correction factor when the microradiographic procedure is applied to tissues with high concentrations of lipids. This is especially required when the lipid fraction is determined by total dry weight determinations before and after lipid extraction, as the error of the weight of the lipid fraction will be between —20 and —25%, if no such correction is made.

D. The Reference System

When soft, continuous X-rays have been used at the quantitative microradiographic analysis, the total dry weight of a biological structure has in general been determined by comparing the photocurrents corresponding to the X-ray absorption image of the structure with those corresponding to the image of the step wedge of nitrocellulose foils (schematically illustrated in Fig. 4). By means of the calibration curve (the photo-current versus the number of foils) the photometer deflections for the structures are transformed to foil equivalents (u_1 and u_2, respectively). In this way the weight per unit area of a reference system with the same absorption as that of the biological structure is determined.

For a continuous X-ray spectrum this principle of comparison can be expressed by the following equation,

$$\sum_{j=1}^{n_1} I_{1j} = \sum_{j=1}^{n_1} I_{0j}e^{-\left(\frac{\mu}{\rho}\right)_{\text{prot}_j}(1+\varepsilon)m_x} = \sum_{j=1}^{n_1} I_{0j}e^{-\left(\frac{\mu}{\rho}\right)_{\text{ref}_j}m_{\text{ref}}} \qquad (31)$$

analogous to Eq. 27. Here, $(\mu/\rho)_{\text{ref}_j}$ is the mass absorption coefficient of the reference system at the wavelength λ_j, and m_{ref} is given by,

$$m_{\text{ref}} = u_x \cdot w \qquad (32)$$

where u_x is the foil equivalent of the biological structure, and w is the weight per unit area of one foil in the reference system. Equation 31 can be transformed to the following expression:

$$\sum_{j=1}^{n_1} I_{0j} \left(\frac{\mu}{\rho}\right)_{\mathrm{prot}_j} (1 + \varepsilon) m_x = \sum_{j=1}^{n_1} I_{0j} \left(\frac{\mu}{\rho}\right)_{\mathrm{ref}_j} m_{\mathrm{ref}} \tag{33}$$

The weight per unit area of the biological structure is then obtained,

$$(1 + \varepsilon) m_x = m_{\mathrm{ref}} \frac{\sum_{j=1}^{n_1} I_{0j} \left(\frac{\mu}{\rho}\right)_{\mathrm{ref}_j}}{\sum_{j=1}^{n_1} I_{0j} \left(\frac{\mu}{\rho}\right)_{\mathrm{prot}_j}} \tag{34}$$

If

$$\left(\frac{\mu}{\rho}\right)_{\mathrm{ref}_j} \Big/ \left(\frac{\mu}{\rho}\right)_{\mathrm{prot}_j}$$

is constant at different wavelengths, Eq. 34 is simplified to

$$(1 + \varepsilon) m_x = m_{\mathrm{ref}} \frac{\left(\frac{\mu}{\rho}\right)_{\mathrm{ref}_j}}{\left(\frac{\mu}{\rho}\right)_{\mathrm{prot}_j}} \tag{34a}$$

Therefore, the reference system ought to have about the same elementary composition or at least be composed of the same elements as the average protein. For a reference system, composed of Parlodion foils, the ratio between

$$\left(\frac{\mu}{\rho}\right)_{\mathrm{ref}_j} \text{ and } \left(\frac{\mu}{\rho}\right)_{\mathrm{prot}_j}$$

is approximately constant at wavelengths longer than 6 Å, but a continuous, small decrease of the ratio is obtained at increasing wavelengths. With the distribution of the intensity of the X-rays according to Fig. 8A, Eq. 34 takes the form

$$(1 + \varepsilon) m_x = 1.315 \cdot m_{\mathrm{ref}} \tag{34b}$$

and according to Fig. 8B the following form:

$$(1 + \varepsilon) m_x = 1.323 m_{\mathrm{ref}} \tag{34c}$$

Hence, the uncertainty regarding the exact shape of the continuous

X-ray spectrum has only a small influence upon the value of the ratio and the mean value 1.32 can be considered as fairly exact.

When utilizing soft, *continuous* X-rays, the introduction of a reference system has three main advantages.

(1) The principle of comparison makes the determination of the density curve of the fine-grained photographic emulsion unnecessary.

(2) The photographic effect of the extremely soft, continuous X-ray spectrum transmitted through the specimen may not be the same as that of an incident continuous X-ray spectrum with the same intensity, because of the filtration of the X-rays in the specimen. This possible source of error is eliminated when utilizing a system of comparison, as the filtration of the continuous X-ray spectrum in the specimen is about the same as that in the reference system with the same total absorption.

(3) It is not necessary to determine the exact absolute value of the resulting mass absorption coefficient of the average animal protein, which is of importance, since the thin filters (X-ray windows) often have to be replaced.

V. DETERMINATION OF WATER IN BIOLOGICAL SAMPLES BY SOFT X-RAY MICRORADIOGRAPHY

Microradiography can be utilized to obtain information about the water content in various cellular structures and tissues. Two methods, outlined by Engström and Glick (1956), will be described briefly in this section.

If the fresh frozen section is microradiographed before and after evaporation of the ice, the amount of water can be determined from two microradiograms according to the following calculations:

$$E_{\mathrm{To}} = \log_e \frac{I_0}{I_{\mathrm{To}}} = \left(\frac{\mu}{\rho}\right)_{\mathrm{Sa}} \cdot m_{\mathrm{Sa}} + \left(\frac{\mu}{\rho}\right)_{\mathrm{aq}} \cdot m_{\mathrm{aq}} = E_{\mathrm{Sa}} + E_{\mathrm{aq}} \qquad (35)$$

In the equalities in Eq. 35 the subscripts "To," "Sa," and "aq" refer to total sample before the dehydration, dry substance in the sample, and water, respectively. Therefore, I_{To} and I_0 represent the X-ray intensities transmitted and incident, and E is the natural extinction. From the two microradiograms E_{To} and E_{Sa} can be calculated, if the densities in the microradiograms are kept within the straight portion of the density versus exposure curve of the photographic emulsion. Then the amount of water per unit area can be calculated:

$$m_{\mathrm{aq}} = \frac{E_{\mathrm{To}} - E_{\mathrm{Sa}}}{\left(\frac{\mu}{\rho}\right)_{\mathrm{aq}}} \qquad (36)$$

If for the same area the amount of dry substance m_{Sa} has been determined as outlined before, the percentage of water in the fresh tissue can be obtained independent of the change of the density caused by the freezing procedure.

As shown in the preceding section a microradiogram of a dried sample registered with 6 Å or softer X-rays can be evaluated in terms of the amount of dry weight. The negative of such a microradiogram represents in a crude way the distribution of water. The effective thickness l_{Sa} of the dry matter in a given structure equals m_{Sa}/ρ_{Sa}, where ρ_{Sa} is the density of the dry substance. Then the thickness of the water present in the fresh material is

$$l_{aq} - l_{To} - m_{Sa}/\rho_{Sa} = m_{aq} \qquad (37)$$

where l_{To} is the measured total thickness of the fresh sample. The weight of the water per unit area m_{aq} is equal to l_{aq}, as the density of water is 1.

Of the two procedures described, the first one is the more unequivocal, but it has the disadvantage of requiring the registration of a microradiogram of the fresh frozen sample in addition to one of the dried sample. Although the second procedure requires only the latter microradiogram, the thickness of the fresh sample l_{To} must be accurately known, since it will usually be between 2 and 5 times the effective thickness of the organic material, m_{Sa}/ρ_{Sa}. Methods for determining the thickness of sections have been reviewed by Lange and Engström (1954), but lately a method has been devised by Hallén (1956) which seems suitable for these purposes. The value of ρ_{Sa} can be taken as 1.3, and only small errors will result from deviations in different tissues from this assumed value. In both procedures it is necessary to consider the morphological effects of the drying process. Although freeze-drying minimizes these, the use of embedding media may involve a loss of substance on the removal of the media. Whether or not such a loss is significant will depend on the nature of the specimen and on the medium. Chemical fixatives ought to be avoided, because they produce structural distortions and also effect the quantity of substance remaining.

VI. EQUIPMENT FOR MICRORADIOGRAPHY WITH POLYCHROMATIC X-RAYS

As indicated in Sections III and IV, X-rays useful for imaging thin biological specimens must have low quantum energies, corresponding to voltages in the range between 500 and 2000 volts. The construction of suitable X-ray sources with high output of radiation therefore presents certain constructional difficulties. The window, screening the light from the hot filament cathode, must be very thin (Fig. 9) in order to transmit the ultrasoft X-rays. Furthermore, the sample must be contained within the high vacuum, as is illustrated in Fig. 10. The X-ray source must also be designed to give as high output of soft X-rays as possible, because the fine-grained photographic emulsions, utilized for contact microradiography, have an extremely low sensitivity. For quantitative microradiographic analyses the incident X-ray intensity must not vary more than about 1% at the plane of the sample in contact with the photographic emulsion. The area in the microradiogram with an

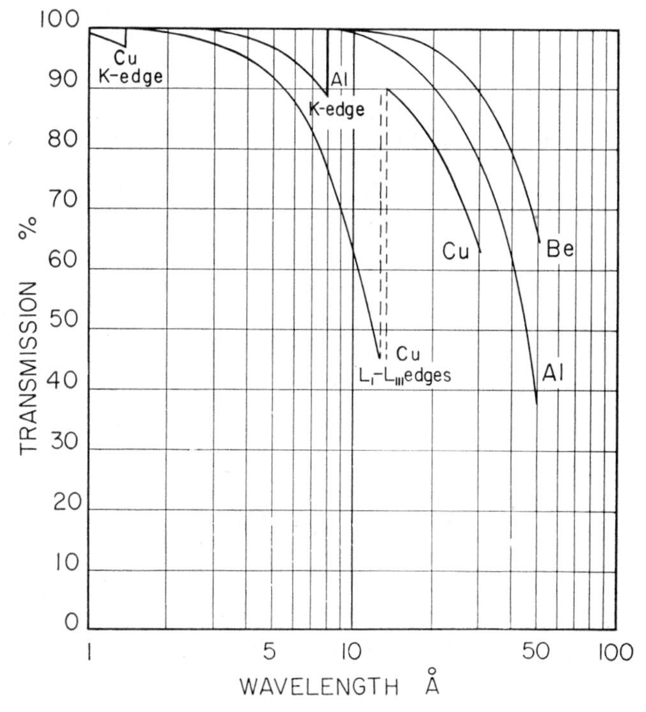

FIG. 9. Transmission of 1.000 Å thick foils of Al, Be, and Cu as function of X-ray wavelength.

even intensity has a linear dimension which is less than one-fifth to one-seventh of the distance between the focal spot at the anode and the image plane. Therefore, if a field with a diameter of 10 mm. is desirable for the particular experiment, the distance between the focal spot in the X-ray tube and the photographic emulsion should be at least 50 to 70 mm. When a high resolution in the microradiographic technique is required, the geometrical unsharpness must be kept below about 0.2 μ in the microradiogram. Therefore, when imaging a 10 μ thick section of a biological sample, the ratio between the size of the focal spot and the focal spot to image distance should be less than 1:50.

Cross-sections of two small X-ray tubes which in practice have given excellent results, are shown in Fig. 11A and 12. For a voltage of about 1000 volts applied to the tube in Fig. 11 the exposure time is between

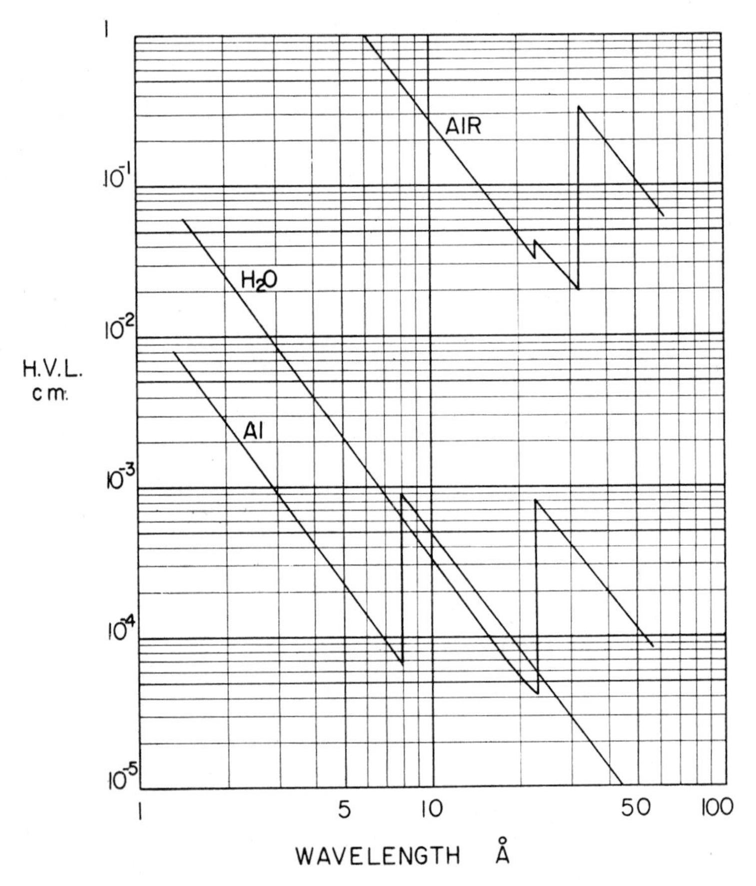

FIG. 10. Half-value layers, i.e. the thickness of absorber which reduces the incident X-ray intensity to half of its original value, for air, water, and aluminum.

15 and 30 minutes, when using an Eastman Kodak No. 649 Spectroscopic Plate and a small focus-to-object distance. But for very low voltages exposure times up to several hours are frequently met. The tube is continuously evacuated and provided with a by-pass valve system. Therefore, when a new sample and photographic plate have been introduced in the X-ray tube, this is ready again for exposure after less than 1 minute of pumping. The principle of such a vacuum system is shown

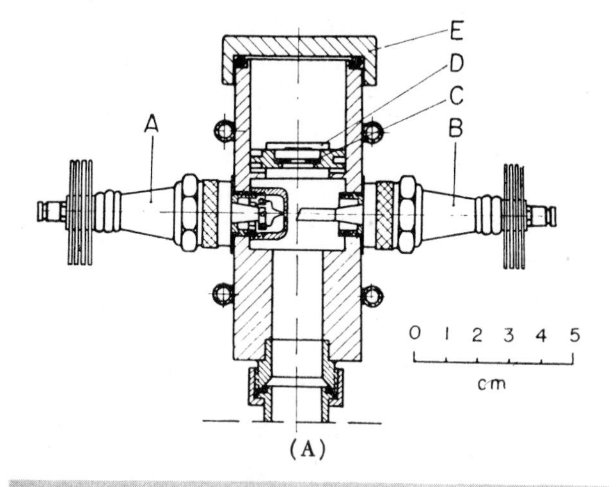

(A)

(B)

FIG. 11. A: Midget X-ray tube (cross-section) for microradiography with soft X-rays. The cathode (A) and anode (B) are attached to automobile spark plugs. The thin window is placed at C and the sample and emulsion at D. Through the top with its removable cap (E) the sample and film is introduced (Engström and Lundberg, 1957). B: The midget X-ray tube shown in A.

in Fig. 13. In order to protect the very thin window from rupturing, it must have a position securing the same pressure on both sides during the evacuation. Typical exposure times for the X-ray tube in Fig. 12 are given in Fig. 14.

Other types of tubes for microradiography with soft X-rays have been published by Engström and Lindström (1950), Clemmons (1955), Brattgård and Hydén (1954), Lindström (1955) (Fig. 15), and Fitzgerald (1957). A commercial, sealed-off unit for microradiography has

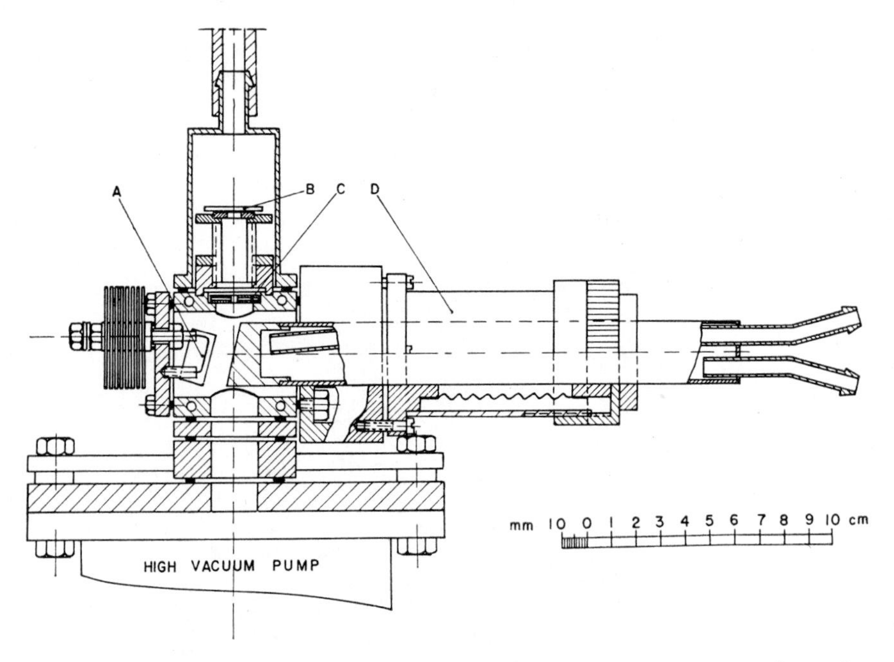

FIG. 12. Cross-section through X-ray tube for ultrasoft microradiography (A) cathode; (B) sample and emulsion; (C) window; (D) anode (Engström *et al.*, 1957).

been marketed by Philips Company, consisting of a small, compact high voltage source with a miniature X-ray tube provided with a 50 μ thick beryllium window. This unit is suitable for microradiography of about 10 μ thick specimens. However, for very thin specimens, such as blood cells, sperms, or 1–2 μ thick microtome sections, the radiation from this tube is too hard, as the extremely soft components in the X-ray spectrum are absorbed in the beryllium window. Specimens with greater values of $(\mu/\rho \cdot m)$ than in soft tissues which is the case for ground sections of bone and teeth, can be microradiographed with considerably harder X-rays. For such purposes the radiation from commercially available X-ray diffraction units have been of great use. Figure 16 shows the principle of microradiography with the aim of determining the distribution of hydroxyapatite in 50 to 100 μ thick bone specimens.

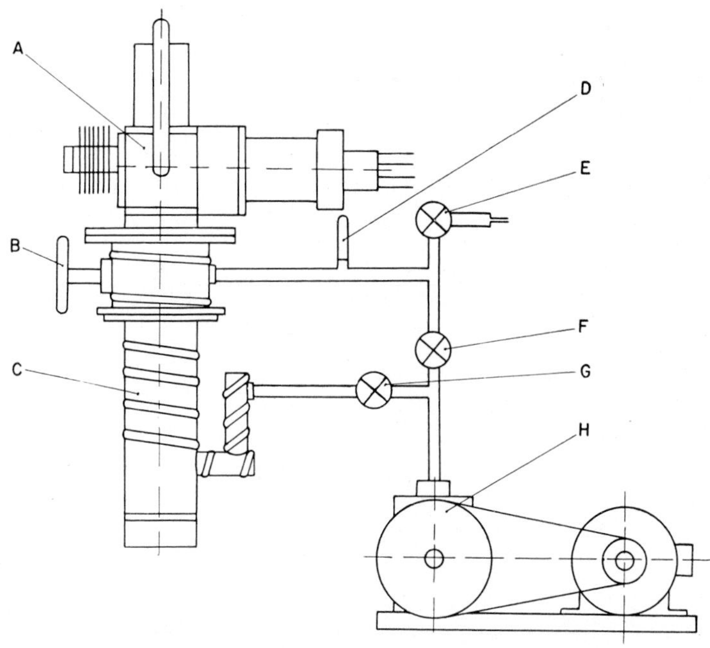

FIG. 13. General outline of the vacuum connections for a soft X-ray tube. A, X-ray tube; B, high-vacuum valve; C, high-vacuum diffusion pump; D, vacuum meter, E, air inlet valve; F and G, bypass valve system; H, forepump.

VII. EQUIPMENT FOR MICRORADIOGRAPHY WITH MONOCHROMATIC X-RAYS

A. HARD TISSUES AND THICK SPECIMENS OF SOFT TISSUES

Due to the low sensitivity of the fine-grained photographic emulsions, the selection of soft monochromatic X-rays by means of crystal reflection leads to extremely long exposure times. Thin sections of bone, however, have a relatively high X-ray contrast, and can therefore be microradiographed with X-rays having as short a wavelength as 1.54 Å, corresponding to the copper Kα lines. Equipment for such microradiography only requires the introduction of a crystal in proper position

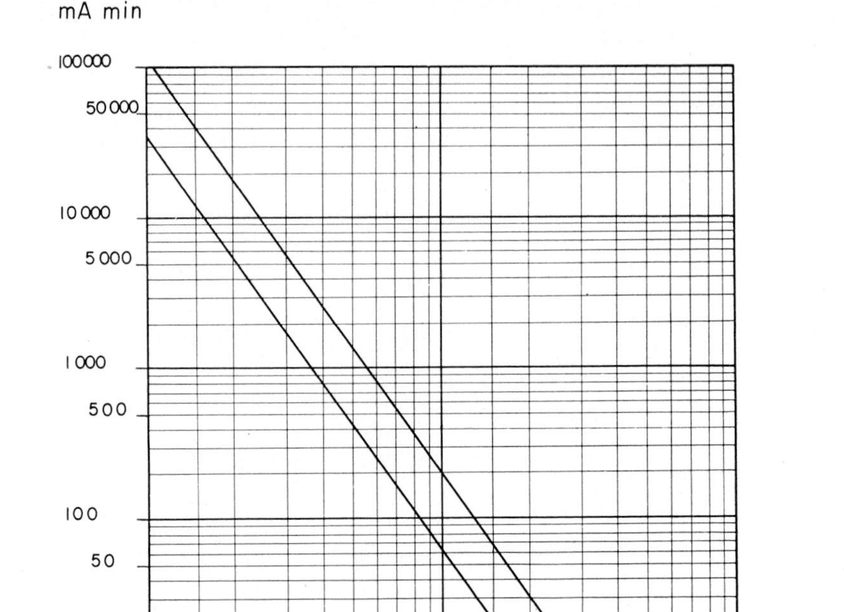

FIG. 14. Exposure time as function of voltage for two different target to sample distances for the fine-grained Eastman Kodak Spectroscopic Plate 649. This diagram was obtained with the X-ray tube shown in Fig. 12.

for reflection in the beam from an ordinary diffraction unit with a copper target tube. This type of microradiography has been extensively used to study quantitatively the mineral distribution in bone (Carlström and Engström, 1956).

If a copper target diffraction tube is run at a low voltage, e.g., 10 kv., and the radiation is filtered in a 0.02 mm. Ni foil, the transmitted radiation is practically equivalent to Cu Kα (1.54 Å) as is clear from the experimental curves in Fig. 17. This technique gives a homogeneous, large field of X-rays of high intensity.

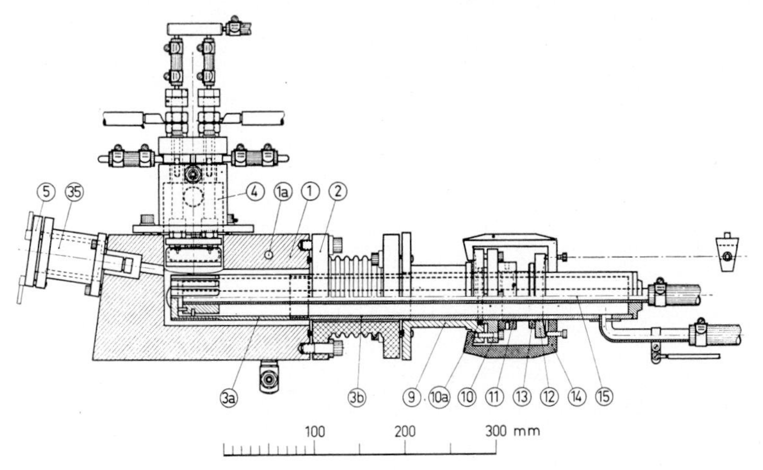

Fig. 15. Longitudinal section through the X-ray tube devised by Lindström (1955). The target has several materials in order to secure output of various emission lines for eventual elementary analysis. For details, the reader is referred to the original publication.

B. Soft Tissues

Microradiography of thin, soft tissues, however, requires much softer radiation than the wavelength of the copper Kα line, and the low voltage ion X-ray tube introduced by Henke (1957) has been found to work satisfactorily. A cross-section of such an ion tube including its principle is presented in Fig. 18A, and a photograph of the tube is shown in Fig. 18B. The X-rays emerging from the target are taken out in a direction 180° from the incidence of the electrons, which secures a minimum of continuous X-rays giving a maximum of the ratio line to white radiation. The radiation undergoes total reflection in a cylindrical mirror, where the hard portion of the continuous radiation is cut off. The angle of total reflection is adjusted for the particular X-ray line wanted, for example copper Lα (13.3 Å), or aluminum Kα (8.34 Å).

The parts of the continuous radiation with wavelengths softer than the required line are cut off by a filter, and Fig. 19 shows the characteristics of such a band pass filter (Henke, 1957). The systematic error of the X-ray weighing method can be calculated more exactly when the microradiograms are registered with monochromatic X-rays; this is evident from the discussion in Section IV. Also, when monochromatic radiation is used, it is not necessary to introduce a reference system, as the mass absorption coefficient at a given wavelength is known (see

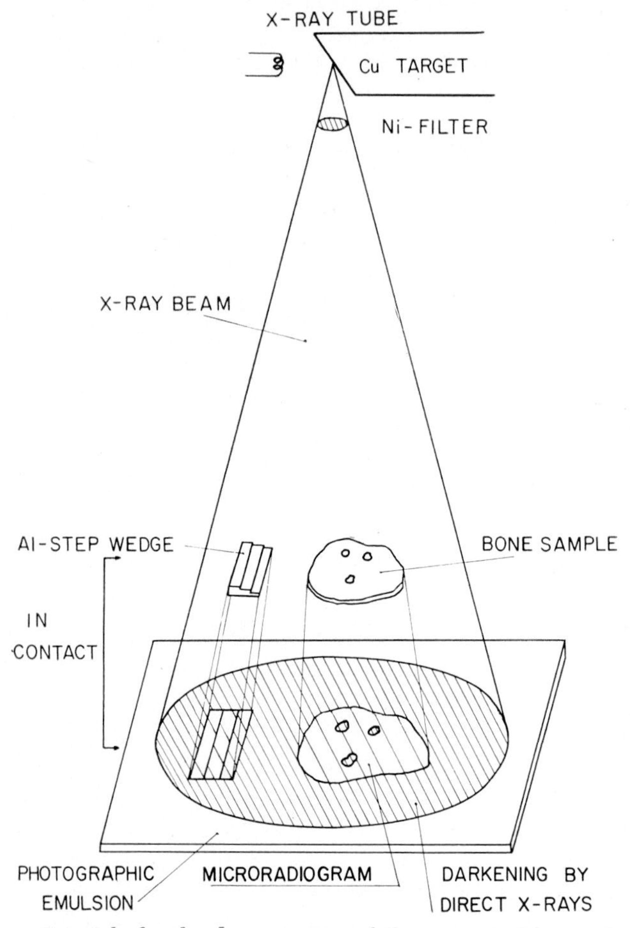

FIG. 16. Principle for the determination of the amount of bone salts by X-rays using the radiation from an ordinary diffraction tube.

Table II). Then the mass can be calculated from the X-ray absorption according to Eq. 2, if only the linear portion of the density curve is utilized.

VIII. THE SPECIMEN

A. GENERAL

The preparation of biological specimens for subsequent cytochemical analyses naturally should utilize procedures which to the greatest possible extent preserve the morphological appearance and the original

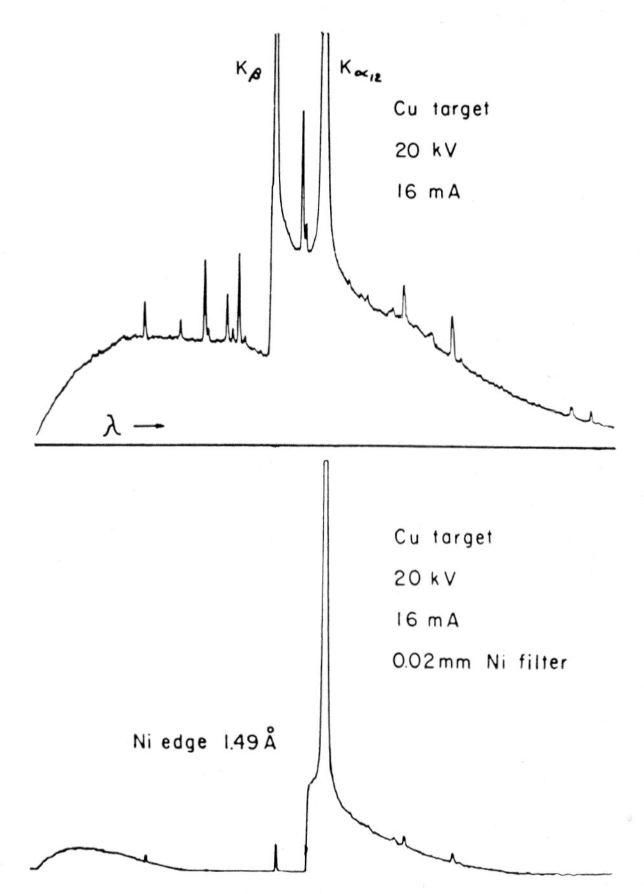

FIG. 17. Distribution of X-ray wavelengths from a copper target X-ray diffraction tube in the Philips PW 1010 X-ray diffraction unit, before and after filtration in Ni. The lower curve shows that good purification of the radiations can be obtained (Wallgren, 1957).

chemical composition of the fresh tissue. Therefore, frozen-sectioned and frozen-dried thin sections of composite tissues are the most appropriate samples for quantitative cytochemical analyses. Frozen-dried specimens and tissues fixed in different solutions, embedded in paraffin, and then sectioned and deparaffinized, can also be used. However, it must be remembered that the solvents, used at these procedures, may introduce severe morphological and chemical artifacts. Therefore, fixatives and treatments which result in removal or addition of compounds should be avoided to the highest possible extent.

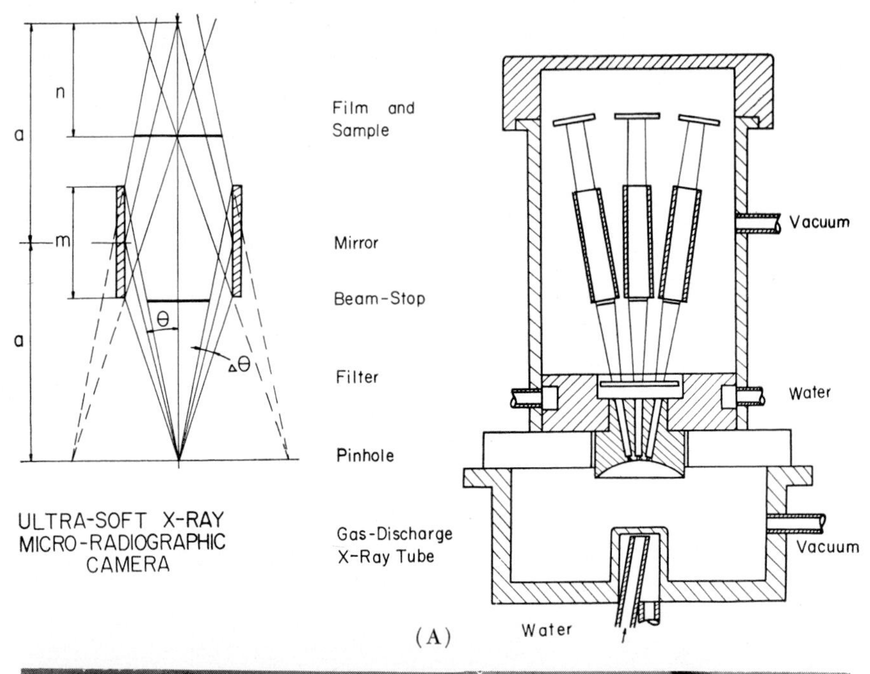

(A)

(B)

Fig. 18. A: Ultrasoft X-ray microradiographic camera. The radiation from the gas discharge X-ray tube is purified by filtration and total reflection (Henke, 1957). B: Mechanical design of the tube and camera shown in A (Henke, 1957).

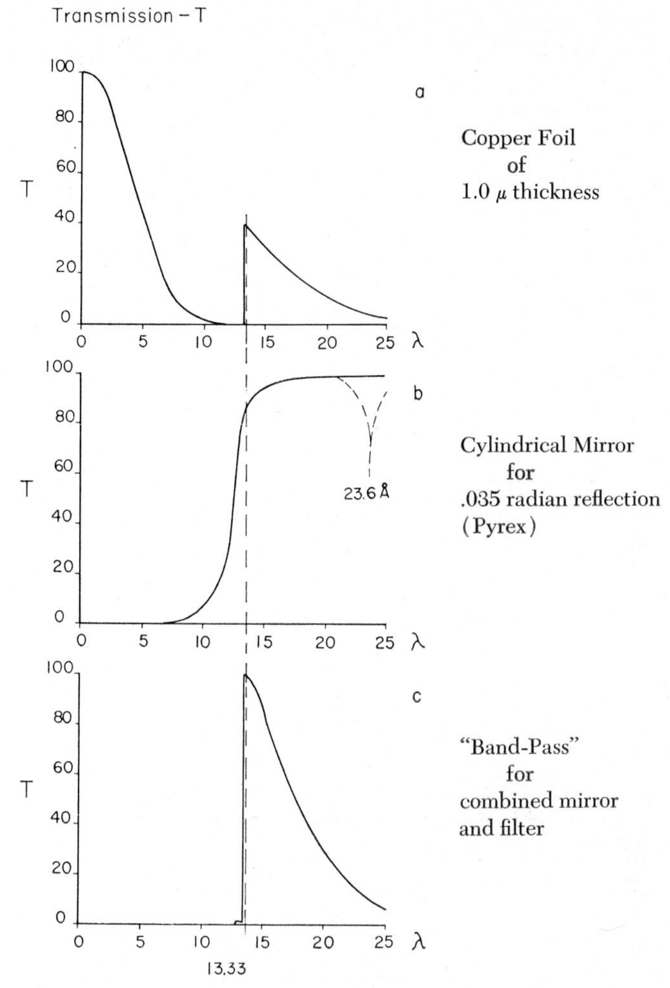

FIG. 19. The characteristics of the band pass filter in the camera in Fig. 18 when adjusted for Cu Lα radiation (Henke, 1957).

B. THE PRACTICAL PROCEDURE OF X-RAY WEIGHING WHEN UTILIZING REFERENCE SYSTEM

The specimen, usually a microtome section of a biological object, is mounted on a metal disc with a slit 0.5–1 by 8 mm., covered with a thin foil of collodion or similar material (see Fig. 4). The supporting foil is made by spreading a few drops of for example diluted, clean Zapon varnish solution on distilled water in a small glass container. Beforehand the metal discs are placed on a perforated metal plate at the bottom of this container. When the Zapon varnish foil on the water surface begins to wrinkle slightly, the plate with the discs is raised out of the water, and the foil is allowed to dry on the discs. In case of paraffin-embedded sections these are stretched on a small drop of cold, distilled water over one-half of the slit. The specimen is then allowed to dry for at least 6 hours at 37° C. and is then deparaffinized in chloroform for about 20 minutes. The reference system is placed over the other half of the slit. Smear preparations of individual cells on thin Parlodion foils on coverslips can either be frozen-dried directly on the supporting discs or on a thin supporting foil which is subsequently placed on the sample holder (Lindström, 1955).

Different methods have been devised to make the reference system and determine its weight per unit area (Engström and Lindström, 1950; Brattgård and Hydén, 1954; Clemmons, 1955). The procedure has been further refined by one of the authors (Lindström, 1955) and is

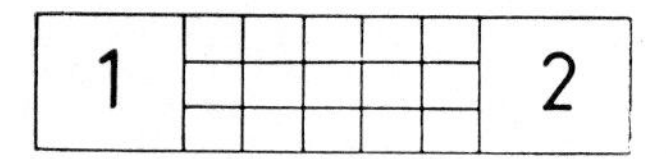

FIG. 20. Cutting of the reference system (Lindström, 1955).

described in the following. A carefully cleaned glass microscope slide is repetitively dipped in a solution of 2.5% Parlodion dissolved in amyl acetate and is allowed to dry in a vertical position, resting with the long edge against a filter paper. An even number of dippings is made, and the slide is allowed to rest alternately on each longside. After four dippings the weight of the foil is about 0.15 mg. • $cm.^{-2}$, an appropriate value for making up a reference system, when measuring 5 to 10 μ thick tissue sections. Figure 20 shows how the reference system is suitably divided on the slide. This is done with a small knife mounted on a special holder attached to a microscope. The Parlodion-covered slide is placed on the microscope stage, the movement of which is controlled by vernier readings. The cut pieces of the foil are floated free by immersing the slide in water, and the large pieces, marked 1 and 2 (area 0.90 $cm.^2$), are dried and weighed on a torsion balance. The small central pieces are used to compose the step wedge on the supporting disc. The supporting disc with sample is placed in intimate contact with the fine-grained photographic emulsion according to Fig. 4. With this procedure a great number of microradiograms of the same specimen can be recorded, for example, with different wavelengths and with no change of the internal relationships of the structures in the sample. The weight per unit area of the reference system can be determined with an accuracy of $\pm$ 2%.

C. Mounting of Specimen Directly on the Photographic Emulsion

Recently Greulich and Engström (1956) have introduced a new method for the mounting of specimens for microradiography. The fine-grained photographic emulsion is coated with a 500–1000 Å thick nitrocellulose layer. Free cells can be spread directly on this coated emulsion and microradiographed. Paraffin sections of the biological object are spread on water and floated onto the coated photographic emulsion. The specimen-film complex is then deparaffinized in benzene and dilute ethanol. After exposure the specimen is removed from the photographic emulsion by immersion and agitation in acetone. The sections become free and can be remounted on glass slides and stained according to routine methods. The photographic emulsion is developed and fixed in the usual way. This procedure has the advantage that the supporting foil mentioned in Section VIII B can be omitted and only the very thin coating on the photographic emulsion has to be penetrated by the ultrasoft X-rays (involving less absorption). Also by this technique the useful field can be larger, as the other field (see Section VIII B) is limited by the mechanical properties of the thin carrier foil.

IX. PHOTOMETRY OF THE MICRORADIOGRAM FROM A QUANTITATIVE POINT OF VIEW

When discussing the accuracy and resolution of a cytochemical microabsorptiometric technique two features must be clearly distinguished. The first one is the linear resolution of the method, indicating the smallest structure that can be adequately resolved from a qualitative point of view. The other is the "volume resolution," which is a measure of the smallest volume of the sample that can be analyzed *quantitatively* with a certain accuracy. In the following, the meaning of these two types of resolution will be briefly discussed, but these problems will be subjected to a more thorough analysis in a coming communication.

As mentioned before, the quantitative evaluation of the X-ray transmission data recorded in the contact microradiogram, is performed by microdensitometry. The outstanding advantage of the photometry of the microradiograms is the extremely small thickness of the X-ray image, which is in the order of 0.3 μ for an image produced by ultrasoft X-rays. This means that all of the absorbing material, in this case the silver grains in the developed microradiographic image, lies within the focal depths of even the strongest optical systems. When performing microphotometry by means of a light optical microscope on a cytological specimen it must be remembered that the beam passing through the sample has a biconical shape. An optical resolution of the order of

0.25 μ means that the numerical aperture of the optical system has to be slightly larger than 1. If the objective is focused in the middle of a 10 μ thick sample, the light beam passing through a point in the focal plane has a diameter of approximately the same magnitude as the thickness of the tissue section when entering and leaving the sample. If the linear resolution in a *microradiogram* is about 0.25 μ, quantitative microphotometry of this extremely thin X-ray image can be performed on areas as small as about 1 μ^2, corresponding to an X-ray "volume resolution" of about 10 μ^3 in a 10 μ thick section of tissue, as the incident X-rays are parallel. At direct light optical microphotometry of the same specimen the corresponding "volume resolution" is no less than about 250 μ^3 because of the biconical shape of the light beam even if its crossing point is infinitely small. Therefore, the "volume resolution" of the X-ray technique is considerably higher than that of direct optical microspectrophotometry of, e.g., a stained tissue section or specific light-absorbing substances. Even if the linear resolution of the microradiographic technique is only about 1 μ, the quantitative "volume resolution" of this method is considerably higher than for direct microspectrophotometry performed in an optical microscope with highest possible resolution for equal sample thickness. In this discussion the absorbing substance is assumed to be homogeneously distributed in the sample. However, the conditions become much more complicated at direct light optical photometry, when the absorbing material is inhomogeneously distributed. Thus the advantages of the X-ray absorption image obtained by means of contact microradiography are that the absorption of a narrow beam of parallel X-rays is measured, and that all the structures are projected into an extremely thin plane, suitable for accurate microdensitometry.

Various types of commercially available photometers can be used for the photometry of the microradiograms, and for the description of such instruments the reader is referred to the booklets issued by the various instrument manufacturers. The recently developed automatic scanning densitometer developed by Bourghardt *et al.* (1955) and Hydén and Larsson (1956), is of particular value for the microphotometry of microradiograms, if a great number of data must be collected. In this equipment the optical transmission values from closely situated small points in the microradiogram are recorded and printed on a paper. In this way it is possible to obtain a great number of precise transmission values within an optically enlarged projection of a microradiographic image of a biological structure. The photometric measurements can then be referred to particular tissue structures, and can also, if necessary, be summed up to yield the total dry weight of a cellular structure. Such a collection of printed transmission values for a single nerve cell is shown in Fig. 21. This type of densitometry thus combines accurate microphotometry with precise localization which is of greatest importance in quantitative cytochemistry.

It has been shown that the fine-grained photographic emulsions are sufficiently homogeneous to allow a precise densitometric evaluation in

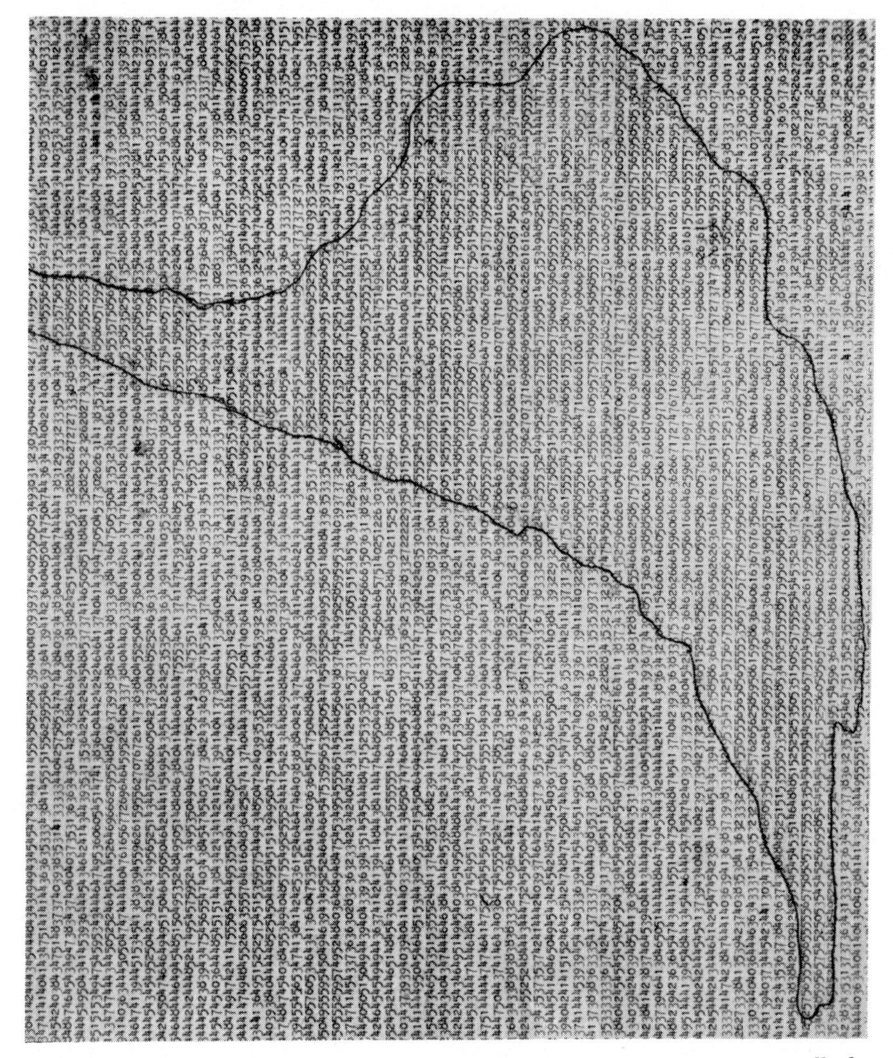

FIG. 21. Transmission data from a microradiogram of a single nerve cell obtained in the automatic scanning and printing densitometer (courtesy of Prof. H. Hydén).

extremely small areas. In a series of experiments Lindström (1955) found that the error was about 5% when measuring areas of the order of 1 μ^2. Recently Wallgren and Holmstrand (1957) tested the reproducibility of the microphotometric technique applied to microradiograms of bone tissue. It was possible to achieve a photometric accuracy of a few per cent, when determining the X-ray absorption characteristics on areas of the order of 900 μ^2, and under certain experimental conditions, the reproducibility could be kept as low as about 1%.

X. RESULTS OF THE WEIGHING OF CELLS BY X-RAYS

In this section only a few applications of the X-ray weighing technique will be given, and for details the reader is referred to the original papers and certain reviews (Engström, 1955a; 1956). The quantitative

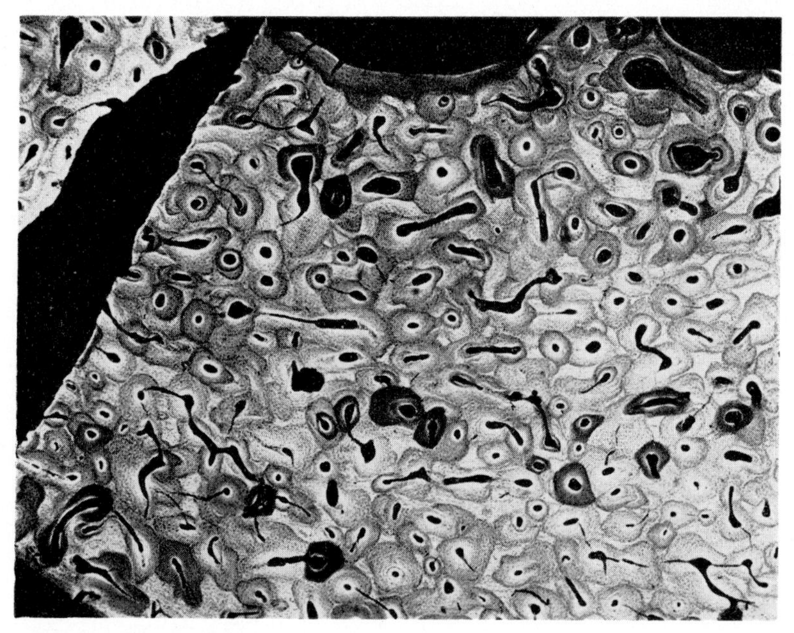

FIG. 22. Microradiogram of cross section from compact bone showing the distribution of mineral salts.

X-ray weighing technique has with particular advantage been applied to the study of the amount of mineral salts in normal bone tissue as well as the mineral fluctuations during a great number of physiological and pathological conditions, *inter alia,* Amprino and Engström (1952), Bergman and Engfeldt (1954), Carlström and Engström (1956). Figure 22 is a microradiogram taken with 1.54 Å X-rays and showing the distribution of mineral salts in the various structures of a cross section of compact adult bone.

Ultrasoft X-ray microradiography has been used to study among other things the distribution of dry substance within nerve fibers (Eng-

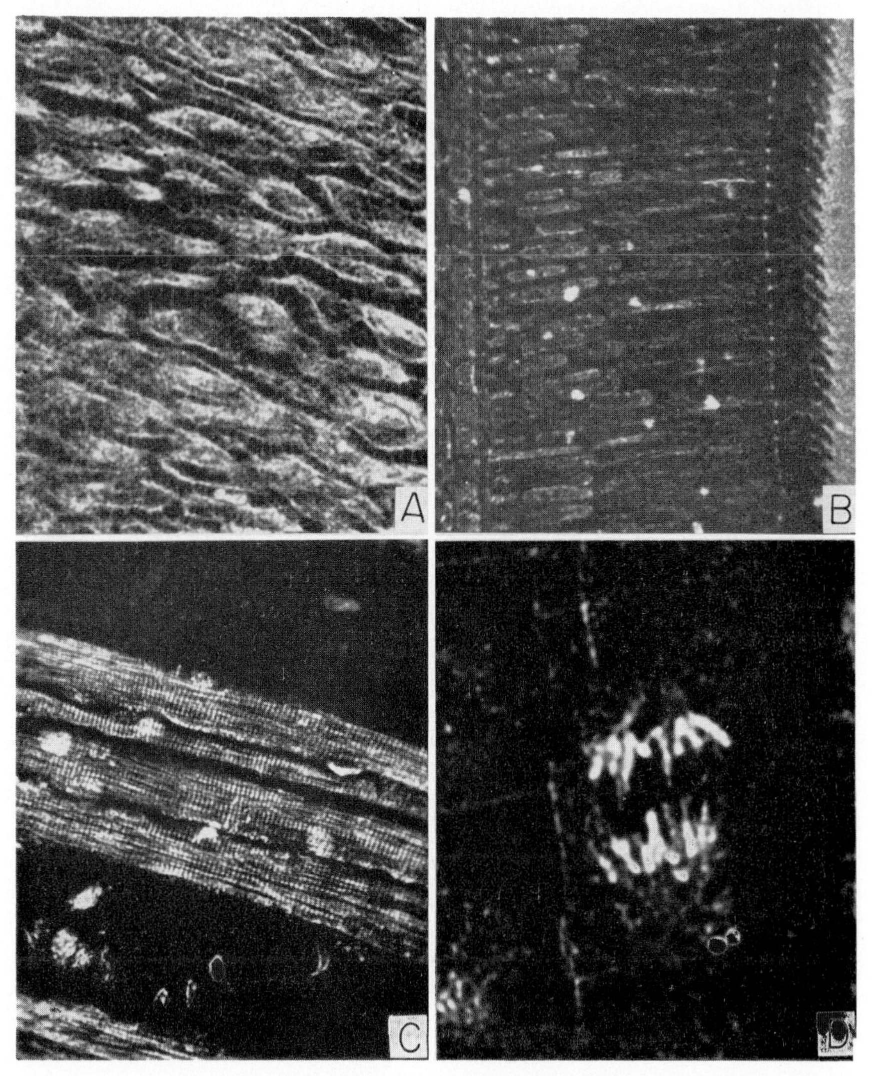

FIG. 23. Microradiograms of various soft tissues showing the distribution of mass in sections of, A: skin (note the tonofibrils), magnification: ×900; B: decalcified tooth, magnification: ×720; C: striated rat muscle, magnification: ×880; and D: chromosomes in mitosis (*Allium cepa*), magnification: ×1300 (Engström *et al.*, 1957).

ström and Lüthy, 1950), the weight of gastric mucosa cells (Engström and Glick, 1950), ascites tumor cells (Lindström, 1955), the amount of organic substance in developing bone and teeth (Greulich, 1957), nerve cells (Nürnberger *et al.*, 1952; Hydén, 1955), cartilage (Clemmons, 1955), and many other types of cells and tissues. The set of photographs in Fig. 23 may illustrate the possibilities of the X-ray weighing technique. Figure 24 shows a high-resolution microradiogram of the cells in a section of root tip from *Allium cepa.*

For the quantitative evaluation of the total amount of dry substance in ascites tumor cells, Lindström (1955) obtained an accuracy of ± 5%,

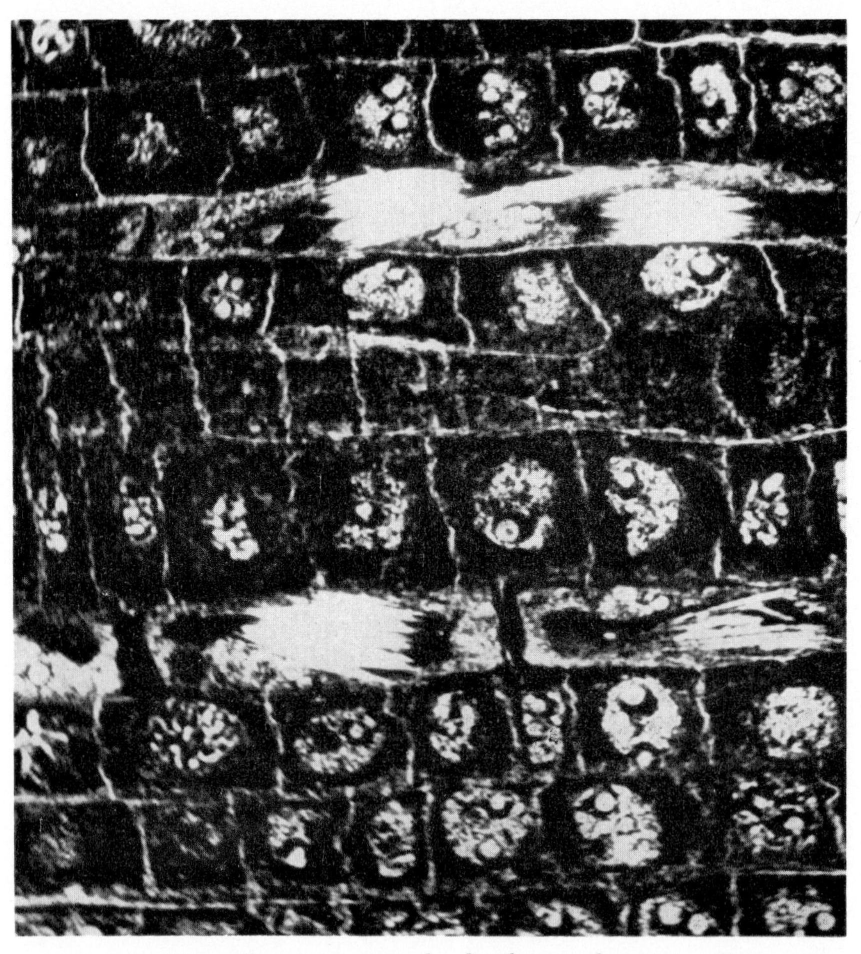

FIG. 24. Microradiogram showing the distribution of mass in cells in various stages of mitosis (*Allium cepa*), magnification: ×700 (Engström and Lundberg, 1957).

and the weights obtained were in good agreement with the results obtained on homogeneous bulk samples analyzed by straight forward microchemical techniques. A more thorough review of the applications of the X-ray weighing technique is beyond the scope of this presentation, which is to deal mainly with the technical aspects of cytochemical techniques, and therefore the discussion of the application has been considerably restricted.

XI. GENERAL DISCUSSION OF THE VALUE OF X-RAY WEIGHING OF CELLS AND TISSUES

The direct contact microradiographic technique reported in this survey has a linear resolution limited by the resolving power of the optical system utilized to inspect the microradiograms. It is not too unrealistic to believe that the new developments in X-ray microscopy now in progress, i.e., point focus X-ray microscopes and ellipsoidal mirror systems working with total reflection, in the future will provide us with X-ray microscopes having a linear resolution in the order of 500 Å. At present, however, these latter two types of X-ray microscopy utilize X-rays somewhat harder than those suitable for imaging thin biological samples, but on the other hand it can be expected that the mirror microscope will work better the softer are the X-rays.

The ultrasoft X-ray weighing technique has the unique feature that it can be methodologically applied in such a way that it gives the same relative accuracy for various amounts of cellular mass. This is accomplished in such a way that, when a very thin specimen with a low amount of dry mass per unit area is microradiographed, the voltage is lowered to give the same contrast in the image of the specimen, as when a thick specimen with high mass is examined (higher voltage). The possibility of taking X-ray absorption images of frozen-sectioned and frozen-dried biological specimens gives an opportunity to record the absolute amount of dry matter present in the tissue, as the specimen has not been subjected to any kind of solvent present in common fixatives, besides the fact that deparaffinizing is avoided. It appears that the X-ray method, combined with this preparation technique, is one of the very few histochemical methods capable of recording total mass with no loss, due to solvents, whatsoever.

The disadvantage of the X-ray method in comparison with the microinterferometric weighing technique is the difficulty in investigating average living cells, as the specimen must be contained within the high vacuum in the X-ray tube. Furthermore, water, the major component in soft tissues, has a slightly higher X-ray absorption than the common organic substances, which obscures the X-ray contrast of organic material in a fresh specimen. On the other hand the theoretical basis of quantitative microinterferometry of high resolution is not yet fully clarified.

The weighing of cellular structures by means of monochromatic ultrasoft X-rays also yields information about the elementary composition of the sample (Engström, 1946; Lindström, 1955). The registration of microradiograms with X-rays of wavelengths lying on the top of the oxygen, nitrogen, and carbon absorption edges for example may give quantitative information of the amount of these elements. This technique has not yet been worked out in detail, but it seems possible to arrive at a method capable of determining the ratio between carbon, nitrogen, and oxygen, as well as the total dry weight of the sample together with its water contents. Besides the possibility of weighing soft tissues, it is relatively easy to determine other major components in tissues, for example, the amount of apatites in sections of bone and teeth.

The versatility of X-ray microanalytical techniques goes beyond their use for various types of microradiography. It ought to be emphasized in this discussion that X-ray microdiffraction procedures have recently been developed to an extent where it is possible to register the diffraction patterns of cellular details.

This technique will certainly find many applications, when studying the crystallographic structure and also the chemical composition of biological specimens. The X-ray microfluorescence method, i.e., the use of a very fine X-ray or electron beam to excite the emission of characteristic X-rays in the specimen, has a great sensitivity. This technique will become a highly welcomed development in addition to the X-ray absorption method for elementary analysis, because when determining elements by X-ray absorption, the various elements must have a minimum concentration per unit area. In general, the concentration of element must be 1% or more in thin tissue sections, in order to produce any measureable difference of the transmitted X-ray intensities on both sides of the absorption discontinuity. However, in order to detect elements by X-ray microfluorescence procedures, the detectable concentration is much smaller, and therefore this method will be most valuable for the analysis of various types of trace elements within biological tissues.

The use of soft X-rays in cytochemistry has given us much new information of the composition of cells and tissues in various stages of development and function, but these methods will most probably be technically further advanced. In the future, X-ray absorption, X-ray emission, and X-ray diffraction procedures may be applied to the analysis of small (submicroscopic) cellular structures to determine their dry mass, water content, elementary composition, and content of defined biochemical substances, as well as their ultrastructural organization.

REFERENCES

Amprino, R., and Engström, A. (1952). *Acta Anat.* **15,** 1.
Bergman, G., and Engfeldt, B. (1954). *Acta Odontol. Scand.* **12,** 133.
Bourghardt, S., Hydén, H., and Nyquist, B. (1955). *J. Sci. Instr.* **32,** 186.
Brattgård, S. O., and Hydén, H. (1954). *Intern. Rev. Cytol.* **3,** 455.
Carlström, D., and Engström, A. (1956). *In* "Biochemistry and Physiology of Bone" (G. Bourne, ed.), p. 149. Academic Press, New York.
Clemmons, J. J. (1955). *Biochim. et Biophys. Acta* **17,** 297.
Engström, A. (1946). *Acta Radiol. Suppl.* **63.**
Engström, A. (1955a). *In* "Analytical Cytology" (R. Mellors, ed.), Chapter 8, McGraw-Hill, New York.
Engström, A. (1955b). *Medica Mundi* **1,** 119.
Engström A. (1956). *In* "Physical Techniques in Biological Research" (G. Oster and A. W. Pollister, eds.), Vol. III, p. 489. Academic Press, New York.
Engström, A., and Glick, D. (1950). *Science* **111,** 379.
Engström, A., and Glick, D. (1956). *Science* **124,** 27.
Engström, A., and Lindström, B. (1950). *Biochim. et Biophys. Acta* **4,** 351.
Engström, A., and Lundberg, B. (1957). *Exptl. Cell Research* **12,** 198.
Engström, A., and Lüthy, H. (1950). *Exptl. Cell Research* **1,** 81.
Engström, A., Greulich, R., Henke, B. L., and Lundberg, B. (1957). *In* "X-Ray Microscopy and Microradiography" (V. E. Cosslett, A. Engström, and H. H. Pattee, eds.), p. 218. Academic Press, New York.
Fitzgerald, P. (1957). *In* "X-ray Microscopy and Microradiography" (V. E. Cosslett, A. Engström, and H. H. Pattee, eds.), p. 49. Academic Press, New York.
Greulich, R. (1957). Personal communication.
Greulich, R., and Engström, A. (1956). *Exptl. Cell Research* **10,** 251.
Hallén, O. (1956). *Acta Anat. Suppl.* **26.**
Henke, B. L. (1957). *In* "X-Ray Microscopy and Microradiography" (V. E. Cosslett, A. Engström, and H. H. Pattee, eds.), p. 72. Academic Press, New York.
Henke, B. L., Lundberg, B., and Engström, A. (1957). *In* "X-Ray Microscopy and Microradiography" (V. E. Cosslett, A. Engström, and H. H. Pattee, eds.), p. 240. Academic Press, New York.
Henke, B. L., White, R., and Lundberg, B. (1956). *J. Appl. Phys.* **28,** 98.
Hydén, H. (1955). *In* "Biochemistry of the Developing Nervous System" (H. Waelsch, ed.), p. 358. Academic Press, New York.
Hydén, H., and Larsson, S. (1956). *J. Neurochem.* **1,** 134.
Lange, P., and Engström, A. (1954). *Lab. Invest.* **3,** 116.
Lindström, B. (1955). *Acta Radiol. Suppl.* **125.**
Nürnberger, J., Engström, A., and Lindström, B. (1952). *J. Cellular Comp. Physiol.* **39,** 215.
Wallgren, G. (1957). *In* "X-Ray Microscopy and Microradiography" (V. E. Cosslett, A. Engström, and H. H. Pattee, eds.), p. 448. Academic Press, New York.
Wallgren, G., and Holmstrand, K. (1957). *Exptl. Cell Research* **12,** 188.

An authoritative and complete account of the various types of X-ray microscopy including several types of applications are found in the proceedings of "Symposium on X-Ray Microscopy and Microradiography" held in Cambridge, England, 1956 and edited by V. E. Cosslett, A. Engström and H. H. Pattee, Academic Press, New York, 1957.

THE DETERMINATION OF MASS AND CONCENTRATION BY MICROSCOPE INTERFEROMETRY

By HOWARD G. DAVIES

Medical Research Council, Biophysics Research Unit, King's College, London, England

I. INTRODUCTION

Mass is a fundamental parameter in biology just as it is in physics and chemistry. The interferometer (interference) microscope method for measuring the mass of single cells and other tissue elements has an *empirical basis.* Stated simply, it depends only on the fact that, equal masses of any of the chemical substances of biological importance when dissolved in water to the same final volume of solution give rise to approximately the same increase in refractive index (Davies and Wilkins, 1951, 1952). What is measured is the mass of substance other than water, the so-called dry mass. A similar method was also proposed independently by R. Barer (1952). Determination of concentration of dry substance and thickness of cells can also be made. The essential physical measurements are made with an interference microscope.

The method is simple to apply but it has, in common with other cytochemical techniques certain limitations, and there are several ways in which erroneous results can be produced. The method finds increasing application, alone, and in combination with the techniques for measuring specific fractions of the total mass. To indicate the scope results to date will be summarized. A proper understanding of microscope interferometry requires an appreciation of the physical optics of light interference; it was thought that biologists might find it helpful if the relevant facts were briefly outlined and this has been done in one of the appendices to this chapter.

II. THEORY OF THE METHOD

A. Relationship between Dry Mass, Concentration, and Optical Path Difference

As a simple model for the cell we consider a gel or solution confined to a plate of thickness t(cm.) and area A(cm.^{2}) (Fig. 1). The plate has a refractive index μ_c and is supposed situated in water of refractive index

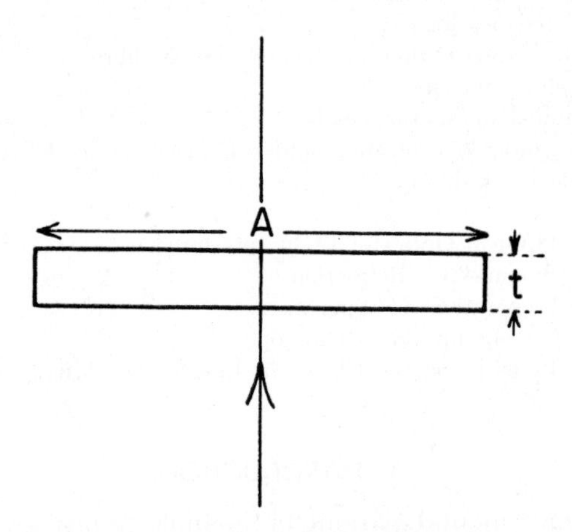

Fig. 1. Cell model, a homogeneous plate of thickness t and area A.

μ_w. If M(gm.) is the mass of substance other than water contained in the plate (i.e., the so-called *dry mass*), then the concentration c (gm. cc.$^{-1}$) is given by

$$c = \frac{M}{At} \tag{1}$$

The difference between the refractive index of the plate (μ_c) and that of water (μ_w) divided by the concentration of dry substances is, by definition, the specific refraction (or refractive) increment α, i.e.,

$$= \frac{\mu_c - \mu_w}{100c} \tag{2}$$

Eliminating c from Eqs. 1 and 2 we obtain

$$M = (\mu_c - \mu_w)t \frac{A}{100\alpha} \text{ gm.} \tag{3}$$

$$= \frac{\Delta_w A}{100\alpha} \text{ gm.} \tag{4}$$

where Δ_w is equal to $(\mu_c - \mu_w)t$(cm.) and will be recognized as the optical path difference (o.p.d.) due to the plate for light rays parallel to the optic axis (Fig. 1). In our previous publications (e.g., Davies and Wilkins, 1952) the symbol ϕ was used for the o.p.d. (Δ) in cm. ϕ now denotes the angular measure of o.p.d., i.e., the phase difference in radians or degrees ($\phi = 2\pi/\lambda . \Delta$; cf. Appendix 1).

It is also useful to deduce a formula which applies not only to solutions or gels but also to substances which contain no water, e.g., the lipid droplets or crystalline deposits which may occur in cells or tissues. It may be easily shown by considering a plate of substance (water-free) of refractive index μ_s and density ρ (gm. cc.$^{-1}$) that its mass M(gm.) is given by

$$M = \frac{\Delta_w A}{(\mu_s - \mu_w)/\rho} \text{ gm.} \tag{5}$$

Hence, from Eqs. 4 and 5, we may write, in general

$$M = \frac{\Delta_w A}{\chi} \text{ gm.} \tag{6}$$

where χ for solutions or gels is 100α and for water-free solids is $(\mu_s - \mu_w)/\rho$.

Equation 6 is the basic formula which we need. It shows that the optical path difference at a point is proportional to *mass per unit area* of dry substance. *Total dry mass* may be calculated from measurements of area and o.p.d. provided χ is known. Optical path difference is determined by microscope interferometry. No measurement of cell thickness is required. If the specimen thickness is known then the *concentration* of dry substance (gm. cc.$^{-1}$) can be calculated from the equation:

$$c = \frac{\Delta_w}{t\chi} \text{ gm. cc.}^{-1} \tag{7}$$

Cell thickness may be difficult to determine unless the cell is for example spherical. Methods for measuring cell thickness by interferometry are discussed in Section II, C and D.

B. The Parameter χ

a. χ for Biochemical Substances.

Table I which is discussed below lists values of χ for a variety of substances of biological importance. The two important points which emerge are (a) χ does not vary very much for the different chemical substances which may be found in cells, and (b) χ is approximately

TABLE I
VALUE OF χ FOR BIOLOGICALLY IMPORTANT MOLECULES

Compound		Physical State	Density	Refractive Index	λ mμ	χ
Class	Example					
Unconjugated Proteins	Ovalbumin [a]	1.61% sol.				0.187
	Ovalbumin [a]	6.45% sol.				0.188
	Bovine plasma albumin [b]	0–50 gm./100 cc. solution				0.182
	Gelatin [c]	Dil. solution				0.18
	Gelatin [d]	Dry solid	1.27	1.525 av.		0.151
Amino acids	Glycine [e]	Dil. solution				0.179
	Alanine [e]	Dil. solution				0.171
	Valine [e]	Dil. solution				0.175
	Tryptophan [e]	Dil. solution				0.25
Nucleic acids	DNA [f]	Dil. solution				0.16
	DNA [g]	Dil. solution				0.20
	DNA [h]	Dil. solution			436	0.20
	DNA [i]	Dil. solution			546	0.175
	DNA [i]	Dil. solution			436	0.181
	DNA [j]	Dil. solution			436	0.188
	RNA [k]	Dil. solution			436	0.194
	RNA [l]	Dil. solution				0.168
Carbohydrates	Starch [d]	Solid	1.50	1.53		0.133
	Sucrose [d]	2% solution				0.141
	Sucrose [d]	Crystalline	1.588	1.558		0.141
Lipid	Fats [d]	Natural	0.93 av.	1.46 av.		0.14 av.
Salts	Sodium chloride [d]	5.25% solution				0.163
	Sodium chloride [d]	Crystalline	2.165	1.544		0.097
	Calcium chloride [a]	1.7% solution				0.21
	Calcium chloride [a]	Crystalline	2.512	1.52		0.075
	Potassium chloride [a]	10% solution				0.115
	Potassium chloride [a]	Crystalline	1.98	1.490		0.079

Conjugated Proteins	β_1-lipoprotein [m]	Dil. solution			0.17
	Hemoglobin [n]	Dil. solution			0.193
	Turnip yellow mosaic virus [o]	Dil. solution			0.171
	Tobacco mosaic virus [p]	Dil. solution			0.17
	Tobacco mosaic virus	Dried gel	1.335 av. [q]	1.534 av. [r]	0.151

[a] Perlman and Longsworth (1948).
[b] Barer and Tkaczyk (1954).
[c] Kenchengton, in Barer and Joseph (1954).
[d] Handbook of Chemistry and Physics.
[e] Adair and Robinson (1930).
[f] Tennent and Villbrandt (1943).
[g] Vincent, in Barer (1952).
[h] Northrop *et al.*, (1953).
[i] Brown *et al.* (1955).
[j] Reichmann *et al.* (1954).
[k] Northrop and Sinsheimer (1954).
[l] Davies and Wilkins (unpublished).
[m] Armstrong *et al.* (1947).
[n] Adair *et al.* (1946).
[o] Markham (private communication).
[p] Oster (private communication).
[q] Bawden and Pirie (1937).
[r] Bernal and Fankuchen (1941).

From Davies and Wilkins (1952) and Davies *et al.* (1954b), slightly modified.

independent of concentration over a certain range of values which depends on the nature of the substance.

χ *for nonconjugated proteins.* The term specific refraction (refractive) increment (α) was first used in connection with dilute solutions of protein by Reiss (1903) (see also Adair and Robinson, 1930). Its determination has been of practical importance. Thus if α is known then a rapid determination of protein concentration can be made from the measurement of refractive index of solution and solvent (by Eq. 2). The parameter α also occurs in the light-scattering formulae for determining the relative molecular weight of large molecules.

Partly as a result of these uses there exists extensive data on the values of α for dilute solutions ($<$ about 0.10 gm. cc.$^{-1}$) of a variety of proteins. Data on the unconjugated proteins shows that α lies between about 0.0018 and 0.0019. ($\chi = 0.18 \rightarrow 0.19$). The values for ovalbumin are shown as an example in Table I, the variation with concentration in the range investigated being negligibly small. It is also generally stated in the literature that α varies only slightly with temperature and pH. The dispersion effect, that is the dependence on wavelength, is also small in the visible region of the spectrum. Thus χ for serum albumin varies from 0.190 in violet light (436 mμ) to 0.183 in yellow light (580 mμ) (quoted in McFarlane, 1935). These points are dealt with at length in the review by Barer and Joseph (1954). Barer and Tkaczyk (1954) have also shown experimentally that the value of α for solutions of bovine serum albumin remains constant up to concentrations of 50 gm. per 100 cc. Measurements at higher concentrations could not be made due to difficulties in dissolving the protein.

χ *for nonproteins.* Values of χ for dilute solutions of deoxyribose nucleic acid ranging from 0.16 to 0.20 have been reported by different workers. Using the dispersion data of Brown *et al.* (1955) to correct the experimental values at 436 mμ given by Northrop *et al.* (1953) and Reichman *et al.* (1954) the average value of χ for DNA at 546 mμ is 0.187, neglecting the earlier value of Tennent and Villbrandt (1943). The value of 0.194 at 436 mμ (Northrop and Sinsheimer, 1954) for ribonucleic acid is judged to be more reliable than the value of 0.168 at 546 mμ given by Davies and Wilkins (1952). Applying the dispersion data for DNA the value of χ for RNA becomes 0.188 at 546 mμ.

Values of χ given in Table I for salts, lipids (~ 0.14), and carbohydrates (~ 0.14) are on the whole somewhat lower than those found for the proteins and nucleic acids. Salts (electrolytes) show a marked variation of χ with concentration, while sugars (nonelectrolytes) exhibit no variation up to the crystalline state. This interesting difference between charged and uncharged molecules is depicted in Fig. 2. The values of χ for amino acids which exist in solution as dipolar ions also decrease with increasing concentration. Values of χ for amino acids are also somewhat lower than for the proteins, with the exception of tryptophan. Such an increased refraction is found to occur in molecules which contain unsaturated bonds.

χ *for conjugated proteins.* Values of χ for the conjugated proteins partly depend, as might be expected, on the nature of the prosthetic group. Thus the value of χ for β_1-lipoprotein, a plasma protein contain-

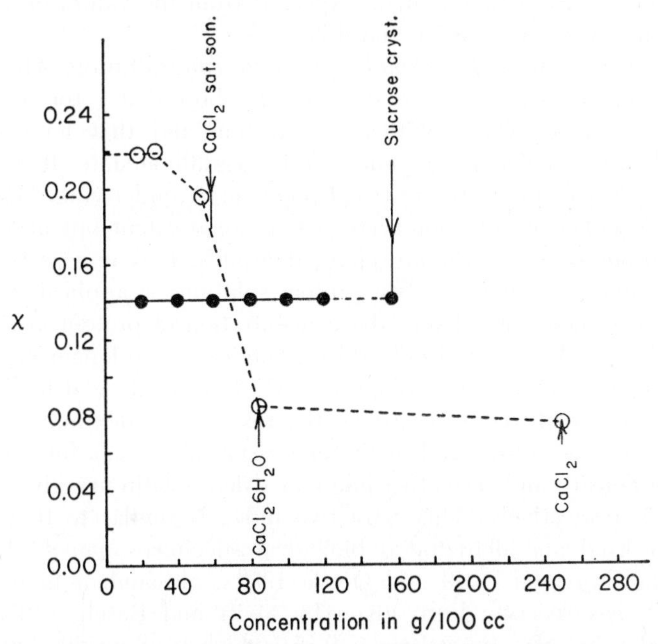

FIG. 2. Indicates the variation of χ with concentration (gm./100 cc.) in solid state and aqueous solutions; calcium chloride and sucrose.

ing about 75% lipid, is given by Armstrong *et al.* (1947) as 0.17, a value lower than the average value for the proteins. Hemoglobin in dilute solution has the slightly higher value 0.193 (Adair *et al.*, 1946). Dilute solutions of tobacco mosaic virus and turnip yellow mosaic virus which contain 5% and 40% RNA, respectively, have a value of 0.17 for χ (Oster and Markham, private communication).

The data on χ for chemical combinations of two substances raises an interesting point. It is to be expected that if x_1 and x_2 are the mass fractions of substances with χ values of χ_1 and χ_2 that χ_{av} for the combination will be given by the weight average, i.e.,

$$\chi_{av} = x_1\chi_1 + x_2\chi_2 \tag{8}$$

This formula may be expected to hold provided that the substances do not interact so as to alter χ for the individual components. An evaluation of the data on α- and β-lipoproteins in Armstrong *et al.* (1947) indicates an approximate additivity. The value of χ for protein-free plasma lipid may be deduced from data on refractive index given by Armstrong *et al.* (1947). The refractive index of plasma lipids is apparently somewhat higher than that of the natural fats listed in Table I. However the data on dilute solutions of virus shows a somewhat different value for χ than might be expected from the values of χ for the constituents, namely protein and RNA.

(i) *Remarks on χ for proteins at high concentration.* The experimental data given above (see also Fig. 2) show that χ for solution of charged molecules varies with concentration and that large charges occur when the molecules assume a solid crystalline state. It would not be entirely surprising if the physical phenomena underlying these variations of χ with concentration were not to some extent operating in the case of proteins which contain charged groups. It is unwise to assume that the value of χ measured in protein solutions is applicable to fixed cells. In fixed cells and tissues the concentration of protein may greatly exceed 50 gm./100 cc. and indeed the proteins are no longer in solution.

In attempts (Davies and Wilkins, 1952) to estimate χ at high protein concentration, such data as exist on the refractive index and density of dry protein were substituted in the expression $\mu_s - \mu_w/\rho$ for χ (Eq. 5). Thus the density and refractive index of dried gelatin are given as 1.27 and 1.525, respectively. This refractive index is similar to that usually given for fixed and dehydrated biological specimens namely 1.53–1.54 (see, e.g., Crossmon, 1949 and Oettlé 1950), although a higher value 1.56–1.572 has recently been given by Swift and Rasch, (1956). The calculated value of χ for gelatin is 0.151 which is about 20% lower than that for dilute gelatin solution. Calculation based on the refractive index and density of dried tobacco mosaic virus (Table I) give the value of χ as 0.151, compared to the value of 0.17 for a dilute solution. However according to Dr. A. Klug (private communication) this dried virus was not completely water free and hence the calculated value should be considered as a lower limit. For a similar reason we may regard the value for gelatin as a lower limit. The presence of small amounts of bound water may possibly account for the range in refractive index values for dried cells given above.

Another method for estimating χ was to use the value for the refractive index of water-free protein calculated with the Lorentz-Lorenz mixture rule (see, e.g., Armstrong *et al.* 1947, also Appendix 2) from experimental data on the specific refraction increment ($\sim$ 0.00185) and partial specific volume ($\sim$ 0.74) for dilute protein solutions. A value of about 1.60 is obtained for most proteins (Armstrong *et al.* 1947; Putzuys and Brosteaux, 1936). The density of dry protein was taken as either 1.27 or 1.35. The lower value is that obtained by direct experiment on dried proteins (Chick and Martin, 1913). The upper value is the reciprocal of the partial specific volume obtained from experiments on dilute solutions of protein (see, e.g., Cohn and Edsall, 1943). (For a discussion of the significance of this difference see Cohn and Edsall, 1943 and Haurowitz, 1950). Combining these data in the formula $(\mu_s - \mu_w)/\rho$ (cf. Eq. 5) we obtain χ as $0.21 \rightarrow 0.19$. This value is somewhat higher than that obtained experimentally for dilute solutions. These estimates of χ must also be regarded as approximate since there is no theoretical or experimental justification for assuming that the Lorentz-Lorenz mixture rule is valid over the extreme range of concentrations from dilute solution to pure solute (Böttcher, 1952).

To summarize, estimates of χ for dry protein indicate that it is not grossly different from χ for dilute solutions of protein. Calculations based partly on the Lorentz-Lorenz mixture rule indicate that χ is somewhat higher, while calculations based on experimental data on dried protein indicate that χ is about 20% lower than χ for dilute protein solutions.

b. χ for Cells and Other Tissue Elements.

Many cells are composed mainly of protein and hence the appropriate choice of χ will be that for protein. In living cells the concentration of protein is often less than about 50 gm./100 cc. (a summary in Barer, 1956) and hence χ will be about 0.185. However in fixed cells the concentration of protein may be higher and for the reasons discussed in the last section there is uncertainty in the value of χ. The writer has chosen to use the value of 0.18 for χ in calculating the dry mass of both living and fixed cells. The error in dry mass of living cells will be small, but there may be a systematic error in the calculated dry mass of fixed cells. According to the calculation of χ based on experimental data on dry protein the error should be less than about 20%, the calculated mass being too low. When comparative measurements of mass of fixed cells are required the systematic error may be ignored.

Due to the assumption of a value of 0.18 for χ in cells, error due to the presence of DNA (0.187 average in dilute solution) and RNA (0.168–0.188 in dilute solution) will be small. When appreciable quantities of carbohydrates or lipids are present, the error will increase, but often the composition of the cell or cell component will be roughly known, and the value of χ can be selected accordingly (cf. Eq. 8). Thus lipids, for example, often occur as recognizable droplets in the cell and allowance can be made for them if a measurement of the total cell mass is required. When measurements of inorganic components of cells and tissues are being made the value of χ must be chosen accordingly. For example, for the inorganic component of bone, hydroxyapatite, χ is about 0.1, calculated from the data on refractive index and density (Davies and Engström, 1954).

C. Measurement in Aqueous Media of Refractive Index Higher than Water

There are several reasons for investigating the possibility of making measurements of o.p.d. in aqueous media other than water. For example, when living cells are measured in physiological media (dilute solutions of salts and solutions of protein) Eq. 6 must be modified. Except in the case of spherical cells, cell thickness is notoriously difficult to measure accurately. By purposely varying the refractive index of the mounting medium and measuring the o.p.d. of the cell, calculations can be made of thickness and concentration of dry substances in cells. An important problem when making the latter measurements on living cells is the possibility of concomitant changes in cell volume, and, or total dry mass due to diffusion of substances into and out of the cells.

1. *Living Cells*

a. Dry Mass in Physiological Media.

Living cells are often immersed in a physiological medium which has a refractive index greater than that of water on account of dissolved substances. The refractive index can easily be deduced from the concentration and value of χ for the dissolved substance. Such media include (a) Tyrode solution with a salt concentration of about 1% and refractive index about $1.333 + 0.0016$, i.e., 1.3346; (b) 5% sucrose with a refractive index of $1.333 + 5 \times 0.0014$, i.e., 1.340; (c) the liquid exudate from blood serum and embryo extract sometimes used for growing animal tissue cultures, and containing dissolved protein and salts in varying amounts.

A formula can be deduced which relates Δ_m the optical path difference in the medium of refractive index μ_m with the dry mass M. If Δ_w is the optical path difference in water and the cell has a refractive index μ_c and thickness t then

$$\Delta_m = (\mu_c - \mu_m)t \tag{9}$$

and

$$\Delta_w = (\mu_c - \mu_w)t \tag{10}$$

$$= (\mu_c - \mu_m)t + (\mu_m - \mu_w)t \tag{11}$$

Substituting (9) in (11)

$$\Delta_w = \Delta_m + (\mu_m - \mu_w)t \tag{12}$$

Substituting for Δ_w from (12) in (6) we obtain

$$M = \frac{A\Delta_m}{\chi} + (\mu_m - \mu_w)\frac{At}{\chi} \tag{13}$$

The second term on the right in Eq. 13 shows that both the refractive index of the medium in which the cell is immersed, and also the cell thickness must be known. The refractive index μ_m can be measured accurately on a refractometer. In the case of flattened cells the thickness will not be known. However the second term will be small provided that the refractive index of the medium is close to that of water, and then rough estimates of cell thickness may give its magnitude with the desired accuracy. As an example (see Davies *et al.*, 1954b) the dry mass of the nucleus of a chick heart fibroblast in Tyrode solution, was 28×10^{-12} gm. neglecting the second term. The correction term was 1.5 to 3×10^{-12} gm. depending on whether the cell thickness was assumed to be 2 or 4 μ.

Whether or not an accurate measurement of thickness will be required must be decided by preliminary calculations. Clearly, however, as the refractive index of the chosen physiological medium increases there will be an increasing uncertainty in the value of M calculated from Eq. 13, unless thickness is accurately known. In measurements of high molecular weight substance, (protein, nucleic acid, etc.) in cells immersed in Tyrode solution it may be more accurate to neglect this correction term since it may represent the mass of low molecular weight material (salts, etc.) already present in the nucleus.

b. Concentration, Thickness and Dry Mass.

(i) *Matching the refractive index.* An examination of Eq. 9 shows that when the refractive index of the external medium is equal to that of the cell, the optical path difference becomes zero. This is the basis of the well-known method for measuring the refractive index of microscopic particles; by immersing them in a series of liquids of known refractive index, the particles disappear from the field of view when the refractive index of the medium is equal to that of the particles. Instead of the conventional Becke line method used by mineralogists, Oettlé (1950) used the phase contrast microscope for detecting the condition of zero o.p.d. He measured the refractive indices of dried blood cells, bacteria, and sperm, using a series of oily refractive index mixtures.

The application of the immersion method to measure the refractive index of living cells is due to the interesting observations of Barer and Ross (1952) who have described the use of solutions of protein, bovine plasma albumin, as immersion media. If μ_c is the refractive index of the living cell determined by matching, then the concentration c in gm. cc.$^{-1}$ is given by

$$c = \frac{\mu_c - \mu_w}{\chi} \tag{14}$$

A detailed description of the technique given by Barer and Joseph (1954, 1955a, b) includes methods for preparing isotonic protein solutions and obtaining statistical data on cell number versus cell concentration. A basic requirement of the immersion method is that the cell volume should remain constant. According to experiments by Ross (1953) the volumes of living spermatocytes of grasshopper remain constant in solutions of protein and salt if the proportions are 0.1% salt to 10% protein. In several applications Barer and co-workers have used the phase microscope to detect the condition of zero o.p.d. A useful discussion of the possible ambiguities arising on account of the mode of operation of the phase microscope has been given by Barer and Joseph (1955b).

The interference microscope can also, of course, be used to detect the position of zero o.p.d., and there is no possibility of ambiguity with this instrument. Points in favor of the use of the phase microscope include its more general availability, and the adjustment is somewhat easier.

In the method for measuring concentration by matching, no measurement of o.p.d. is required. Hence the method is well suited to the determination of concentration of small cells of uniform refractive index, but with irregular shape and indeterminate dimensions. The inner regions of nonuniform cells cannot be measured quantitatively by this method.

(ii) *Measurement of o.p.d. in two media.* It is possible to deduce values for the total mass, thickness, and concentration of dry substance in a cell from two measurements of optical path difference in media of different refractive index. If Δ_1 and Δ_2 are the o.p.d. in media of refractive index μ_1 and μ_2 then if μ_c is the refractive index of the cell

$$\Delta_1 = (\mu_c - \mu_1)t \tag{15}$$
$$\Delta_2 = (\mu_c - \mu_2)t \tag{16}$$

The two equations contain two unknowns, μ_c and t, and hence can be solved for them. We can also evaluate Δ_w and hence dry mass per unit area. The following equations can be derived simply.

$$\mu_c = \frac{\Delta_1\mu_2 - \Delta_2\mu_1}{\Delta_1 - \Delta_2} \tag{17}$$

$$t = \frac{\Delta_1 - \Delta_2}{\mu_2 - \mu_1} \tag{18}$$

$$M/A = \frac{\Delta_1}{\chi} \frac{\mu_2 - \mu_w}{\mu_2 - \mu_1} - \frac{\Delta_2}{\chi} \frac{\mu_1 - \mu_w}{\mu_2 - \mu_1} \tag{19}$$

$$c = \frac{\Delta_1}{\chi} \frac{\mu_2 - \mu_w}{\Delta_1 - \Delta_2} - \frac{\Delta_2}{\chi} \frac{\mu_1 - \mu_w}{\Delta_1 - \Delta_2} \tag{20}$$

In applying the method the accuracy is increased if the media are chosen with refractive indices which are well separated, for example, near that of water and near that of the cell.

2. *Fixed Cells*

a. Thickness and Concentration.

When fixed cells are measured in aqueous solutions it is necessary to enquire a little more closely into the meaning of cell thickness, refractive index, and concentration. We take as a model for the fixed cell a three-dimensional network of denatured substance. In many fixed cells this substance will be mainly nucleoprotein but in some tissues this will not be so. The substance within the cell membrane will be more or less clumped leaving microscopic and submicroscopic spaces. Let μ_p be the refractive index of this substance.

The meaning of μ_p is not as clear as it would be if the substance was, for example, a network of glass fibers. However μ_p may be interpreted here as follows. If measurements of o.p.d. are made in aqueous solutions of low molecular weight substance, then the molecules can be considered to penetrate the network; the refractive index μ_p is equal to that of the immersion solution in which the o.p.d. of the cell substance becomes zero.

In solutions of sucrose the substance in mammalian sperm heads ($\sim$ 60% protein, 40% DNA) has a refractive index of about 1.54, obtained by extrapolation (see Section VI, C). It should be noted that the refractive index of the fixed cell varies somewhat depending on what medium is being used. Thus sperm heads have a refractive index of about 1.56 in alcoholic media (Section II, G). Another point of interest is that these refractive indices, 1.54–1.56 obtained from immersion in aqueous solutions are very similar to those given for dried cells in oily media.

(i) *Thickness, geometrical and effective.* As a result of the above considerations we see that when measuring cell thickness, according to Eq. 18, if the molecules in the medium penetrate the cell, it is the effective thickness t_e which is being measured. The effective thickness is the geometrical thickness minus that occupied by the immersion medium.

On account of the similarity of μ_p of wet cells in aqueous media (sucrose solution) and dried cells in oily media, we may regard the effective thickness as referring approximately to the dry cell substance.

If measurements can be made in an aqueous solution in which the solute does not penetrate the fixed cell then the geometrical thickness t_g is measured. If F is the fraction of the geometric volume occupied by cell substance then (cf. Davies *et al.*, 1954b, and Section VI, C)

$$t_e = t_g F \tag{21}$$

The penetration of the molecules of the aqueous immersion medium into the geometrical volume of the fixed cell may not be an all or nothing process. Depending on the extent of penetration the value obtained for the thickness (from Eq. 18) will lie between the geometrical thickness (t_g) and the thickness of dry substance (cf. t_e).

(ii) *Concentration.* The meaning of concentration of dry substance when applied to living cells is quite clear. The meaning to be ascribed to concentration of dry substances in fixed cells will obviously depend on whether the mass is referred to the geometrical volume or effective volume. The meaning of dry mass per unit geometric volume is quite clear although the significance of such data on fixed cells may not be very great on account of possible changes in volume due to fixation. In practice the geometric thickness of fixed cells is not easy to determine since most media penetrate the cell.

When the concentration refers to the ratio of total mass to effective volume then what is measured is the density of the cell substance. On account of the similarity between refractive index of wet cells in aqueous media and dried cells in oil, the density refers to that of the dry substance. This is not of much interest since it is constant for all cells with the same composition, just as the refractive index μ_p is approximately constant and equal to 1.54. We have already remarked on difficulties in interpreting absolute values of mass in fixed cells on account of uncertainty in χ and similar considerations apply to concentration.

b. Dry Mass.

A useful relationship is that connecting the dry mass of the fixed cell with its refractive index μ_p and the refractive index μ_m of the medium which surrounds and permeates the cell. If Δ_w and Δ_m are the o.p.ds in water (μ_w) and medium (μ_m), respectively, then

$$\Delta_w = (\mu_p - \mu_w)t_e \tag{22}$$

$$\Delta_m = (\mu_p - \mu_m)t_e \tag{23}$$

By using Eq. 6

$$
\begin{aligned}
M &= \frac{\Delta_w A}{\chi} \\
&= \frac{A}{\chi}(\mu_p - \mu_w)t_e \frac{\mu_p - \mu_m}{\mu_p - \mu_m} \\
&= \frac{A\Delta_m}{\chi} \cdot \frac{\mu_p - \mu_w}{\mu_p - \mu_m} \qquad (24)
\end{aligned}
$$

D. The Bubble Method for Cell Thickness and Refractive Index

There is an ingenious method for measuring cell thickness and refractive index used by Ambrose (private communication). The method is applicable to objects (e.g., some living cells) which can be deformed so that their thickness is equal to that of an air bubble, artificially introduced, and located adjacent to the object. An advantage is that the physiological medium around the cell need not be changed. The pressure of coverslip onto cell must be controlled. If μ_a, μ_m, and μ_c are the refractive indices of air, medium, and cell, respectively, then if t is the cell thickness

$$\Delta_1 = (\mu_m - \mu_a)t \qquad (25)$$

$$\Delta_2 = (\mu_c - \mu_m)t \qquad (26)$$

Δ_1 and Δ_2 are determined by experiment. Since μ_a and μ_m will be known, Eq. 25 can be solved to give cell thickness t. Substituting for t in Eq. 26 we obtain the cell refractive index μ_c. Hence concentration and dry mass per unit area can also be calculated.

E. The Water Content of Living Cells

The reader interested in concentration of water, total wet mass and density, and concentrations expressed in terms of weight of wet tissue may refer to Barer and Joseph (1955b). Expressions have been deduced for these quantities in terms of the concentration of dry substance per 100 cc. of protoplasm. To illustrate their method of calculation expressions for the weight of water in the cell and density of the cell will be derived, in terms of the cell volume V (i.e., $A \cdot t$). The method for deducing water content assumes the cell to consist of protein. The calculation rests on the fact that the volume v occupied by 1 gm. of protein in solution (i.e., the partial specific volume) is about 0.75 cc.

If M is the mass of dry substance in the cell then the mass of water w (specific volume assumed to be unity) is given by

$$w = V - 0.75M \text{ gm.} \qquad (27)$$

The total mass W is given by

$$
\begin{aligned}
W &= V - 0.75M + M \\
&= V + 0.25M \text{ gm.}
\end{aligned}
\tag{28}
$$

Hence the density of wet cell is $(V + 0.25M)/V$; i.e., $1 + 0.25\ M/V$ gm. $cc.^{-1}$

F. Measurement of Specific Substances

The interference microscope method being basically nonspecific gives no indication of the nature of the substances measured. The composition of cells and tissues in terms of the major groups of chemical substances must be obtained from biochemical experiments and specific cytochemical tests. It is obvious, in principle, how to measure the mass of a particular substance. This can be done if the substance can be removed specifically, measurements of mass being made before and after removal. It is also clear that, in general this is a very tricky problem, since specificity of extraction for any one substance may vary from tissue to tissue and depend on nature of fixation, etc. The substance must also be present in amounts large enough for accurate measurement. The techniques, however, appear worth developing on account of the paucity of specific tests.

In the majority of soft tissues we may put total dry mass equal to total protein plus nonprotein where nonprotein consists chiefly of nucleic acid, lipid, carbohydrate, and a fraction 1-m. This last fraction 1-m refers to relatively low molecular weight substances which include salts, nucleotides, and amino acids which are present in variable and often unknown amounts in living cells. In osseous tissue for example the fractional content of salts, inorganic substance, is high.

a. Total Protein.

Measurement of total protein content of cells and tissues is an important aspect of microscope interferometry. For various reasons the method appears to be more reliable than either ultraviolet light absorption or available staining methods.

Many living cells and tissues consist predominantly of protein and hence protein content may be equated approximately to total dry mass. In order to increase the accuracy with which protein content may be estimated, living cells may be fixed so as to render the protein insoluble and then treated so as to remove nonprotein. Making use of the insolubility of proteins in absolute alcohol, initial fixation can be by alcohol freeze-substitution (Simpson, 1941; Davies, 1954), a procedure which avoids the localized increases in concentration caused by fixing in absolute alcohol at room temperature. Tissue may also be frozen-dried and fixed in alcohol or sometimes fixed in 8% trichloroacetic acid (TCA) as in the Schneider (1945) procedure for assay of nucleic acid.

It seems probable that during the initial fixation and subsequent immersion in water that most of the low molecular weight fraction will be removed from the cell. Fixation in alcohol will also remove some cell lipid and this can be further extracted in alcohol-ether. The extent to which proteins may be removed during these treatments is not known; according to Folch and Lees (1951) proteolipids, i.e., a new type of lipoprotein soluble in absolute alcohol, are present in small amounts in certain tissues. The material remaining will consist mainly of protein, nucleic acid, and carbohydrate. The total mass can be measured and the mass of nucleic acid determined by some specific technique (e.g., Feulgen staining or ultraviolet light absorption) subtracted to give total protein. This is a very useful procedure which provides data on the ratio of protein to nucleic acid. Alternatively the nucleic acid can be removed with hot trichloroacetic acid (Schneider, 1945) and the mass of the remainder, total protein, measured (Davies *et al.*, 1957).

The procedures will require modification depending on the tissue. When appreciable quantities of carbohydrate are present as in some plant cells, it may be difficult to accurately assess the mass fraction due to protein. In thin sections of bone tissue the total protein may be estimated after removal of inorganic substance (Davies and Engström, 1954).

It may sometimes be possible to determine certain of the major protein fractions of the total protein content of cells and tissues by combining measurements of mass with specific extraction procedures. An example is the assay of myosin in myofibrils (Huxley and Hanson, 1957).

b. Other Substances.

For an attempt at measuring nucleic acid content of sperm heads by mass measurements before and after extraction with TCA at 90° C. see Davies *et al.* (1957).

Measurement of lipid content of mammalian sperm heads are discussed in Section VI, C, e. The appropriate value of χ for lipid which is combination with protein can be deduced from the data of Armstrong *et al.* (1947) on plasma lipoproteins, together with Eq. 8. This yields a value of 0.165 for lipid.

Mineral salts in bone can be estimated by mass measurements before and after decalcification (Section VI, C, a, (ii)).

A useful principle for increasing the sensitivity of the method is to make use of increase in mass due to enzyme activity. This principle has been employed in determining a tissue enzyme, alkaline phosphatase (Section VI, C, d).

G. Measurements in Nonaqueous Media

It is useful to consider the possibility of measuring dry mass of cells mounted in nonaqueous media. Under these conditions and in suitably prepared material water-soluble substance may be retained and meas-

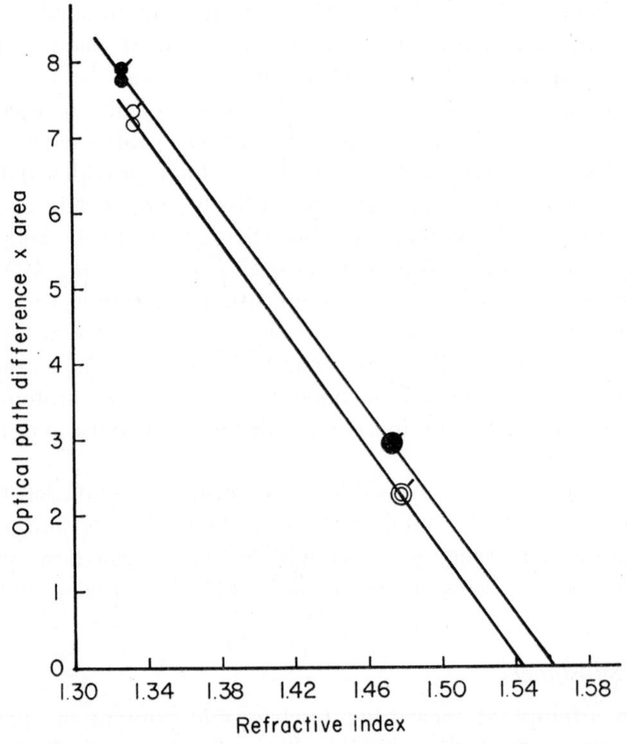

Fig. 3. Shows the variation in total refraction (optical path difference $\times$ area) in units of $k \cdot 10^{-12}$ cm.^{3} ($k = 0.18$) with refractive index. The total refraction in two sperm heads from bull (sperm heads distinguished by line attached to points), measured in methanol (full circles), water (empty circles), glycerol (full circles), and finally in a solution of sucrose (empty circle). The lines represent averages between data on the two cells in nonaqueous media, and in aqueous media (from Davies *et al.,* 1957).

ured by interferometry. The value of χ for dilute protein solutions is well established. As was pointed out the exact value of χ for precipitated protein is in question. It would also not be entirely surprising if the total refraction (defined here as o.p.d. × area) of water-free protoplasm in a nonaqueous medium was different from that of protoplasm in an aqueous solution of the same refractive index, the electrical properties of the media being different. For this reason an experimental approach is required.

In experiments (Davies *et al.*, 1957) on alcohol-fixed lipid-extracted sperm heads of various mammals the refraction was measured in a series of aqueous solutions of sucrose and in alcohol media, namely, methanol, ethanol, and glycerol. Figure 3 shows that the total refraction is apparently somewhat greater in alcoholic than in aqueous media. The difference is about 5% at a refractive index near that of water and the effect is reversible. The refractive index obtained by extrapolation is about 1.54 in aqueous solution and 1.56 in alcoholic medium.

The main conclusion is that dry mass can be measured in a medium such as methanol with the same value of χ as in water without a large error.

The interpretation of the small changes in refraction is not clear. Whether or not it occurs in other cell types which may contain different protein needs investigation. The effects are relevant to the further study of the value of χ for fixed cells. One suggested explanation is that there is water bound to the denatured protein which is not available as water of solution for the sucrose molecules. This water is removed by the alcohols and hence the refractive index obtained by extrapolation in that medium will be higher. The increased refraction in methanol over that in water could be due to an increase in volume in the methanol; it may only be coincidental that the fractional increase in total refraction is nearly equal to the fractional decrease in density which has been reported by some for dried protein (see discussion Section II, B, a, (i)). Further study including measurements in nonpolar media are required.

III. MICROSCOPES AND INTERPRETATION OF IMAGE

A. Remarks on Interference and Phase Contrast

The interference (interferometer) microscope like the phase contrast microscope renders visible transparent objects, that is objects which do not absorb light but differ only in refractive index from their surrounds. Both microscopes rely on interference phenomena to change phase difference into amplitude difference. The essential difference lies in the way in which the two coherent light beams, necessary for interference, are produced and allowed to interact. In the phase microscope light which has passed straight through the specimen is caused to interfere with light diffracted sideways by it. In the interference microscope the light splitting and combining is carried out by devices external to the specimen itself and hence under the control of the experimenter.

The practical results are that the phase microscope is efficient only at showing up diffracting structures in the specimen, that is, edges and abrupt changes of refractive index. In addition the way in which the beam splitting is carried out results in the well-known "halo" around images, which is more or less objectionable. On the other hand in the interferometer microscope, the halo effect can be eliminated entirely and the image gives an easily interpreted picture of the variations in optical path in the object. The o.p.d. due to the specimen can be easily measured with the interference microscope and therein lies its great advantage.

The phase contrast microscope was invented by Zernike in about 1935 and is now in routine use for the inspection of transparent specimens. The interference microscope, on the other hand, was invented much earlier. Dr. W. E. William in his textbook on interferometry (Williams, 1948) refers to the interference microscope devised by Sirks and Pringsheim in about 1890 (Sirks, 1893; Pringsheim, 1898). An interesting description of the various forms of interference microscope has recently been given by F. H. Smith (1955), (see also Barer, 1955). Interference microscopes employing multiple beams have for various fundamental physical reasons a limited application in the study of biological material and will not be discussed here. In this chapter the general principles of two-beam interference microscope are discussed with special reference to the two instruments now commercially available (Smith, 1950, manufactured by C. Baker, London; and Dyson, 1950, manufactured by Cooke, Troughton and Simms, York).

B. General Principles of Two-Beam Interference Microscopy

The general principles upon which the design of any double-beam interference microscope are based may be made clear by briefly considering a purely schematic version (Fig. 4). A ray of light from the source (S) is split into two parts, of equal intensity, by some device known as a *beam splitter,* the nature of which we need not for the moment consider. One ray passes through the specimen (O) and another, the comparison ray (C), some distance to the side. These rays are then combined by another device, the *beam combiner.* Since the two combined rays have arisen from a point on the source, any phase difference between them remains constant in time, i.e., they are coherent and hence can interfere to give light or dark. A refractile object in one beam causes a change in light intensity (cf. Eq. 43, Appendix 1) which is a measure of the optical retardation (o.p.d.) due to it. The o.p.d. could, in principle, be measured by introducing a glass wedge the so-called compensator so as to change the light intensity to its original value. If a linear wedge is moved into the comparison ray then the light intensity fluctuates through maxima and minima and in this way wedge movement could be calibrated against o.p.d. in wavelengths of the monochromatic light used. The main difference between the various forms of interference microscope lies in the way in which the beam splitting and combining are carried out. In the Cooke-Dyson microscope this is done by using semireflecting surfaces. In the Baker-Smith microscope use is made of the phenomenon of double refraction, according to which a plane-polarized light ray incident on a nonisotopic crystal is split into two rays each polarized at right angles, the rays in general following different paths. Such devices for beam splitting had, of course, long been known and used in interferometers for macroscopic studies.

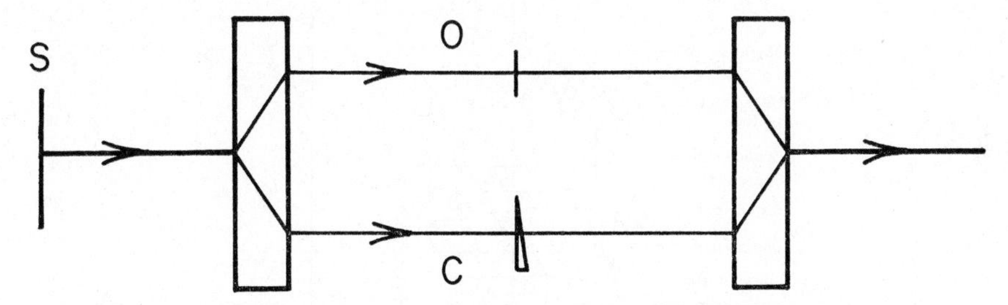

FIG. 4. Purely schematic illustration of the double-beam interference microscope. A ray from the source (S) is divided into two, the specimen ray (O) and the comparison ray (C), and these are then recombined.

1. *The Cooke-Dyson Interference Microscope*

The beam splitting and recombining is carried out in a somewhat similar manner in the macro-interferometer of Jamin (Jenkins and White, 1957) and the low-power interference microscope of Sirks and Pringsheim. In the schematic ray diagram (Fig. 5A) the specimen is located between two identical glass plates, not quite plane-parallel but with a small wedge angle. The surfaces of the plates are semisilvered and the upper plate bears a small opaque spot. Part of the illuminating cone traverses the lower plate (only a single ray is shown) and illuminates the object (full line) while part (broken line) is reflected down and then upwards so as to form a comparison beam which passes through an annular area around, and much larger than the field of view of the microscope objective. The two beams are recombined by the second plate situated above the object. The angle subtended at the object by the opaque spot and the thickness of the plates determines the separation of the direct and comparison beams. The reflectivity of the surfaces is controlled during manufacture so as to make the two beams which finally interfere as nearly as possible equal in intensity.

The mode of operation of the microscope interferometer is best understood by considering the images of the light source (S) formed in the various optical components (see the similar explanation of the Michelson interferometer in Jenkins and White, 1951). It can be shown

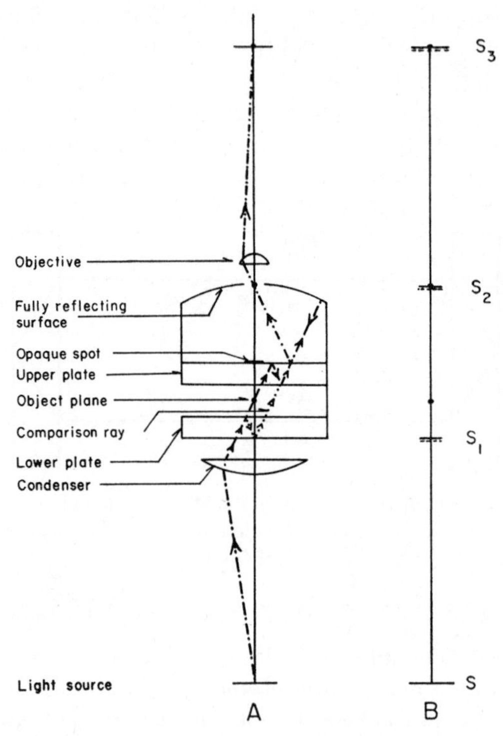

FIG. 5. The Cooke-Dyson interference microscope (schematic). A: Optical paths; the upper and lower plates have semialuminized surfaces. B: Two images of the light source (S), one (full line) containing an image of the object, are formed at S_1, and also at S_2 and S_3.

that the action of the condenser and lower wedge is to form an image of the source at or near the lower surface (S_1 in Fig. 5B). The action of the condenser plus upper wedge is to form a second image of the source, containing also an image of the specimen, at the same site. A hemispherical glass block with a silvered spherical surface forms a real image at S_2 which can then be viewed at a central transparent hole by a conventional high-power objective. This glass block is often referred to as the long working distance attachment. Provision is made for tilting the lower wedge and when in correct alignment the two images overlap precisely so that interference may occur.

As a result of the wedge shape of the plates the two images of the source in monochromatic light will in general overlap at an angle (Fig. 6A). The separation of the two images (parallel to the optic axis), and hence the o.p.d., varies linearly with distance across the field. Consequently the field is crossed by a set of parallel equidistant interference bands. These bands have a minimum spacing when the axes of the glass wedges are anti-parallel. This minimum spacing is determined by the angle at which the wedges are cut and is therefore constant for any one instrument.

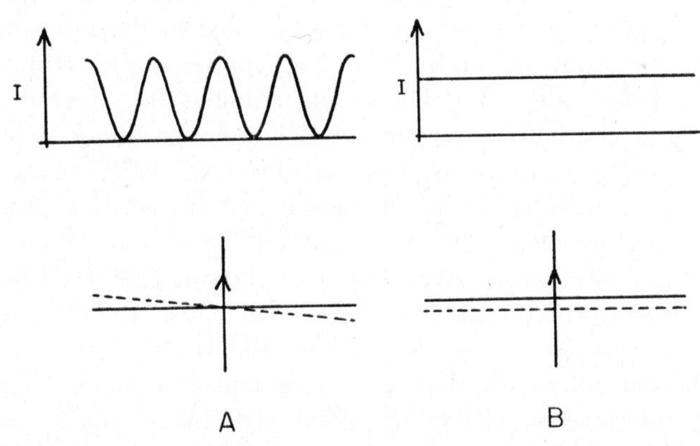

FIG. 6. A: Shows two images of the light source formed in the Dyson microscope, lying at an angle, with linear variation of o.p.d. Shows also the resulting variation in light intensity (I) in the image plane. B: Shows the two images of the light source with wedge axes parallel, the separation being equal to the optical path difference introduced; as a result there is no variation of intensity (I) in the image plane.

The lower wedge plate can be moved mechanically at right angles to the optical axis by a fine calibrated screw. This has the effect of shifting one image of the light source (not the one containing the object) parallel to the optical axis. As a result the point of intersection moves either to the right or left (in Fig. 6A) and the interference bands apparently move across the field of view.

As the upper glass wedge is rotated about the optic axis, and provision is made for doing this, a condition is reached when the wedge axis are parallel. If now the lower wedge is readjusted so that the two images of the source overlap, it can be shown that there is no variation in separation and hence o.p.d. between the two images. Hence the field of view is evenly illuminated, either light or dark depending on the separation (Fig. 6B). As the micrometer phase screw is adjusted the separation of the two interfering fields varies as before and the light intensity goes through maxima and minima (cf. Eq. 43, Appendix 1).

The introduction of an object causes a separation of the two interfering fields by an amount equal to the o.p.d. introduced. As a result of interference the object shows up in interference contrast.

a. Effect of Finite Condenser Aperture.

In Fig. 6 the images of the same point on the source lying one above the other have been assumed to emit light parallel to the optical axis. In practice the rays from any one point will be at angles to the optic axis, determined by the angular aperture of the condenser. Without going into details the effect will be to reduce contrast in the interference pattern by an amount depending on the condenser aperture and the separation of the interfering fields. Surprisingly the effect is not large except at high condenser apertures and large separations. Unpublished calculations referred to in Davies and Deeley (1956) show that the reduction in contrast is negligible in the first few orders provided the condenser aperture does not exceed about 0.5 to 0.6. What is apparently a similar effect was investigated by Dyson (1950), who pointed out that the introduction of an object will cause the two interfering images to become separated, this being the basis of the interference contrast. Dyson calculates that, for a condenser numerical aperture of about 0.6, interference effects of "good visibility" will be obtained if the o.p.d. due to the object does not exceed about four wavelengths.

b. Interference Pattern in White Light.

When white light illumination is employed a system of colored bands replaces the system of equidistant bands seen in monochromatic light (Fig. 6A). The band system has a white central band indicating that the optical path differences between light of all wavelengths is zero along that line. The symmetrical system of colored bands on either side of the central zero-order band exhibits the so-called Newton's sequence of colors.

The way in which this band system arises is shown in Fig. 7A where interference patterns are drawn for three monochromatic wavelengths. In a continuous spectrum due to the overlapping several characteristically colored bands appear, shown schematically in Fig. 7B. For large o.p.d. the bands so overlap as to give the appearance of white light. Each colored interference band is quite distinct from any other and this fact is made use of in a manner to be indicated in Section IV.

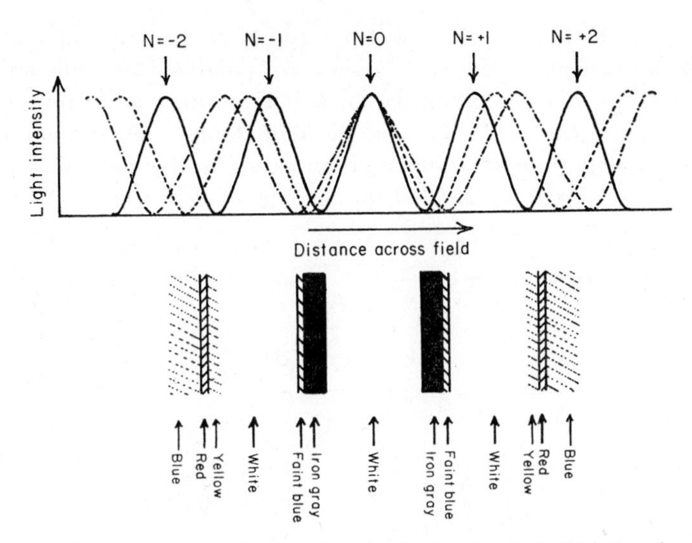

FIG. 7. Light intensity distribution across the banded field for three monochromatic wavelengths, blue ————, green – – – – –, red — · — · — · The order of interference (N) for blue light is given together with approximate indication of the colors in white light as seen in the Cooke-Dyson interference microscope. In the Baker-Smith interference microscope the analyzer may be set so as to make the zero-order band appear black, with a symmetrical colored pattern on either side.

When the glass wedges are parallel the entire field of view is evenly illuminated with light of one color, which changes to others through a defined sequence as the phase screw is rotated.

2. *The Baker-Smith Interference Microscope*

This microscope (Smith, 1950) makes use of the phenomenon of double refraction to produce the beam separation. A low-power instrument was introduced by Lebedeff (1930). A. F. Huxley (1952, 1954) has also designed and built interference microscopes with double-refracting components. The commercially available microscopes exist in two forms. In the first the two images of the light source are formed by the beam splitter, one at the level of the specimen and one some distance below it. This *double-focus* arrangement is not entirely free from halo effect; measurements of o.p.d. with it will not be so accurate as in the second form and it will not be described here.

In the second shearing form the double refracting plates are attached to the condenser and objective and act as beam splitter and combiner respectively. Light from the source (S) (Fig. 8) plane-polarized by polaroid is focused by the condenser and lower plate to form two laterally displaced (sheared) images of the source in the plane of the object (at S_1 in Fig. 8). It is a property of the birefringent plate that the light vibrations in the two images are polarized at right angles to each other. As a result of the equal and opposite double refraction in the upper plate the two images of the light source, vibrations mutually polarized at right angles, are superimposed in the final image. Two images of the object are formed on account of double refraction in the

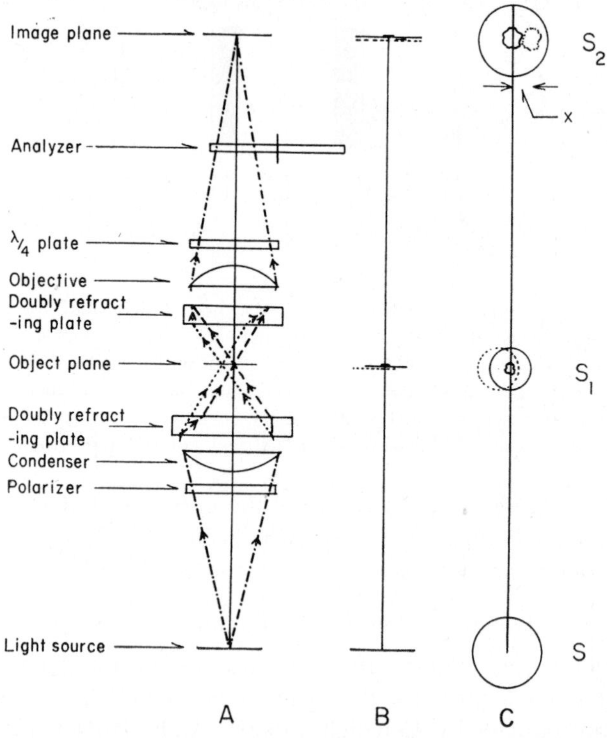

FIG. 8. The Baker-Smith shearing interference microscope (schematic). A: Optical paths. B: Shows the two laterally separated images (S_1) of the light source (S) formed in the plane of the object; shows also the two superimposed images of the light source formed at S_2 the image plane, both containing an image of the object. C: a plan view of B showing the two images of the object separated by a distance x.

upper plate, one of which is defocused (due to astigmatism) so as to contain no detail. This is usually referred to as the "ghost" image. The two images of the source each of which contains an image of the object can be superimposed, one above the other, by tilting the microscope condenser and are then in a condition to interfere when an analyzer (polaroid) is located above them. The analyzer produces two components vibrating in the same direction. On account of the optical path difference, depicted in Fig. 9 as the vertical separation, the object viewed through the analyzer will be seen in interference contrast. The contrast is maximum when the analyzer is located at 45° to the planes of vibration.

There is a restriction on the size of object which can be measured, imposed by the ghost image. Only objects which have a dimension (diameter) less than x the shear distance in Fig. 8 will be entirely free from the ghost image.

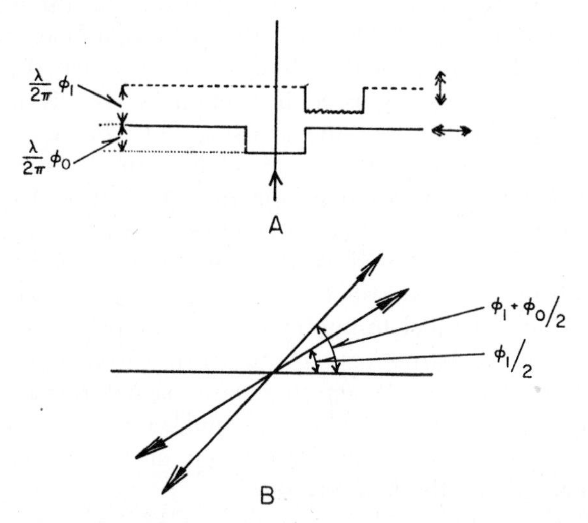

FIG. 9. A: Shows the two images of the light source in the Baker-Smith shearing microscope separated by an amount $\lambda\phi_1/2\pi$ where ϕ_1 is the phase difference between the two beams. The separation is altered by the plate-like object by an amount equal to the optical path difference $\lambda\phi_0/2\pi$ where ϕ_0 is the phase difference due to the object. The vibrations in the two images are polarized at right angles. No $\lambda/4$ plate or analyzer present. B: After the two vibrations have passed through the $\lambda/4$ plate the resultant is a single plane polarized vibration, the plane of polarization being rotated through an angle which is equal to half the phase difference between the two beams. Rotation of background is $\phi_1/2$ and for one image of object is $(\phi_1 + \phi_0)/2$ from arbitrarily chosen horizontal base line.

For the purpose of measuring phase difference a birefringent compensator consisting of a combination of quarter-wave plate and analyzer is used, which functions, briefly as follows. When two vibrations polarized at right angles are passed through the suitably oriented quarter-wave plate they emerge circularly polarized in opposite senses. These combine to form a single plane-polarized beam the orientation of polarization depending on the original phase difference between the vibrations. It may be shown that increasing the phase difference between the beams by say ϕ causes the plane of polarization to rotate through $\phi/2$. The phase differences indicated by the separation in Fig. 9A are shown as affecting the plane of polarization in Fig. 9B.

When the analyzer (polaroid) is rotated the light intensity fluctuations in the surround and object are proportional to $\cos^2 \Psi$ and $\cos^2 \left(\Psi + \frac{\phi_0}{2}\right)$ where Ψ is the angle between the vibration direction in the analyzer and in the background beam. The variations in light intensity are thus essentially similar to those observed in the Cooke-Dyson microscope set for even field when the path difference between the beams is varied by means of the phase screw. The actual measurement of o.p.d. of an object is discussed in Section IV. Briefly it is made by noting the angle $(\phi_0/2)$ that the analyzer must be moved to reduce to minimum intensity first the object and then the background.

In the Baker-Smith interference microscope a special eyepiece, the so-called fringe eyepiece, is used to produce an image field crossed by interference bands. This is used with the quarter-wave plate and analyzer removed. This eyepiece consists of a birefringent quartz wedge oriented with its optic axis parallel to one of the vibration directions. Owing to the wedge shape this introduces a continuously variable phase difference between the two beams. When these are passed through a suitably oriented analyzer a system of parallel equidistant interference bands is formed in monochromatic light. Movement of the wedge causes the bands to move across the field of view.

3. *Note on the Distribution of Energy in Interference Contrast*

It is useful to keep in mind the basic physical principle of conservation of energy when interpreting image formation in the phase and interference microscope. When two light beams interfere to form an interference pattern, energy is not destroyed but merely redistributed so that either the bright parts of the image are due to energy being removed from the dark parts, or a complementary pattern is formed, the bright parts of which correspond to the dark parts of the observed pattern. In the phase contrast microscope the light energy is redistributed in the field of view, i.e., the energy drawn from the "image" of the object is redistributed as a bright halo round it. In the interference microscopes complementary patterns are formed. In the Baker-Smith microscope the complementary pattern is absorbed by the polaroid. In the Cooke-Dyson microscope the complementary pattern is to be found in a reflected pattern which is not seen in the field of view of the microscope.

C. Interpretation of Image in Interference Contrast

The interpretation of the image formed in an interference microscope will be made clear by considering the appearance of some objects of simple geometry, first in a field of view crossed by bands and second in an evenly illuminated field of view, that is zero path difference over the field. We have previously noted that the effect of introducing an object is physically to separate the two interfering images. The observations in Section III, C, 1 and 2 apply equally to the Cooke-Dyson or Baker-Smith microscopes. It is assumed that the objects, large with respect to the wavelength, are illuminated with parallel monochromatic light and that the bending of the light rays by the object due to refraction or diffraction can be ignored.

1. *Background Field Crossed by Interference Bands*

When the image plane is crossed by parallel equidistant interference bands, the path difference between the two interfering beams varies linearly in a direction at right angles to the bands. The path difference changes by one wavelength (λ) from one minimum to the next. Due to an object of thickness t being present, the path difference is modified by an amount $(\mu_o - \mu_m)t$ where μ_o and μ_m are the refractive indices of object and medium, respectively. Hence the o.p.d. at a point in the image of the object becomes altered from what it was, and becomes equal to that at some other point in the surrounding clear field. The object is said to cause a shift in the interference pattern. The points in the image of the object and surround at which the o.p.d.'s between the two interfering beams are equal are defined as *corresponding points*. The distance separating corresponding points is a measure of the o.p.d. The o.p.d. between two interfering beams in terms of the wavelength is usually referred to as the order (N) (See Fig. 7).

In monochromatic light all the interference bands look alike and it is not possible to tell which are corresponding bands (or points). It is possible to decide this by using white light when corresponding bands (or points) have the same color, allowance being made for the fact that the colored band system has a center of symmetry.

a. Large Uniform Object.

Figure 10 shows the appearance of a thin strip object of uniform refractive index and thickness, and also the light intensity in it and the background field (Fig. 10B). The bands are displaced uniformly by an amount d_1. Since the path difference changes by one wavelength (λ) from one band to the next, distance d, the o.p.d. due to the object

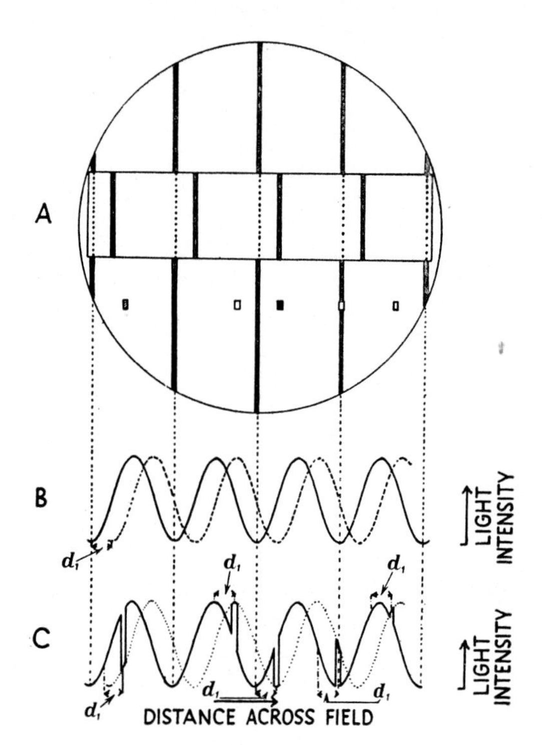

FIG. 10. Simple objects of known geometrical form in the field of view crossed by interference bands. A: In the upper section interference bands are seen displaced in the large strip object of constant optical thickness; in the lower section five small objects of the same optical thickness are shown situated on a line at right angles to the bands, and only part of a band will be seen in them. B: Interference pattern in clear background field (full line) with band separation *d;* the light intensity distribution in the object (broken line) is similar in shape but displaced through a distance d_1. C: Light intensity distribution (full line) across a line containing the five similar objects is simply constructed from the light intensity in the clear background field (dotted line); all distances d_1 are equal (from Davies *et al.*, 1954b).

is $d_1\lambda/d$ cm. (λ in cm.). The necessity for using white will be clear: all bands look alike and the displacement could be d_1, $d \pm d_1$ $2d \pm d_1$, etc. Whether the refractive index of the object is higher or lower than its surround can be ascertained by determining, once and for all, the direction of displacement in an object the refractive index of which is greater than that of its surround.

b. Small Uniform Object.

When the size of the image is small compared with the distance between band minima only a part of an interference band will be seen in it, and the object will appear to be approximately evenly illuminated.

In small plates (Fig. 10A) with an o.p.d. equal to that of the large strip the light intensity can be derived as shown in Fig. 10C. The image intensity depends on the position of the object in the field. If we imagine one object to be moved across the field in a direction at right angles to the bands then the light intensity in it will alternately be greater and less than the surround. There are positions in which its intensity is equal to the surround and hence it apparently disappears. Alternatively, with a stationary object a similar variation of light intensity is seen in its image if the bands are caused to move across the field of view.

This shows how an object (or part of an object) can be viewed in variable contrast by operating the phase screw (or analyzer). Hence by a simple adjustment any detail can be set to minimum visibility.

c. Large Nonuniform Object.

In the image of a large object of varying o.p.d. the interference bands are no longer straight. For example, the image of a homogeneous lens-shaped object of central o.p.d. slightly less than one wavelength is shown in Fig. 11. The o.p.d. at any point on a band is proportional to

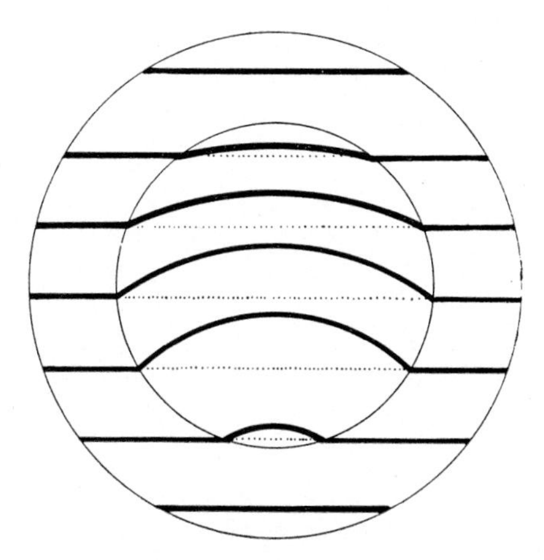

FIG. 11. Shows the interference bands in a homogeneous lens-shaped object of optical path difference about 0.9 λ (from Davies *et al.*, 1954b).

its displacement from the clear-field band (interpolated). Distances are measured in a direction at right angles to the clear-field bands.

To summarize, when an object is viewed in a banded field an examination of the white light interference pattern in it, and especially a study of the way in which the pattern changes inside the image when the bands are moved relative to it (or vice versa) enables corresponding points and hence the o.p.d. throughout the image to be mapped out.

2. *Background Field Evenly Illuminated*

Under these conditions the o.p.d. between the interfering beams does not vary across the field. When the object is several wavelengths thick the interference bands seen in the image are contour bands of constant optical thickness (μt). When the thickness is constant, as in sections, variations in optical thickness are indicative of variation in refractive index and hence of concentration of dry substance. As examples, the images of three model objects are shown in Fig. 12.

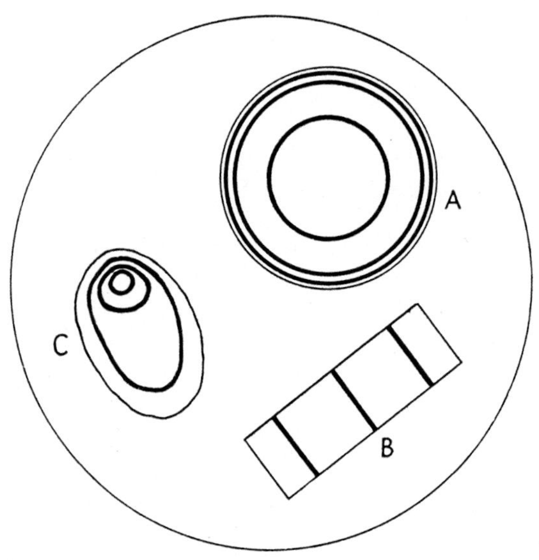

FIG. 12. The appearance of objects in even-field interference contrast and maximum brightness. A: A homogeneous sphere of o.p.d. 3 λ. B: A wedge increasing in o.p.d. from 0 to 3 λ. C: an irregular object of maximum o.p.d. 3 λ. The darker bands in the images are contour lines of equal o.p.d. (from Davies *et al.*, 1954b).

An important special case is that of objects in which the optical path difference is less than half a wavelength. If the background field is set to its brightest then the higher the o.p.d. of an area the darker will it appear. Hence in such objects, the distribution of o.p.d., or dry mass per unit area can be seen at a glance. Contrariwise when the background field is set near minimum light intensity, the regions of greater mass per unit area will appear brighter. For an example see Fig. 15.

To further illustrate the last two sections several photomicrographs with keys (Figs. 13, 14, 35) are included. Ideally these should have been in color when corresponding bands (or points) would be the same color.

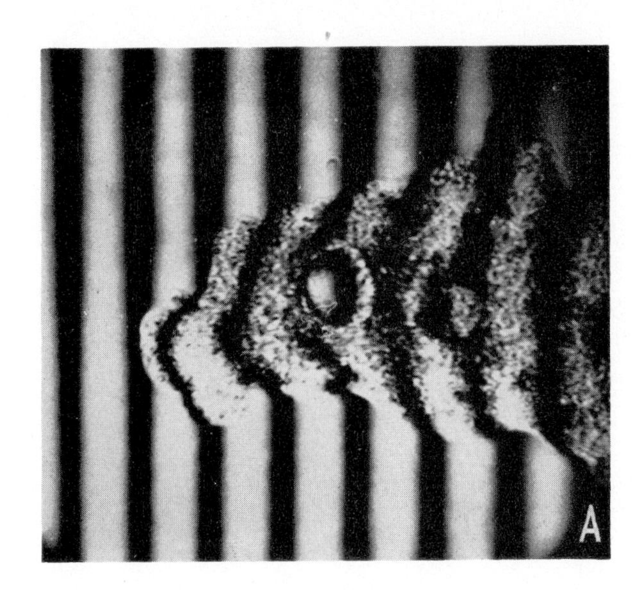

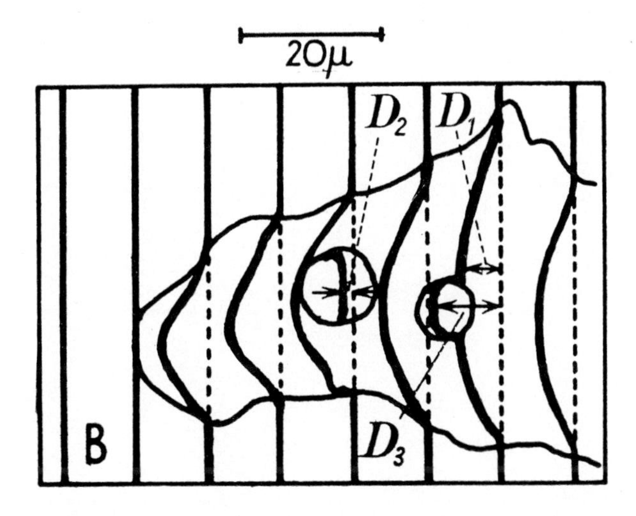

FIG. 13. A: part of a living *Ameba* in Chalkley's solution. Ilford Pan F film. The photomicrograph was taken with a Huxley (1952) type interference microscope, with objective of NA 0.2, and a primary magnification of ×24. B: schematic diagram of interference bands seen in background field and image of *Ameba*. The o.p.d. varies from zero at the cell edge to a maximum near the cell center, e.g., the o.p.d. D_1/d is about 0.6 where d is the band separation in the surround. In the contractile vacuole the o.p.d. is about 0.1 λ (D_2/d), while in the food vacuole it is about 1.0 λ (D_3/d) (from Davies *et al.*, 1954b).

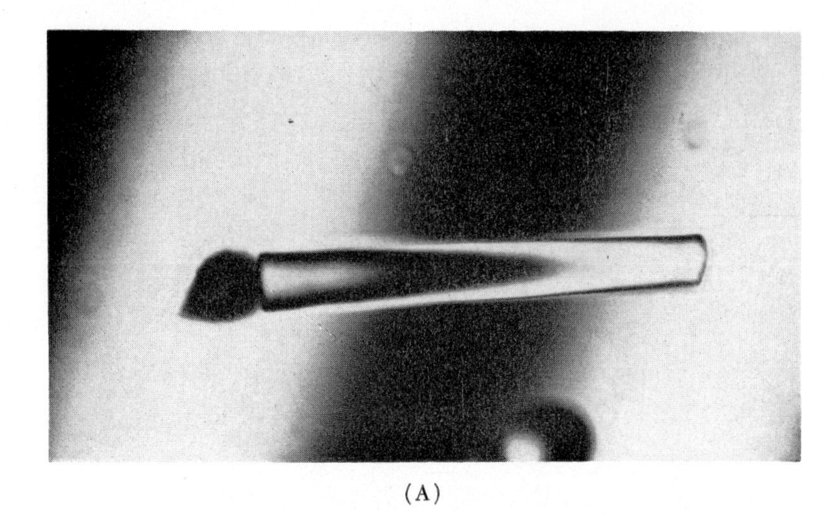

(A)

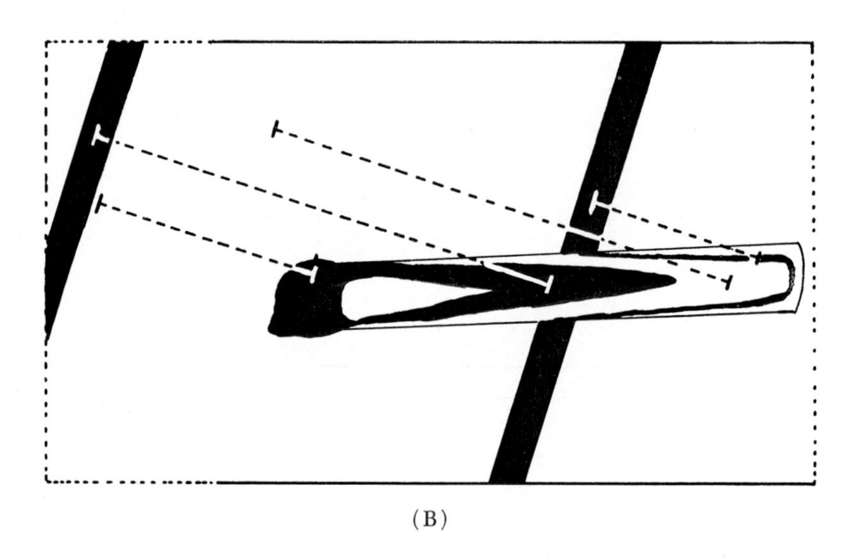

(B)

FIG. 14. A: Living retinal rod from frog. Taken with a Cooke-Dyson interference microscope, with objective of NA 1.3, condenser NA $\sim$ 0.5 and is at magnification $\times$1050. B: Schematic diagram of interference bands in image and surround. The lines (broken) join corresponding points, that is points of equal order N (Davies, Fernández-Moran, from C. T. S. booklet).

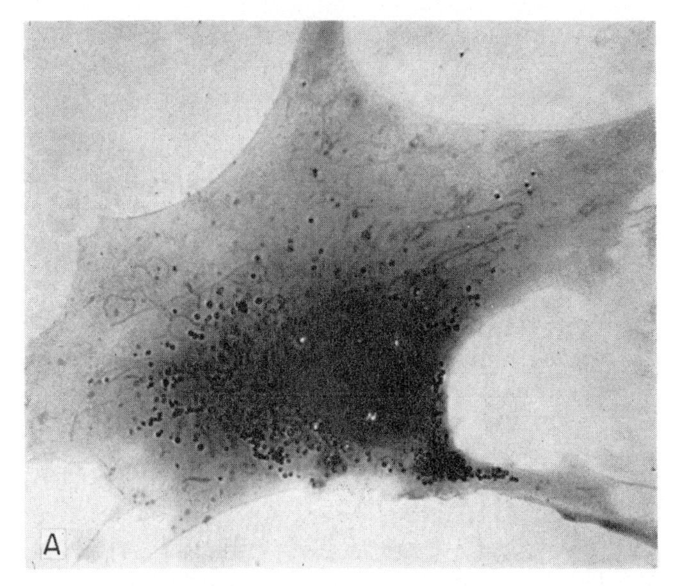

FIG. 15. A: Living fibroblast from mouse heart in tissue culture in even-field interference contrast. Taken with Cooke-Dyson interference microscope with objective of NA 1.3, condenser NA $\sim$ 0.5, and is at magnification $\times$1275. The darker parts have a greater o.p.d. and hence mass per unit area (Davies and Richards, from C. T. S. booklet). B: Shows the relationship between light intensity and optical path difference, with the various regions of the cell marked; the o.p.d. is equal to the horizontal separation between the marked site and the background.

IV. METHODS OF MEASUREMENT

A. Measurement of Optical Path Difference (Δ)

The writer recommends examining an object of unknown optical path in white light illumination in a banded field. The o.p.d. will be n wavelengths ($n\lambda$), in general, where n consists of a *whole number* plus a *fractional part.* By moving the bands relative to the object, or the object relative to the bands, corresponding points (Section III) in image and surround can be found. Hence the whole number part of n can be determined; by changing over to monochromatic light the fractional part of n can be ascertained, with ease, to about 0.2 λ. When the object is complicated draw a rough diagram and map out the o.p.d. of the parts. A characteristic and hence useful white light band is one of the two bluish black minima (order N, $\pm \frac{1}{2}$ in Fig. 7) but any band will do. By carefully observing the sequence of bands in the image and surround no confusion will arise due to the presence of the two minima; in any case they are not identical, one having a blue edge to its right and the other to its left.

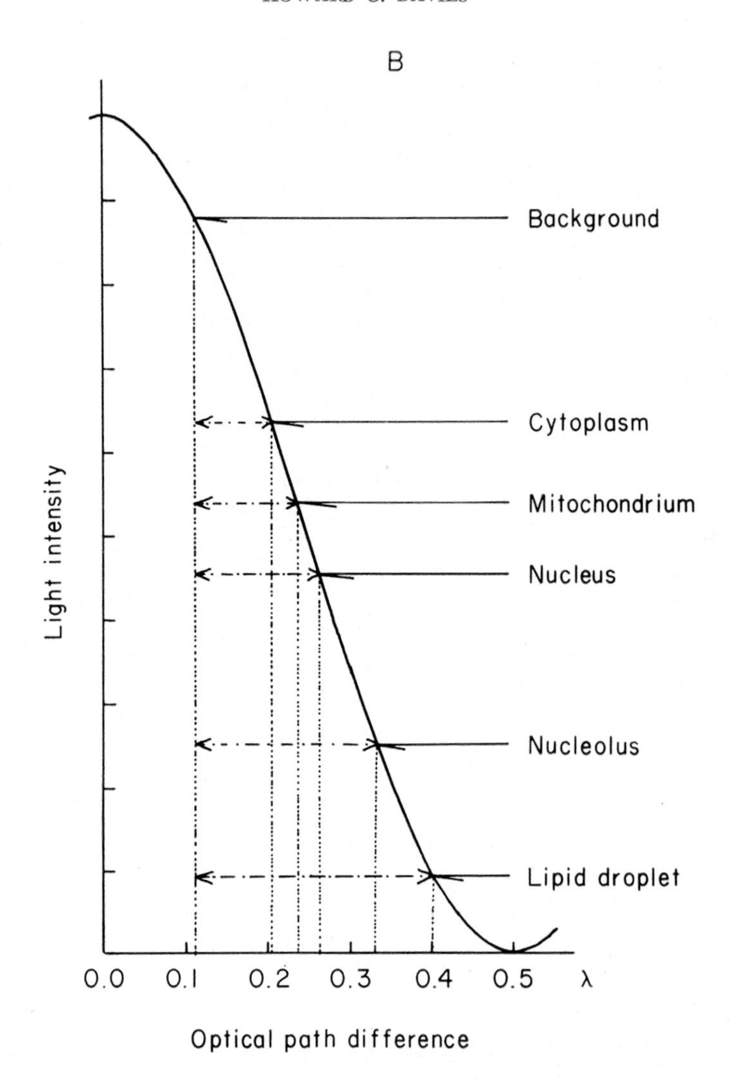

The estimation of n can also be carried out when the field is evenly illuminated but it is more difficult to remember the sequence of colors.

1. *Measurement of Fractional Part of Optical Path Difference*

a. Visual Photometry.

(i) *Setting on a minimum in a banded field.* The chosen site in the image is visually set to a minimum light intensity by either moving the object or the interference bands. Monochromatic light must be used so that λ can be specified. The distance (d_1 in Fig. 10) to the corresponding background band minimum and the separation of the bands in the surround (d in Fig. 10) can be measured on an eyepiece scale or a photomicrograph. Alternatively by using an eyepiece with a cross-hair the distances d and d_1 can be read off on the micrometer phase screw in the Cooke-Dyson.

It has been found that this method works well when the site chosen is large so that an appreciable part of a band can be seen in it. The accuracy of setting an interference minimum on a cross-hair is about 1/10 to 1/50 λ depending on the contrast. However, when the size of the object is small compared with the band separation, so that only a part of an interference band is seen in it, the accuracy in setting on a minimum is less. This setting is a subjective matter, judgment being impaired by the variations in light intensity of the surround. Thus the object may look as if it is getting somewhat darker merely because the surround is rapidly getting brighter. It has been found that although the settings are fairly reproducible the mean setting varies from one individual to the next. Comparisons with an objective method have indicated that large systematic errors may be present. The accuracy of setting could be improved, for example by enclosing the measured area in a diaphragm so as to exclude the surround.

(ii) *Setting on a minimum in even-field contrast.* Measurements can be made equally as well with even-field interference contrast. By setting the field at successive positions of minimum light intensity the scale reading on the micrometer screw in the Cooke-Dyson can be calibrated. In the Baker-Smith the analyzer is moved through 180° from one minimum to the next. The reading (cf. d_1) for transference of the condition of minimum light intensity from background to object area when divided by the reading between successive minima (cf. d) gives the o.p.d. as a fraction of the wavelength of the light used. It will not be obvious which way to rotate the phase screw or analyzer. In the case of the analyzer in the Baker-Smith microscope reference can be made to the instruction booklet. The appropriate direction can simply be ascertained by finding which direction gives a value similar to the rough reading obtained first in a banded field. The direction of rotation will always be the same provided that the measurements are always made in the same sequence, e.g., surround first, then object. Although this is simple to do it is also easy to read, in error, an o.p.d. of 0.56 wavelengths as 0.44 wavelengths. Care must be exercised.

Subjective errors in setting to a minimum of the type described in (i) are also possible here. The reproducibility is about 1/10 to 1/50 λ.

(iii) *The Cooke Photometer Eyepiece.* The method makes use of two facts. First, the effect of a refractile object is merely to shift the interference pattern without altering its shape; all distances d_1 in Fig. 10 are equal. Hence instead of measuring the distance between corresponding points of minimum light intensity we measure the distance between corresponding points of equal light intensity the latter being chosen to lie about half way between minimum and maximum, i.e., on the relatively straight part. At this position the rate of change of light intensity with o.p.d. is a maximum and it is clear that the accuracy of measurement will thereby be markedly increased. Second, the positions

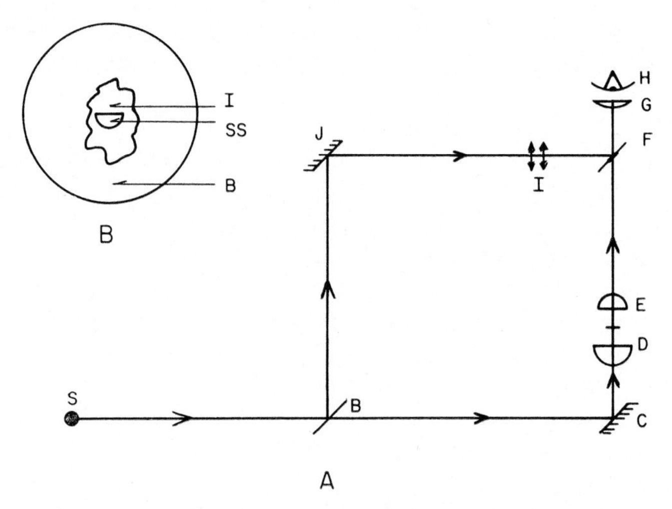

FIG. 16. Schematic diagrams of the Cooke photometer eyepiece for accurate measurement of o.p.d. A: Optical paths. S_1 light source; B, beam splitter; C, microscope mirror; D, E, condenser and objectives of the interference microscope; F, transparent plate with semisilvered spot; G, positive eyepiece; H, eye; I, polaroids for adjusting light intensity in comparison beam; J, reflecting plate. B: Appearance of image plane. I, image of object; SS, silvered semicircular spot; B, image of surround.

of equal light intensity in object and background are determined by successively matching them with a spot of light maintained at a constant light intensity. This kind of matching is well known in photometry and can be made by the eye with precision.

The eyepiece attachment manufactured by Cooke, Troughton and Simms Ltd. is described in detail elsewhere (a C.T.S. booklet) and is for use with an evenly illuminated field (i.e., band separation infinite). Briefly, the primary image plane contains a plate, inclined at 45° to the microscope axis, on which is a fully reflecting semicircular spot (Fig. 16A). A positive eyepiece is focused on this spot and the primary image. Part of the light from the source is deflected and again reflected so as to fall on this spot. This provides an area of constant light intensity in the image plane for use as a comparison spot (Fig. 16B). The intensity can be adjusted by polaroids so as to lie on the relatively straight part of the interference pattern.

The micrometer phase screw is first calibrated by successively matching the field with the spot. The object detail and surround are successively matched against the comparison spot and from a combination of these reading (cf. d_1, d) the o.p.d. is computed. The screw must be turned in the right direction and the readings made in the correct order in a manner similar to that described above (ii). The reproducibility of the setting is about $\pm \frac{1}{200} \lambda$.

There are three critical adjustments to be made. (1) The light must be strictly monochromatic; slight color differences between the two fields upsets the matching. (2) The exit pupils from the two light trains must coincide; otherwise small movements of the head may alter the light intensities. (3) The positive eyepiece must be carefully focused on the straight edge of the comparison spot so as to produce as sharp as possible a division between the two fields. Due to the tilt only this edge can be focused. The condition is essential for accurate visual photometry. A limit to the accuracy of matching is set by the presence of a demarcation line between the two areas, present due to diffraction (i.e., point not imaged as a point). As a result the spot cannot be made to completely disappear against its background.

The photometer eyepiece, although designed for use with the Cooke-Dyson microscope, can be used with the Baker-Smith microscope. Dyson (private communication) has recently described a device which reduces the diffraction effect and he claims an increase in the accuracy with which a match can be made. This has apparently been done by increasing the angular cone of light onto the mirror causing the diffraction pattern to become narrower.

(iv) *The Baker Half-Shade Eyepiece.* This is a device for use only with the Baker-Smith interference microscope set for even-field conditions. With it measurements of o.p.d. can be made with an accuracy similar to that obtained with the photometer eyepiece (iii); the technique of measurement is similar. The construction and mode of operation are described in a booklet by C. Baker (1955) and by Smith (1954).

Briefly, the primary image is formed on one face of a totally reflecting glass prism which has on it a metallized strip (Fig. 17A). Owing to the different nature of the reflection from the dielectric and metal surfaces the perpendicularly polarized light beams ($\lambda/4$ plate and analyzer out) suffer relative phase shifts which are different at the two surfaces. As a result of this difference the rays after passing through a $\lambda/4$ plate are plane-polarized and rotated through about 60° as shown in Fig. 17B. The metal strip may be regarded as exactly equivalent to an object of phase difference about 120°. The light intensity in the different regions of the image plane vary with position of analyzer as shown in Fig. 17C. The o.p.d. of the object is obtained by successive matching, first the image of surround inside and outside the metal strip (points 1 or 1′ in Fig. 17C), second the image of object inside and outside the metal

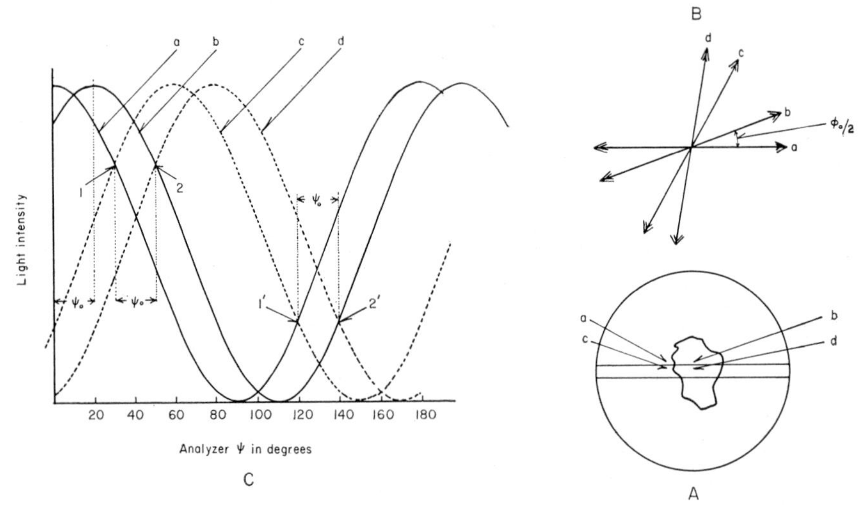

FIG. 17. Diagrams illustrating the Baker Half-Shade Eyepiece. A: The field of view contains an image of the object and that of the narrow metallized strip. B: After passing through the analyzer the plane of polarization in surround and object are rotated through an amount $\phi_0/2$ (cf. Fig. 9B) where ϕ_0 is the phase difference due to the object. The images of object and surround in metal strip are further rotated through about 60° due to relative phase changes at reflection from glass and metal. C: Shows the light intensity curves obtained by rotating analyzer; curves a, b, c, d correspond to points a, b, c, d in 17A. The angle ψ_0 between positions 1′ and 2′ is half the phase difference due to the object, i.e., $2\,\psi_0 = \phi_0$.

strip (points 2 or 2′ in Fig. 17C). For reasons which cannot be discussed here the positions of lower intensity (1′ and 2′) give the best match. The accuracy of this device is also limited by edge effects due to diffraction.

(v) *Other methods.* In the routine use of the Cooke and Baker microscopes the above methods have proved to be adequate for visual spot measurements of o.p.d. However, it is of interest to mention some other methods, which have been proposed and sometimes employed, for measuring of o.p.d. F. H. Smith has suggested (private communication and C. Baker, 1955) a method for small isolated particles, which makes use of the fact that in an evenly illuminated field there are positions (two in number) between adjacent interference minima (see also Fig. 10C) where isolated plate-like particles will disappear against the background. The setting, from which the fractional part of n can be determined, is very critical since the brightness of object and surround vary in opposite directions. It might be expected that difficulty would arise in measurements on isolated spheres and rods in which the o.p.d. is varying.

Dyson (1953) has proposed an interference microscope, relying on double refraction, basically similar to that of Lebedeff (1930), with a special "checker board" eyepiece similar in function to the half-shade device (iv). This eyepiece, suitable for use with the Baker-Smith microscope has no special advantage over the half-shade device, and may indeed be somewhat confusing to use.

Finally in the interesting interference microscope, relying on double refraction, recently described by Johansson and Afzelius (1956) the "sensitive color" method, plus a special compensator, is used for measuring o.p.d. The "sensitive color" method relies on the fact that there is one special region of the white light (tungsten lamp) interference pattern (purple-violet of the first order) where an abrupt change in color occurs, for a small change in phase screw or analyzer. The accuracy claimed under ideal conditions is about $\lambda/500$. The "sensitive color" method has not been much used with the Cooke-Dyson and Baker-Smith microscopes. One suspects that the color saturation and hence band visibility may be rather better in the most recently described microscope.

b. Photographic Photometry.

The methods are based partly on the above principles, with the visual detector replaced by a photographic film. This provides a permanent record of the object from which measurements of o.p.d. and also of area can be made at leisure. This is an important advantage when studying living and moving cells. However a microphotometer, preferably recording, which can read in either transmission or log-transmission units is required.

(i) *The trace displacement method.* This method, used with a banded field, relies on the fact that the effect of the object is to shift the bands without altering their shape (cf. Fig. 10B, C). The method can conveniently be applied to both the Cooke-Dyson and Baker-Smith microscopes.

The method will be made clear by studying Fig. 18. Briefly the

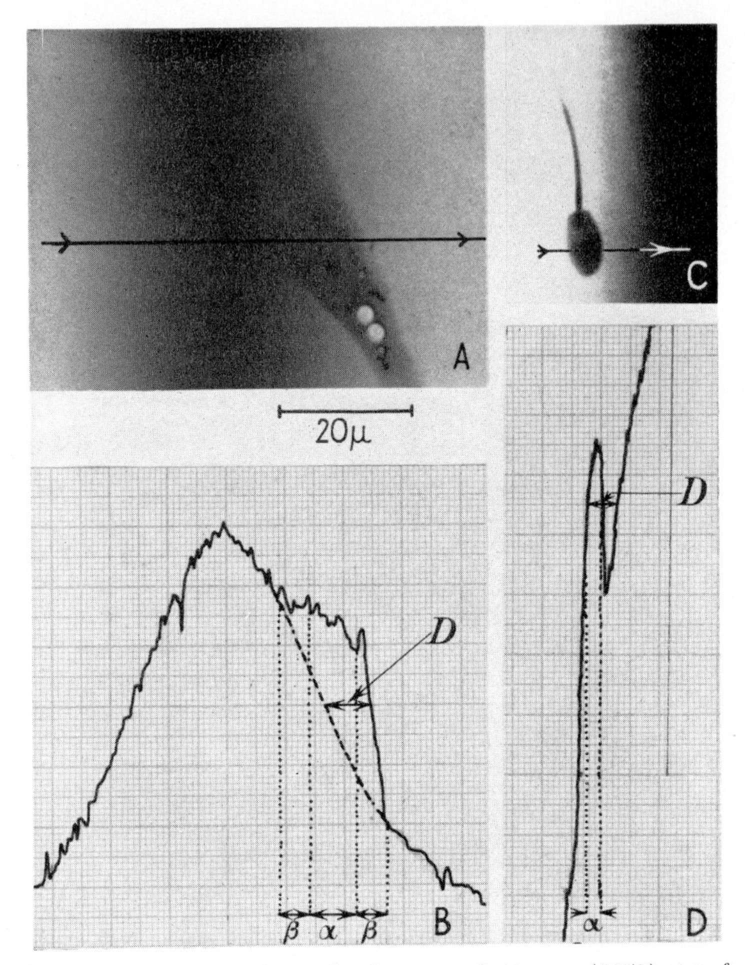

FIG. 18. Photomicrographs with the original Dyson (1950) interference microscope objective NA 0.7, condenser NA ~ 0.5, magnification ×880. A: living interphase chick heart fibroblast mounted in Tyrode solution. B: film density (ordinate) of negative, recorded along line indicated in A. The dotted curve is filled in from a suitable "blank" field, nuclear dimensions α, cytoplasmic dimensions β. The o.p.d. at any chosen point on the line in A is determined from the distance D. C: ram sperm head, fixed in 10% neutral formalin, mounted in water. D: film density (ordinate) from negative along the line indicated in C; cell dimensions α. The o.p.d. at any chosen point on the line is determined from the appropriate distance D (from Davies *et al.*, 1954b).

object occupies only a fraction of the space between bands, and the position of the bands is adjusted so that the light intensity in the image of the specimen lies on the linear part of the interference pattern. A record of film transmission (or density) is made along any line. From this the distance d and d_1 (Fig. 18 and Fig. 10) can be measured and hence o.p.d. at any point in the object can be computed.

This method is very suitable for making absolute measurements of o.p.d. on small isolated objects with an accuracy of about $\pm \lambda/200$. An evenly illuminated field, in absence of interference, is required. The method is not so well suited to measure objects of a size which partly obscure the reference band. The interpolation of the background (cf. Fig. 18B) relies on the interference pattern being perfectly symmetrical.

(ii) *Measurements in an even field.* In a photomicrograph of an object of o.p.d. less than 0.5 λ, darker (or lighter) areas will have in general a greater o.p.d. (cf. Section III, C, 2). This fact can be usefully employed to measure o.p.d. if the relationship between blackening of film negative and o.p.d. of object can be simply deduced. Two aspects of this will be discussed in the next two paragraphs.

Part of the interference pattern (i.e., light intensity versus o.p.d.) is approximately linear over a range of about 0.15 λ (see also Section IV, B, c). Simple theoretical considerations lead to the conclusion that for a film of gamma unity, the intensity of transmitted light is proportional to the light intensity incident during exposure. That is, for a limited range of o.p.d. and on the straight part of the gamma curve, a simple linear relationship is to be expected between o.p.d. and film blackening as measured by transmitted intensity. However experiments indicate that under conditions far removed from the ideal, namely with films of gamma between two and three and with a photometer recording log-intensity (density), there is a region of linearity between o.p.d. and film density. This can be deduced from Fig. 18D and the fact has been made use of in measurements of muscle fibrils in even-field contrast (Huxley and Hanson, 1957). This method requires considerable care in application and does not readily yield absolute values for o.p.d., and is applicable only when the o.p.d. is small ($< 0.2\,\lambda$).

We suggest another method which may be very useful in photographic photometry with the doubly refracting type of interference microscope. Adjacent to the image of the object on the film, it should be possible to image in a birefringent step wedge of effectively variable o.p.d. This should provide a calibration relating film blackening against o.p.d., from which the unknown o.p.d. of the object can be read off.

c. Photoelectric Photometry.

It is simple to conceive of procedures for measuring o.p.d., similar to those outlined in Section IV, A, 1, a in which the eye is replaced by a photosensitive surface. The required apparatus consists basically of an opaque screen at the image plane, containing a small aperture behind which is located a photomultiplier.

It may be expected that the accuracy in setting on a minimum (cf. Section IV, A, 1, a (ii)) will be increased when employing photoelectric detection. A constant light source is not needed. A second method, with even-field contrast, has the increased sensitivity of the method described in Section IV, A, 1, a (iii). Instead of employing a spot of constant brightness, we determine the rotation of the phase screw (or analyzer) needed to transfer a condition of constant light intensity from the background to the object detail. The light is adjusted to lie about half way between maximum and minimum. A constant light source is required. This procedure has been used successfully by Leuchtenberger *et al.* (1956).

The above methods of photoelectric photometry are suitable for measuring the o.p.d. at selected points in the cell. Apparatus constructed by Caspersson *et al.* (1954, 1955) permits the measurement of o.p.d. at points on line-scans across the object. The scanning is achieved by a combination of mechanical movement of the object in one direction (x) and optical movement (a lens device) of the light in the other (y). The measurements are made in even-field interference contrast and the output of the photomultiplier is fed into a function transformer with a resulting linear variation between a pen reading and o.p.d. (Lamakka, 1954, 1955). The feed may be direct, or the record on paper may be followed by hand and the output transformed. The apparatus of Caspersson and co-workers has particular application to the measurement of the product, o.p.d. times area, as discussed in Section IV, B, b.

B. Measurement of Optical Path Difference times Area ($\Sigma\Delta_A\, dA$)

The *total dry mass* of some biological objects may be computed from measurements of o.p.d. and area in an image plane of the microscope (cf. Eq. 6). The problems of measurement are analogous with those occurring in the determination of total amount of absorbing substance by absorption spectophotometry (Caspersson 1936, 1950, and review by Davies and Walker, 1953). We ignore, as we have done hitherto, the special difficulties associated with the image formation in the microscope, and due to the form, shape, and size of the biological material.

We assume as before that the object is illuminated with parallel light, and ignore the bending of light rays by refraction, or diffraction at the object or objective aperture. A single light ray through the object parallel to the optical axis corresponds to a single point in the image plane (Fig. 19), and the interpretation of the optical path difference due to the object is quite clear. The area occupied by the image of the object is by definition the magnified *projected area.* It should be noted that wherever the image plane is situated that the distribution of light intensity is always the same apart from a scale factor, i.e., all parts of the object are in focus. Hence the problem of determining total dry mass of the object is merely that of summing (integrating) the varying o.p.d. over the projected area of the object. If Δ_{Ax} is the o.p.d. in an elementary area dA_x of the image then

$$M = \frac{1}{x^2\chi} \sum \Delta_{A_x} \, dA_x \text{ gm.} \tag{29}$$

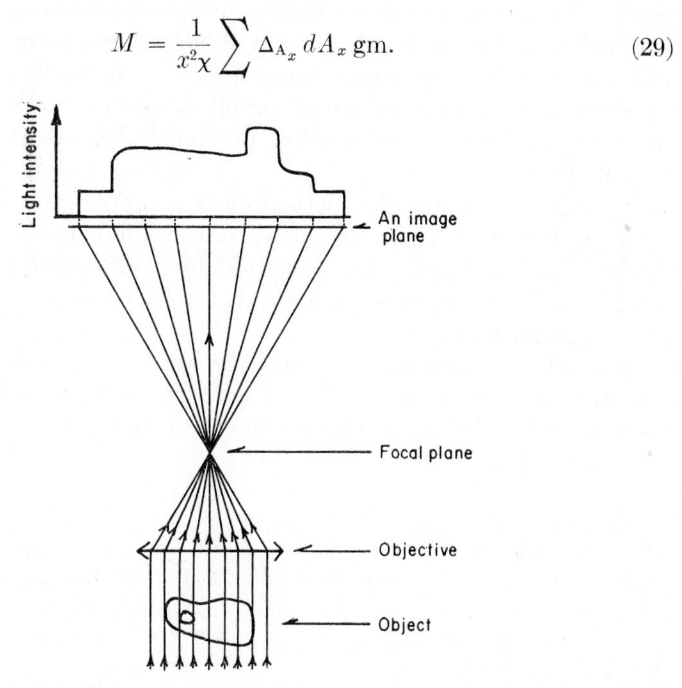

FIG. 19. Diagram illustrating image formation in an objective lens illuminated with parallel light. There is a one-to-one correspondence between rays in the object and points in the image; light intensity shown as if in even-field contrast.

where x is the magnification of the microscope and $\Sigma dA_x = A_x$ the magnified projected area. The unmagnified projected area A is A_x/x^2. Methods for evaluating the product $\Sigma \Delta_A \, dA$, the so-called total refraction, are described in the following three sections.

a. Separate Measurements of Optical Path Difference and Area.

Some objects have a well-defined projected area which can be measured from, say, a photomicrograph. To obtain total mass this area (A) is multiplied by the average o.p.d. measured in a separate experiment. When the o.p.d. does not vary much over the projected area the average may be calculated from several point measurements. When this is not so, an average, weighted with respect to area, is required. Such an average can be calculated from measurements, made photographically or photoelectrically, along a series of lines spread over the projected area. The number of traces will depend on the accuracy required.

When the object has a center of symmetry, measurements along one radius can be used to compute the average o.p.d. If the object is *known* to be a homogeneous sphere, only a measurement of o.p.d. at the center is required. If this is Δ and A is the projected area then it can be shown that the total mass is proportional to $2/3 \cdot \Delta A$. Similar considerations apply to homogeneous cylinders and ellipsoids. The problems are similar to those occurring in absorption photometry discussed in Swift and Rasch (1956).

Example: To make the method clear a simple example is given. To calculate the dry mass in the nuclear area (minus nucleolus) in Fig. 15, the average o.p.d. (Δ) and projected area (A) have been measured, e.g., by the method in Section IV, A, 1, a (iii) and (iv) and planimetry of a photomicrograph.

Δ is 0.19 wavelengths of green light, i.e., $0.19 \times 5460 \times 10^{-8}$ cm. The magnified projected area A_x is 3.37 cm.^{2} and x the total magnification is 1500. With $\chi = 0.18$ substituting the data in Eq. 6 we obtain for the total mass M

$$M = \frac{\Delta \, . \, A_x}{x^2 \chi} \text{ gm.}$$

$$= \frac{0.19 \times 5460 \times 10^{-8} \times 3.37}{(1500)^2 \times 0.18}$$

$$= 86 \times 10^{-12} \text{ gm.}$$

b. Measurement of Area times Optical Path Difference by a Scanning Method.

Some objects do not have a well-defined projected area and hence it is not possible to measure area and o.p.d. separately. The product can be deduced directly from measurements of area under curves (o.p.d. versus length) obtained by a series of closely spaced scans across the image made with a small aperture. A necessary condition for the integration is that the height of the scan is linearly related to o.p.d.; otherwise, as can be simply shown, errors which may be serious will occur.

The required curves can be obtained from photomicrographs when conditions are such that there is a linear relationship between microphotometer reading and o.p.d. (cf. Section IV, A, 1, b (ii)). This is a very laborious procedure and the photoelectric-recording apparatus and auxiliary equipment of Caspersson and co-workers (1954, Section IV, A, 1, c) have been designed specially to increase speed, range of o.p.d., and accuracy. Although this apparatus can be used for the automatic computation of mass in objects of indeterminate area, it is also equally applicable to the rapid assay of mass in objects of well-defined area, no separate measurement of area being made. When the object is surrounded by material the mass of which is not required it is necessary to record each scanning line separately so as to establish exactly the limits of integration. When the object is isolated (e.g., ascites cell) there is no need to do this and total mass can be obtained automatically. This apparatus is necessarily complicated, not simple to construct, and for details the original papers listed in Section IV, A, 1, c should be consulted. The automatic measurement of mass with this apparatus is analogous to the measurement of total stain with the scanning integrator constructed by Deeley (1955).

c. *Measurement of Area Times Optical Path Difference by a Nonscanning Method.*

The above procedures may be exceedingly laborious unless an apparatus of the type due to Caspersson and others is available. Methods have been developed by Davies and Deeley (1956) and Mitchison *et al.* (1956) which enable the product of projected area and o.p.d. to be measured simply with high speed and accuracy. The former employed a Cooke-Dyson and the latter a Baker-Smith interference microscope. The basis of both methods is that over a restricted range of the interference pattern ($< 0.2\,\lambda$) there is an approximately linear relationship between light intensity and o.p.d. A brief description of the Davies-Deeley integrator is given here, so as to bring out its salient features, and in order to compare the possibilities of measurement with it and with the previously described methods.

(i) *The Davies-Deeley integrator.* When the relationship between light intensity and o.p.d. are linear it may be seen, in a general way, that determination of dry mass (o.p.d. $\times$ area) requires measurements of light energy (intensity $\times$ area). As a model we consider a plate of area A and o.p.d. Δ the image of which is formed, in even-field interference contrast, somewhere within an aperture of area a. Two measurements of energy change are required: first E_1, when the object is removed from the field; second E_2, when the phase difference is adjusted to give maximum and minimum illumination. It may be shown that the total refraction is

$$\Delta\,.\,A = \frac{a}{x^2}\frac{E_1}{E_2}\frac{\lambda}{\pi}\ \text{cm.}^3 \tag{30}$$

where λ is the monochromatic wavelength and x the magnification. No measurement of the area of the object is required and the method can be used to "weigh" objects with indeterminable area or of varying optical path difference. There is an important limitation; the o.p.d. of the object must not exceed about 0.2 λ so as to prevent working on the nonlinear part of the interference pattern. When the background field is adjusted to a phase difference ϕ_1 (Fig. 20) somewhere on the relatively straight

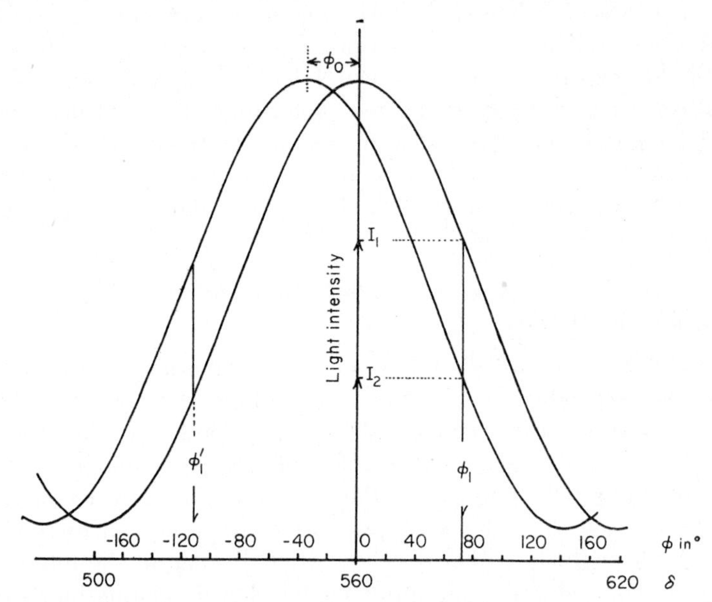

FIG. 20. The light intensity in the Dyson interference microscope as a function of micrometer screw reading δ (typical readings only), and phase difference ϕ (degrees) between the two interfering beams in the background (full line), and in a plate-like object of phase difference ϕ_0 (broken line) where $\phi_0 = 36° = 0.2\pi$ radians. The central maximum corresponds to zero order, $N = 0$, $\phi = 0$. When the micrometer screw is adjusted to $\delta = 584$ ($\phi_1 = 72°$) the light intensities are I_1 and I_2 in background and object, respectively, i.e., the object looks darker than the background. An alternative setting which would give the same value for total refraction in the integrator is $\phi_1' = -108°$, when the object looks lighter than the background (from Davies and Deeley, 1956).

part, the error will be a function of o.p.d. of object. For plate-like objects this has been calculated for various values of ϕ_1 (Fig. 21). Similar calculations hold for spheres. This data enables working conditions and approximate errors to be calculated. It will be seen that errors are small provided the optical path difference is less than about 0.15 λ.

Figure 22 gives an idea of the optical setup. The apparatus has been tested on model spheres and systematic error shown to be small. From

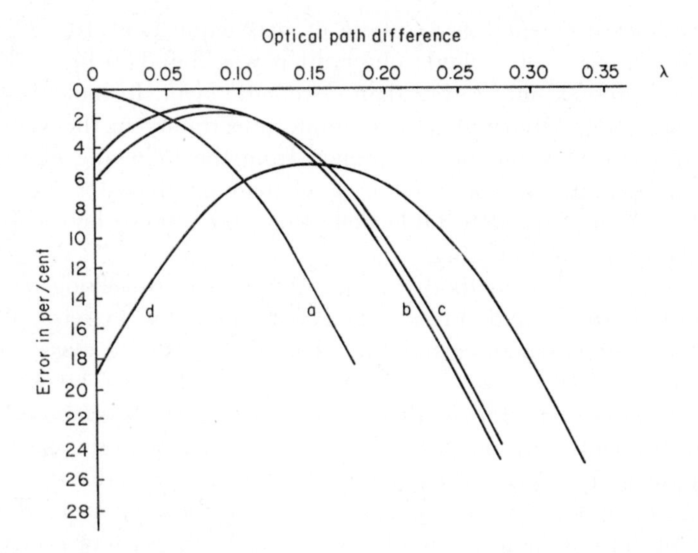

FIG. 21. The percentage error in measuring total refraction (or mass) of plate-like objects in the integrator (Davies-Deeley) as a function of their optical path difference for various values of ϕ_1, the phase difference between the two clear-field interfering beams a, $\phi_1 = 90°$; b, $\phi_1 = 72°$; c, $\phi_1 = 70°$; d, $\phi_1 = 54°$ (from Davies and Deeley, 1956).

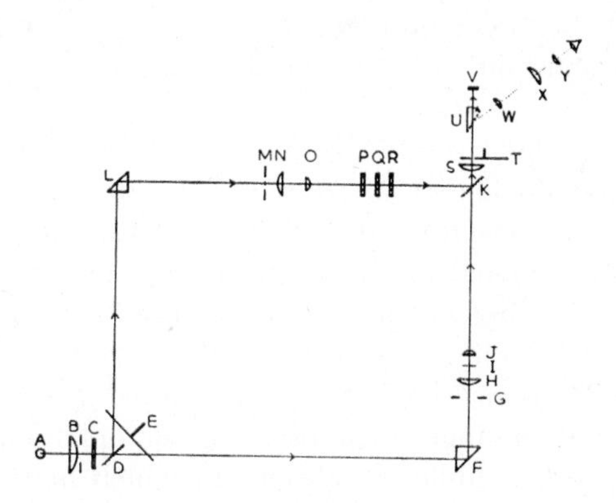

FIG. 22. Schematic diagram of optics of integrator (Davies-Deeley) for measuring total refraction. A, 250-watt compact source mercury arc; B, Köhler lens and iris diaphragm; C, Wratten filters 77A and 58; D, semialuminized beam splitter; E, beam chopper; F, L, prisms; G, aperture stop; H, J, condenser and objective of interferometer microscope; I, specimen; K, beam combiner; M, aperture stop; N, condenser; O, objective; P, Q, R, polaroids; S, lens; T, rotatable disc containing a set of apertures; U, movable mirror; V, photomultiplier cathode; W, X, Y, lens forming an eyepiece (from Davies and Deeley, 1956).

repeated measurements on a sphere of mass-equivalent 10×10^{-12} gm. the probable error of a single observation was found to be $\pm$ 1%. This refers to reproducibility rather than systematic error. In biological material other factors contribute, for example uncertainty in the value of χ. The sensitivity may be roughly judged from the following observation. With an aperture of about 30 μ^2 a change in phase-screw of about 1/1000 λ produced a detectable deflection above noise level under the particular conditions existing.

The range of the method can be extended by squashing objects so as to reduce their o.p.d. Fixed cells can be mounted in refractile solutions, for example sucrose, and Eq. 24 used to compute mass (see also Davies *et al.*, 1957).

(ii) *Comparison with the integrator of Mitchison, Passano and Smith.* Both instruments rely on part of the interference pattern being linear, so that their range of working should be similar. They differ essentially only in the method of compensating for the change in light energy caused by introduction of the object. In the M-P-S system a method of internal compensation is used and as such the method appears to be applicable only to the Baker-Smith interference microscope. In the D-D system a method of external compensation is used and is applicable to both types of interference microscope. Consequent on the different methods of compensation, it follows that in the D-D system the measured product $\Delta . A$ is independent of the relative areas of object and aperture. In the M-P-S system this is not so, altering the size of aperture, with object size constant, resulting in alterations in the calculated value of mass.

(iii) *On measuring from the photographic negative.* The method of integration above (image enclosed by an aperture) can be applied to the photographic negative provided the light intensity transmitted by the film is proportional to o.p.d. and measurements are made with a microphotometer reading in intensity units. For the required conditions see Section IV, A, 1, b (ii).

Huxley and Hanson (1957) have attempted to measure the total amount of material in single myofibrils by microphotometry of the film negative. They worked under conditions in which microphotometer deflection in log units was linearly related to o.p.d. Hence the method in the last section could not be applied. Instead they scanned the negative with a slit oriented at right angles to the fibril, and computed the area under the trace in a manner similar to that suggested in Section IV, B, b (1st paragraph). It can be shown that with a microphotometer reading log units, errors occur if the area under the slit is not of uniform blackening. Huxley and Hanson showed that errors under their particular conditions were small, but point out that their method should not be applied without careful calculation.

d. Summarizing Remarks on Methods.

Measurement of total dry mass is simple if the object is homogeneous and its shape simple, for example, a flat homogeneous plate, or a homogeneous sphere. Separate measurements of area and o.p.d. can be made.

When the distribution of material is nonuniform methods of integration by line scanning of the film negative can be used. These are generally laborious and become exceptionally so when the relationship between deflection and o.p.d. is nonlinear. It would then be necessary to convert to a linear scale before measurements of area under the trace could be made. The other method is to use photoelectric scanning and function transforming as developed by Caspersson and co-workers. This equipment is very complex. This method gives a fairly rapid estimation of mass in isolated cells, but when the mass of one cell, surrounded by others, is required, time taken must increase because limits of integration must be marked on each scan.

The methods developed by Davies and Deeley (1956) and Mitchison *et al.* (1956) have limitations. First the o.p.d. must not exceed about 0.15 λ. Second the method cannot be used to weigh cells surrounded by others because in general cell shape is complex and the required area cannot be outlined by a circular aperture. Apertures of constantly variable outline would solve the problem, but would be difficult to construct. However on account of the comparative simplicity of the apparatus and the excellent accuracy and speed it is worth while attempting to prepare the biological sample so that the above limitations can be removed. For example although irregular nuclei inside cells cannot be measured by nonscanning integration the nuclei can be removed from the cells and then measured. It also so happens that there are, sometimes, fundamental limitations to the measurement of "nonisolated bodies" due to presence of over- and under-lying material.

V. CRITIQUE

A. Introduction

The interference microscope method for measuring dry mass and concentration is capable of yielding results with sufficient accuracy to give an unequivocal solution to many cytological problems. This is only to be expected since the basis of the method is simple, relying on well-estabished values of refraction increments; the method of measuring optical path difference is also standard physical practice. That this expectation is realized in actual measurements with the interference microscope has been established by comparison with independent methods for determining mass (Section V, B).

There are, however, certain limitations to the type of biological object which can be measured with accuracy. For example in objects which do not have a simple shape, or are near the resolving power of the microscope in size, or in which the material is inhomogeneously distributed, the accuracy in measuring optical path difference and in deducing projected area is reduced. It is necessary to make estimates of the magnitude of the errors that arise from the various causes so as to be able to realistically assess the possibility of tackling the various cytological problems. With an awareness of the errors it is sometimes possible to overcome certain of the limitations by special methods of preparing the biological material and by other special techniques. The errors arising in microscope interferometry are closely analogous with those in absorption spectrophotometry which have been much studied (Caspersson, 1950, and review by Davies and Walker, 1953).

Errors, in accordance with general practice, may be divided into two sorts. Accidental errors in measuring o.p.d. (Δ) and area (A) may be reduced to reasonable limits by repeated observations. We will only be concerned with the systematic errors, more insidious in nature, which can arise in the parameters from which mass and concentration are computed. It does not always seem to be sufficiently well appreciated that a mean value given with a minute percentage deviation may be quite incorrect due to a large systematic error.

The uncertainty in χ (Section II) introduces the major uncertainty in microscope interferometry on fixed material. Experiments are in progress to evaluate χ accurately under these conditions. The systematic errors in measuring Δ and A (Section V, C) arise from a combination of effects due to the nature of the object and to the special conditions of image formation in the microscope.

B. Comparison with Other Methods

A comparison between the optical interference and the soft X-ray absorption method (Engström and Lindström, 1949, 1950) on a variety of biological objects gave results in agreement to about $\pm$ 15% (Davies *et al.*, 1953). The objects were of varying inhomogeneity. On homogeneous retinal rods from frog the concentrations of dry substance obtained by X-rays and interferometry were 0.41 and 0.39 $\mu\mu$gm./μ^3 respectively.

Similar good agreement (Davies, Engström, and Lindström, unpublished) was obtained on nerve fibers from frog. A χ value of 0.18 was used throughout this work. For a comparison of the two techniques for measuring mass the original paper should be consulted.

Davies *et al.* (1954b) obtained by interferometry and from a biochemical value for DNA a value of 40% for the ratio of DNA to total dry mass in living sperm heads of ram. The value obtained by Mann (1949) was computed as 45%. A reevaluation, using more extensive and accurate data on lipid-extracted sperm heads, by Mann (1954) and Davies *et al.* (1957) indicates that to obtain agreement a value of χ between 0.15 and 0.16 is needed in this material which has an unusually high content of DNA.

Hale and Kay (1956) compared the dry masses of isolated calf thymus nuclei obtained by interferometry and ordinary weighing procedures and obtained remarkably good agreement with a value for χ of 0.18 (see Section VI).

The above results indicate good agreement when χ has a value 0.18 in fixed material. However this could be due to a balancing out of errors in the methods which are being compared. The possible decrease in χ at high protein concentrations is not large and specially refined methods will be needed to check it accurately.

C. Errors in Measuring Optical Path Difference Δ and $\Sigma\Delta_A dA$

Hitherto we have made certain simplifying assumptions. Referring to Fig. 19 where the microscopic object is illuminated in parallel light, corresponding to a single light ray through the object there is a single point in an image plane. Interference contrast arises as a result of the separation of the two interfering fields by distances equal to the optical path differences of corresponding rays in the object, coherent points lying on a single light ray. Under these simplified conditions, no matter how irregular the object, it is possible to attribute a simple unique meaning to optical path difference and to obtain the true value for the total refraction (or total mass) of the microscopic object. Recording errors due, for example, to unevenness of light intensity within a measuring spot are not fundamental and do not concern us.

Stated simply, the errors in determining Δ and $\Sigma\Delta_A dA$ as defined under the idealized conditions, arise because under the actual conditions of measurement the one-one correspondence no longer holds.

It is convenient to distinguish three effects which upset the one-one correspondence. First, the use of conical illumination which gives rise to the so-called distributional errors; second, the effects of diffraction; third, the effects of glare. Although the errors arise from the same physical causes in interferometry with the Cooke-Dyson and Baker-Smith microscopes their magnitude depends partly on the way in which measurements are made. In what follows the treatment specially refers to the former instrument.

1. *Distributional Error*

Due to the use of a condenser aperture of finite value, in general, rays contributing to the light intensity at a point in the image of a microscopic object will have travelled through varying path lengths in the object. Similar effects are brought about by refraction or diffraction at the object even when parallel light illumination is used. One result is that the light intensity in the image plane varies with the position of the image plane in quite a different way from what it did under the idealized condition with parallel illumination. Only in one position, the so-called in-focus position, is there any simple relationship between o.p.d. (Δ) and total refraction ($\Sigma\Delta_A dA$) for parallel light and the measured o.p.d. (Δ') and total refraction ($\Sigma\Delta'_{A'}dA'$). Choice of focus in the case of objects of simple geometry, homogeneous plates and spheres, will be obvious, but in general parts of irregular objects will be out-of-focus and measurements in any image plane will be in error.

In what follows we assume the objects to be large compared with the wavelength and that ray (geometrical) optics apply. We can neglect the effects noted in Section III, B, 1, a which although also arising from a finite condenser aperture are relatively small.

a. Plate Illuminated with Convergent Light.

When the object is a homogeneous plate at right angles to the optic axis, it is usual to focus on the center so as to reduce distributional error at its edges. The effect of convergent light is to increase the effective optical path. As a result measurements of o.p.d. (Δ') will be higher than with parallel light (Δ). The effect has been calculated theoretically by Davies and Deeley (1956). The results shown in Fig. 23 apply to the Cooke-Dyson interference microscope which has a central obstruction of aperture of about 0.25. For a condenser numerical aperture of 0.6 the error is small, about 5%. The errors should be somewhat less in the Baker-Smith microscope which has no central obstruction.

The vector diagram given by Davies and Deeley (1956) illustrating their calculations appears to be inexact (from considerations of coherence). The following vector theory illustrates that, due to the varying optical paths in the object, a reduction in the contrast of the interference pattern is to be expected, as well as a shift in its position. We divide the light incident on the plate into annular cones of semi-angle θ (Fig. 24A) contrast being even field. In Fig. 24B is shown the individual interference patterns for a cone of semi-angle θ at points in the image plane of surround (Fig. 24B, curve a) and plate (Fig. 24B, curve c). When θ is zero the interference pattern is shown by curve b and for intermediate angles curves lie between b and c. The interference pattern for a solid cone of angle θ is obtained by adding all curves intermediate between and including b and c.

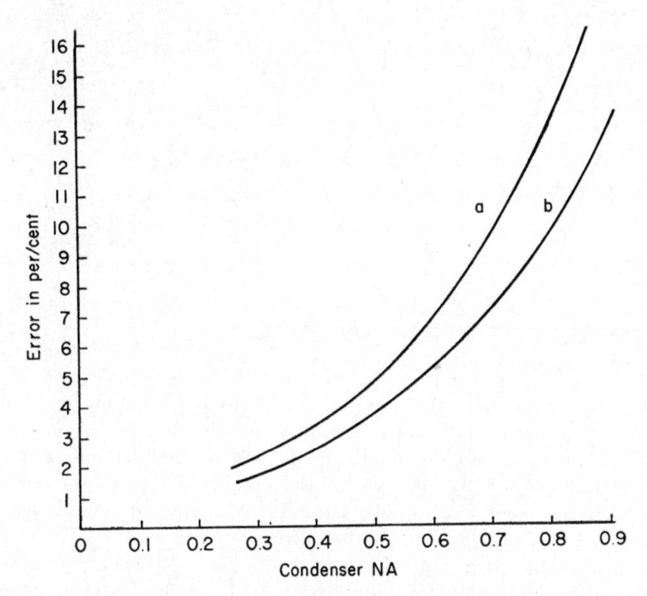

FIG. 23. The percentage error in measuring optical path difference of a flat plate as a function of numerical aperture (NA) of the condenser in the Cooke-Dyson interference microscope. The curves start at 0.265, the NA of the central obstruction in the objective; the angle subtended by this obstruction is 19° 40′. a, refractive index of immersion medium 1.33. b, refractive index of immersion medium 1.52. The refractive index of the microscope immersion medium is also 1.52 (from Davies and Deeley, 1956).

The resulting interference pattern (Fig. 24C) is seen to be not only shifted on account of the increase in effective thickness but also reduced in visibility, i.e., minima no longer zero. Calculations (Davies and Deeley, 1956) indicate that the reduction in contrast in this case is negligibly small. The nature of the effect is discussed in detail since in inhomogeneous objects it may result in serious reduction of contrast (Section V, C, 1, c).

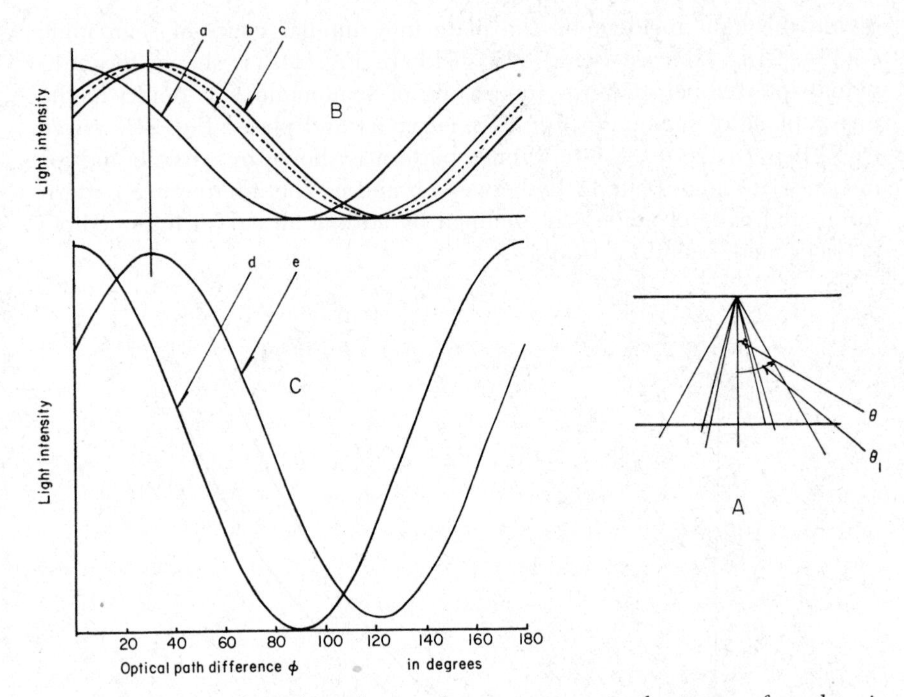

FIG. 24. To show how the interference pattern in the image of a plate is produced in even-field interference contrast in the Cooke-Dyson interference microscope. A: the light incident on the plate can be divided into cones of semi-angle θ, θ_1 being defined by the NA of the condenser. B: shows the individual interference patterns from which light intensity at a point in the image of object and surround can be computed. a, in the surround ($\theta = 0$ or θ_1); b, in the object ($\theta = 0$); c, in the object ($\theta = \theta_1$). The dotted curves between b and c indicate interference patterns for cones of illumination with angles between 0 and θ_1. C: taking into account all the cones ($\theta = 0 \rightarrow \theta_1$) the interference pattern at a point in the image of the object is obtained by summing the interference patterns b $\rightarrow$ c to produce e; d, the interference pattern in the surround, is similarly produced.

b. Other Objects of Simple Geometry Illuminated by Convergent Light.

When the microscope is focused on the center of a homogeneous sphere, measurements at the center with converging light give the same o.p.d. as in parallel light. Corrections for cylinders and ellipsoids will be intermediate between zero and that for a plate. At a distance from the center of a sphere optical path difference in convergent light may differ somewhat from that in parallel light neglecting refraction. No calculations have been made in the way that was done for absorption photometry by Caspersson (1936) and Thorell (1947). There serious distortion occurred only at the edge of the sphere.

c. Out-of-focus Errors.

Irregular three dimensional objects (e.g., a prophase or metaphase chromosome group) are unsuitable for accurate assay of total mass unless they can be flattened into one plane so that all parts can be in focus at once, or unless the optical path differences are everywhere small (see also Section V, C, 1 (d)).

To obtain an estimate of errors due to out-of-focus we consider a plate of o.p.d. Δ and magnified area A_x. We compare the total refraction $\Sigma \Delta_{A_x} dA_x$ with that obtained by integrating over the out-of-focus pattern, i.e., $\Sigma \Delta'_{A'_x} dA'_x$. This integration would in practice be made most easily by the method in Section IV, B, b in which the image plane is scanned, point by point, with a small aperture. We assume that the effect of out-of-focus is merely to smear out the energy difference in the image plane. This is theoretically expected to hold at least to a first approximation. Experiments (Davies and Deeley, 1956) indicate that it is so, negligible change in energy-difference being found with the integrating attachment when objects of o.p.d. 0.1–0.3 λ were put out of focus.

If ϕ_1 (radians) is the uniform phase difference between the two interfering fields (as in Fig. 20), in the absence of the object and ϕ_0 is the phase difference due to the object ($< \pi$) then it can be shown that the error in total refraction is

$$\% \text{ Error} = 100 \frac{\phi_0 - \left[\cos \phi_1 - \dfrac{\cos(\phi_1 + \phi_0)}{\sin \phi_1}\right]}{\phi_0} \tag{31}$$

This is the maximum error, that is when the particle is completely out of focus. By substituting the appropriate parameters in Eq. 31 the error can be calculated. Thus when ϕ_1 is near 0° or 180°, i.e., near light or dark field conditions, the error is negative and may be very large indeed; the reading for total refraction is too high. When ϕ_1 is 90° it may be seen that the error has changed sign so that the reading for total refraction is too low. The equation then reduces to Eq. 10 in Davies and Deeley (1956) and the error is indicated by curve a in Fig. 21. For this value of ϕ_1 the error is small provided the o.p.d. is small ($< 0.1\,\lambda$).

d. Inhomogeneity of Objects.

To simplify the problem we consider the microscope focused on the center of a microscopic sphere. What is the error in measuring optical path difference at the image center when the homogeneous distribution of material is converted into an inhomogeneous one, a collection of plates oriented as shown (Fig. 25), total refraction remaining constant? Light rays passing at various angles through the center of the microscope object traverse varying numbers of plates and hence varying amounts of refractile material. The problem is exactly similar to that in Section V, C, 1, a where it was shown that the interference pattern in the image plane could be obtained by adding a series of curves (Fig. 24B, C). A rough estimate of the magnitude of the effect in the present case can be obtained from a simple treatment kindly provided by A. R.

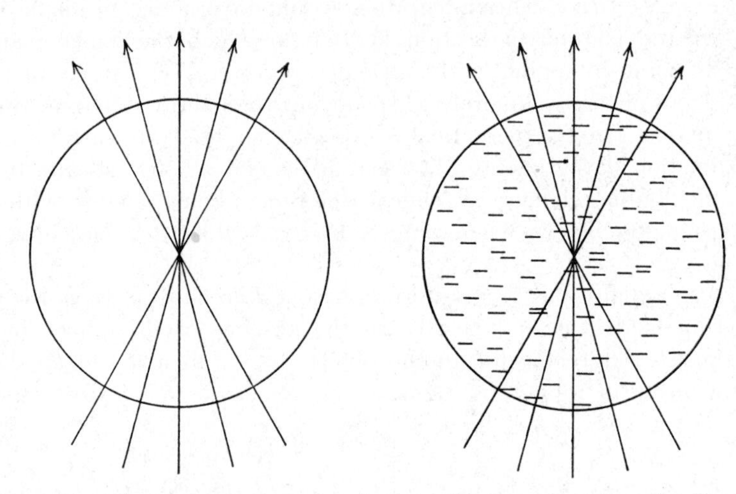

FIG. 25. The material present in the homogeneous sphere is redistributed into plates oriented at right angles to the optical axis of the microscope.

Stokes (unpublished). The plates are assumed to be distributed randomly throughout the sphere, with their planes at right angles to the optical axis. If $d\phi_0$ is the phase difference due to each plate and on average the rays pass through q plates then the average optical path difference is $qd\phi_0$, equal to that for the homogeneous sphere. The path differences of the rays through the center of the sphere are distributed according to a probability function about the value $qd\phi_0$. A mathematical summation corresponding to the graphical summation of curves in Fig. 24 gives for the intensity I^1 in the interference pattern.

$$I^1 = 2r^2[1 + e^{-q(1-\cos d\phi_0)} \cos (\phi + q \sin d\phi_0)] \tag{32}$$

where r is a constant equal to the amplitudes of the interfering beams and ϕ is the phase difference between the beams in the absence of the object. The corresponding intensity I for the homogeneous sphere is

$$I = 2r^2[1 + \cos (\phi + qd\phi_0)] \tag{33}$$

A comparison of Eqs. 32 and 33 shows that the interference pattern is shifted from $qd\phi_0$ to $q \sin d\phi_0$ and that the amplitude (proportional to visibility) is altered by a factor $e^{-q(1-\cos d\phi_0)}$.

TABLE II

SUMMARY OF FORMULAS REQUIRED FOR CELL INTERFEROMETRY [a]

Formula [a]	Eq. no.	Applications and Remarks
$M = \frac{\Delta_w A}{\chi}$	6	For measurement of total dry mass of cell (living or fixed) immersed in water. No measurement of thickness needed.
$M = \frac{A\Delta_m}{\chi} + (\mu_m - \mu_w)\frac{At}{\chi}$	13	For measurement of dry mass of cell (living or fixed) in a medium of refractive index μ_m. For living cells t is geometrical thickness t_g; for fixed cells t is usually effective thickness t_e. Needs a measurement of thickness.
$M = \frac{A\Delta_m}{\chi} + \frac{\chi'}{\chi} m'$		An alternative form of Eq. 13 in which χ' is 100 times the specific refraction increment for the immersion medium and m' is the calculated dry mass of immersion medium in a volume equal to that of the cell.
$M = \frac{\Delta_1 A}{\chi}\frac{(\mu_2 - \mu_w)}{(\mu_2 - \mu_1)} - \frac{\Delta_2 A}{\chi}\frac{(\mu_1 - \mu_w)}{(\mu_2 - \mu_1)}$	19	For calculating dry mass of cell (living or fixed) from measurements of o.p.d. (Δ_1, Δ_2) in two media of refractive index μ_1 and μ_2.
$M = \frac{A\Delta_m}{\chi}\frac{\mu_p - \mu_w}{\mu_p - \mu_m}$	24	For dry mass of fixed cells in a medium of refractive index μ_m. Needs knowledge of μ_p (~ 1.54 in sucrose solutions). Can also be used for computing mass of living cells in a medium (μ_m) if concentration and hence refractive index (μ_c) is known (μ_c replaces μ_p).
$t = \frac{\Delta_1}{\mu - \mu_1}$		For determining thickness t from measured o.p.d. Δ_1, when the refractive indices of specimen (μ) and of medium (μ_1) are known. Useful for determining thickness of dried cells ($\mu \sim 1.54$) mounted in air ($\mu_1 = 1$).

TABLE II (Continued)

Formula [a]	Eq. no.	Applications and Remarks
$t = \frac{\Delta_1 - \Delta_2}{\mu_2 - \mu_1}$	18	Formula for calculating thickness from two measurements of optical path difference in media of refractive index μ_1 and μ_2. Value of t, whether t_g or t_e, depends on penetration of medium.
$\mu_c = \frac{\Delta_1\mu_2 - \Delta_2\mu_1}{\Delta_1 - \Delta_2}$	17	Formula for calculating refractive index (μ_c) from two measurements of optical path difference in media of refractive index μ_1 and μ_2.
$c = \frac{\Delta_w}{t\chi}$	7	For calculating concentration when o.p.d. in water is measured and thickness t is known.
$c = \frac{\mu_c - \mu_w}{\chi}$	14	For calculating concentration from cell refractive index (μ_c) measured by matching method.
$c = \frac{\Delta_1}{\chi}\frac{\mu_2 - \mu_w}{\Delta_1 - \Delta_2} - \frac{\Delta_2}{\chi}\frac{\mu_1 - \mu_w}{\Delta_1 - \Delta_2}$	20	For calculating concentration from measurements of o.p.d. in two media of refractive index μ_1 and μ_2.
$w = V - 0.75M$	24	Formula for calculating mass of water (w) in living cell (only protein) from volume V and dry mass M.

[a] List of symbols. M, total dry mass (gm.); c, concentration of dry substance per unit volume (gm. cc.$^{-1}$); χ, a parameter approximately a constant (see text); A, projected area of cell (cm.^{2}); Δ, the optical path difference (cm.); subscripts w or m refer to water or medium; μ, the refractive index; subscripts c, w, 1, 2 refer to living cell, water, and media 1 and 2, subscript p refers to the fixed cell (see text for interpretation of μ_p); t is the cell thickness.

So as to show the magnitude of the effect a few numerical values are given in Table III. The main effect is to reduce contrast in the interference pattern, the phase shift due to inhomogeneity being small. The reduction in contrast is apparently negligibly small only when the optical path differences ($d\phi_0$) are very small ($< 0.01\ \lambda$) or the over-all o.p.d. is small ($< 0.1\ \lambda$). An estimate of the refraction in microscopic objects can be obtained from size, and refractive index, and density of protein (cf. Table I). Thus particles of solid dry protein in size about equal to the resolving power ($0.25\ \mu$) will have an o.p.d. of about $0.1\ \lambda$.

This suggests that, in some fixed cells, reduction in contrast may be an important effect and it should be tested for.

The extent to which reduction in contrast will produce error in measuring o.p.d. will depend on the method of measurement. This is shown in Fig. 26 which graphically illustrates data in Table III (line 2).

TABLE III

DECREASE IN CONTRAST WHEN A HOMOGENEOUS DISTRIBUTION OF SUBSTANCE BECOMES INHOMOGENEOUS [a]

$qd\,\phi_0$		q	$d\phi_0$ in λ	$q \sin d\phi_0$ in degrees	$e^{-q(1-\cos d\phi_0)}$
In degrees	In λ				
36	0.1	10	0.01	36	0.98
72	0.2	5	0.04	71	0.85
108	0.3	5	0.06	105.4	0.70
108	0.3	10	0.03	107.4	0.84
108	0.3	30	0.01	108	0.94
172.8	0.48	8	0.06	168.6	0.57
172.8	0.48	16	0.03	171.9	0.76

[a] q, number of particles; $d\phi_0$, phase difference particle.

When measuring the shift of the interference pattern due to the object the error is zero provided the background field is set at $\phi = 90°$, or if the distance is measured between interference minima in the object and background. The distortion is symmetrical about a point half-way between maximum and minimum light intensity in the surround. Hence measurements in light and dark contrast with the integrator as at ϕ_1 and ϕ_1' in Fig. 20 will be identical and will not reveal an error. To test for inhomogeneity the interference pattern in the object must be measured, a measure of the error being the extent to which the object does not "black out."

Inhomogeneous objects may be difficult to measure by visual photometry and it may be necessary to use a nonvisual method. Small fluctuations in interference pattern will occur depending on size of recording aperture and whether or not an individual particle is in focus. These fluctuations will usually be small compared with the total effect introduced by inhomogeneity.

2. *Diffraction Effects*

a. Light Loss Due to the Object.

A reduction in the amplitude of the light wave passing through the microscopic object can be brought about by several factors. In homogeneous objects light loss can be caused by reflection of light energy outside the objective aperture (Fresnel's laws apply, see Jenkins and White, 1957). Experiments (Caspersson, 1950; Davies, 1954) on living and fixed cells, however, indicate that by far the most serious cause of amplitude reduction is loss of light due to diffraction (scattering) by particles produced when fixing cells. These particles are in size of the order of the wavelength of light and less. Light loss in visible wavelengths is occasionally brought about by true absorption, due for example to hemoglobin in erythrocytes.

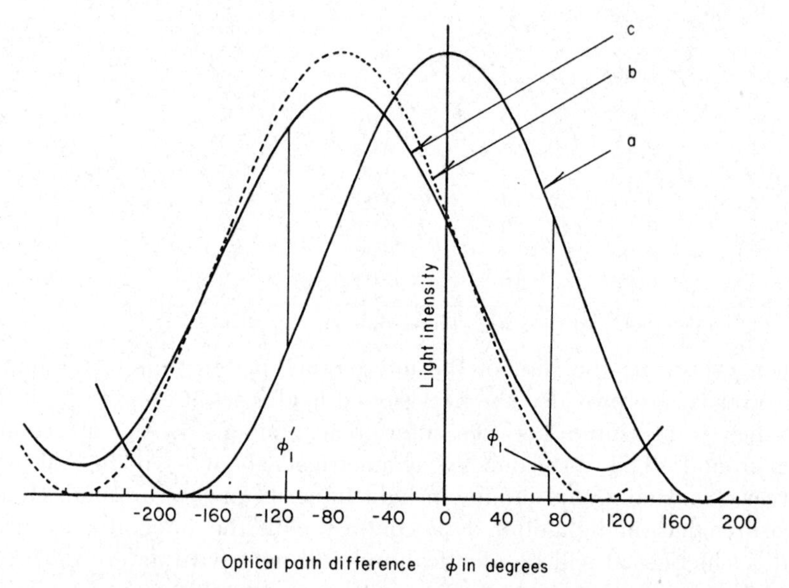

FIG. 26. Shows the effect of inhomogeneity of the object on the interference pattern in the image plane, in even-field interference contrast in the Cooke-Dyson microscope. Shows data in Table III, line 2. a, interference pattern in surround. b, interference pattern in homogeneous object of o.p.d. 0.2 λ. c, interference pattern in inhomogeneous object of same total refraction. Note that the vertical separations at ϕ_1 and ϕ_1' are identical, the curve c being symmetrical.

The effect of light loss due to the microscopic object is to change the shape of the interference pattern in its image. As an example, suppose the light loss is equivalent to an optical extinction (absorbance) of 0.1. Then r_1^2 the square of the amplitude of the beam which has traversed the object is related to that of the comparison beam (r_2^2) by the expression $r_1^2 = 0.8r_2^2$. In the absence of the object the beams are assumed to have equal amplitudes. Figure 27 shows the distortion in the interference pattern obtained by using Eq. 43 (Appendix I); the o.p.d. due to the object is assumed to be 0.2 λ. The change in shape is different from that produced by inhomogeneity the major effect being to depress the interference maximum.

The error introduced by light loss in the object depends on the method of measurement. When setting on a minimum there is no error. If the shift in the pattern is being measured with the photometer eyepiece say, then by setting the surround first to a positive and then to a negative phase value optical path differences greater and less than the true value

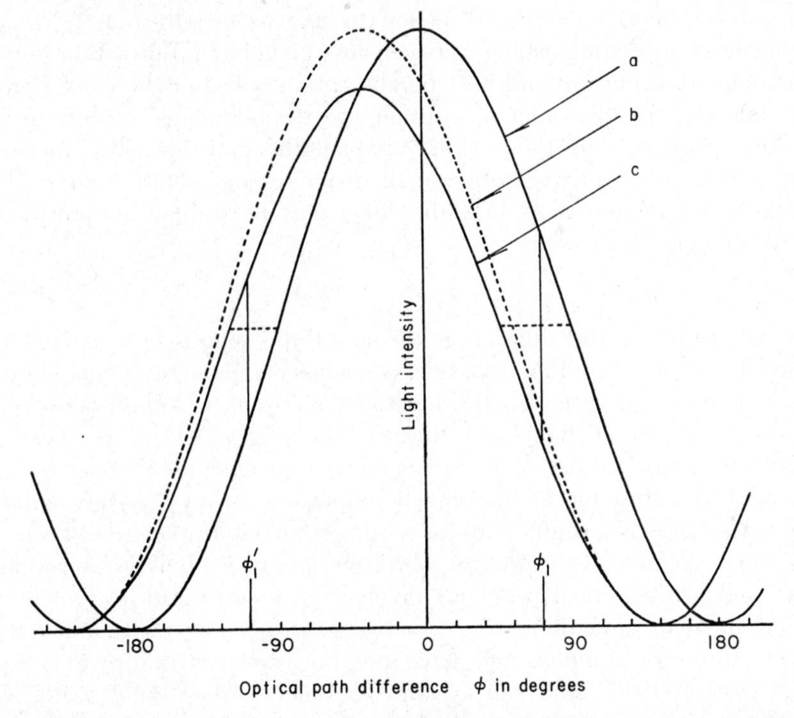

FIG. 27. Shows the effect of light scattering by the object on the interference pattern in the image plane, in even-field interference contrast in the Cooke-Dyson microscope. a, interference pattern in surround. b, interference pattern in an object without scatter, o.p.d. 0.1 λ. c, interference pattern in an object with scattering, equivalent optical extinction (absorbance) being 0.1. Note that the vertical separations at ϕ_1 and ϕ_1' are unequal, while the horizontal separations (dotted) at a background $\phi = \pm 90°$ are less than and greater than the *true* value by an equal amount.

will be obtained. This will be seen from the horizontal dotted lines in Fig. 27. The true value lies precisely midway between the two readings. The magnitude of the effect for $\phi = \pm 90$ in Fig. 27 is $\pm\lambda/30$. The value of 0.1 for the extinction is however unusually high. When the total refraction is measured on the integrator in light and dark field contrast (ϕ_1 and ϕ_1' in Fig. 27), the true o.p.d. lies approximately in between the values obtained (see Davies and Deeley, 1956). In the Baker-Smith interference microscope the effect of light loss due to the object is to introduce elliptical polarization and this distorts the light pattern produced by the rotation of the analyzer following the $\lambda/4$ plate.

We have assumed the amplitude of the interfering waves in the surround to be equal. If this is not so then conceivably light loss due to the object may be in the direction to increase contrast in its image. Difficulties in distinguishing between loss of light by diffraction outside the objective aperture and by true absorption is one of the chief sources of difficulty in ultraviolet absorption spectrophotometry of cells. Light scatter decreases in the visible wavelengths and in many instances errors due to it are negligibly small in microscope interferometry. The data in this section show how the effect may be studied and correction made if necessary.

b. Size of Object.

According to the diffraction theory of the microscope the one-one correspondence (cf. Fig. 19) between object and image plane derived from geometrical optics breaks down as the size of the object approaches the wavelength of light and towards the edges of larger objects. A luminous point in the object plane is not imaged as a point but, on account of diffraction at the objective aperture, as an Airy disc pattern. The Airy disc is a light pattern arising from interference between the secondary light wavelets at the objective aperture (Jenkins and White, 1957). The theoretical problems involved in assessing measurements of o.p.d. in small three-dimensional bodies, as a function of size and refractive index, are complex and have not been solved. In previous work (Caspersson 1936, 1950) an attempt was made to set limits to the size of microscopic object below which the image would cease to correspond to the object. For an objective numerical aperture of about 1.2 the size of object was about three to four times the wavelength of light. An aberration free objective was assumed. Other calculations indicate a lower limit under certain conditions (Wilkins, 1950). In the absence of exact data on errors in microscope interferometry measurements of o.p.d. in the center of bodies below the limit ($3\,\lambda \rightarrow 4\,\lambda$) should be regarded with caution.

There are two types of problem involved in measurement on small microscopic objects which may be illustrated by reference to the photomicrograph of a myofibril (see Fig. 37). First there is the problem of the accurate localization of mass. The Z and H lines (located in isotropic

(I) and anisotropic (A) segments) are of the order of the resolving power and accurate measurements of area and optical path difference in them cannot be made. Hence the mass of substance localized in the Z and H regions cannot be measured. The second problem is the measurement of total refraction (or mass) in the entire fibril. We may approximately regard the effect of diffraction at the objective aperture as causing a redistribution of the energy in the interference pattern in the same way as an out-of-focus effect. Hence similar calculations apply as those in Section V, C, 1, c where it was shown that errors are small provided that the o.p.d. is small (< 0.1) and measurements are made on the linear portion of the interference pattern. Further research is needed into the magnitude of the errors involved in microscope interferometry on bodies of the order of the wavelength.

3. *Glare*

As a result of glare, that is light reflection and scattering by lens surfaces in the objective, etc., a fraction of the light energy from each ray through the object is randomly distributed over the image plane. In microabsorption measurements glare can cause serious errors at high values of optical extinction. In interferometry when the image plane is crossed by interference bands the effect of glare is to remove energy from the bright to the dark part of the pattern, that is to reduce contrast. The contrast is reduced equally in surround and object and no systematic error is introduced.

However in even-field interference contrast and with alteration of the phase difference the intensity of the glare light may also vary and hence introduce errors into the measurement of o.p.d. The effect can be studied by measuring the change in light intensity inside the image of an opaque body (e.g., carbon black) large compared with the wavelength. When glare is absent or constant there will be no change in intensity. In one experiment referred to in Davies and Deeley (1956) the effect of glare was found to be small. However the effect needs further study and glare should be tested in every microscope since it can be expected to vary. The precautions usual in microscopy for reducing glare should be exercised, namely, small field stop, clean surfaces, and as small a condenser aperture as is commensurate with other requirements.

4. *Objects in the Comparison Beam*

In the Baker-Smith microscope the presence of objects in the comparison beam will produce a ghost image overlying the object to be measured. This can easily be seen and hence avoided.

In the Cooke-Dyson microscope objects in the comparison beam cannot be so easily detected, but their effect is as follows. Cones of light through points in the field of view have associated with them comparison areas which are annular in shape and which approximately overlap. If the objects in the comparison areas are distributed at random (with regard to lines radiating from the center of the field), it may be seen that the effect is to reduce the contrast equally at points in the field of view. Hence no systematic error is introduced. If, for example, 30% of the comparison area is occupied by objects of o.p.d. 0.1 λ then it can be shown that the ratio of light to dark field is reduced to 12:1. Measurements can still be made with this contrast.

Alterations in contrast which may be noted as the specimen is moved about in the microscope can be caused by variations in the amount of refractile material in the comparison area. Precautions can be taken during preparation of material to minimize loss of contrast from this cause.

5. *Other Factors*

When the microscopic object is known to be birefringent this should be taken into account.

Possibility of error in optical path difference, when determining the whole number of wavelengths with white light, has recently been discussed, with examples, by Faust and Marrinan (1955). This is a well-known difficulty in interferometry and arises since the achromatic band in white light, that is the position where light of all wavelengths coincide, does not necessarily coincide with the position of zero path difference. The effect of the object is to cause the interference pattern to be shifted by an amount depending on the path difference. If the variation of o.p.d. with wavelength (dispersion) is different in object and surround the band system is not displaced by an equal amount for all wavelengths and the white band first becomes blurred and with increasing dispersion may coincide with a position of finite path difference. Although evidently important in some applications of the interference microscope, the difficulty has not yet been encountered in biological practice due mainly to the thinness of the specimens and the small path differences.

D. Choice of Condenser Aperture and Order (N)

A finite condenser aperture is used, generally, in microscopy for several reasons, primarily because the resolving power increases as the sum of the numerical apertures of objective and condenser. Another reason is that increasing the condenser aperture increases the light intensity in the image. Factors particular to microscope interferometry have been discussed in Sections III, B, 1, a and V, C, 1, a. In practice certain features of microscope design mainly decide the choice of condenser aperture. Thus in the Baker-Smith microscope it is not possible to increase the condenser numerical aperture above 0.5. Experiments (Davies and Deeley, 1956) with the Cooke-Dyson microscope indicate that increasing the condenser numerical aperture much above 0.5 to 0.6 (objective NA 1.3) results in decrease in contrast, as a result of glare, presumably due partly to multiple reflections in the interferometer plates.

A condenser aperture of 0.5 to 0.6 appears to be a reasonable choice for most applications. It is desirable to keep the path difference (order N) of the interfering rays in the image plane small, on account of the effect discussed in Section III, B, 1, a. It is convenient to work at or near the zero order in the surround.

E. Limitations in Measuring Bodies Inside Cells

There is a well known "geometrical" limit to optical measurements on bodies inside cells due to the fact that light rays must also pass through over- and under-lying layers. Unless the thickness of these layers and also that of the body itself can be measured neither the mass nor concentration of dry substance in the body can be determined. An expression for the dry mass of the body may be obtained as follows.

Let μ_b, μ_c and μ_m be the refractive indices of the intracellular body, the cell, and the immersion medium, respectively (Fig. 28). Let t and d

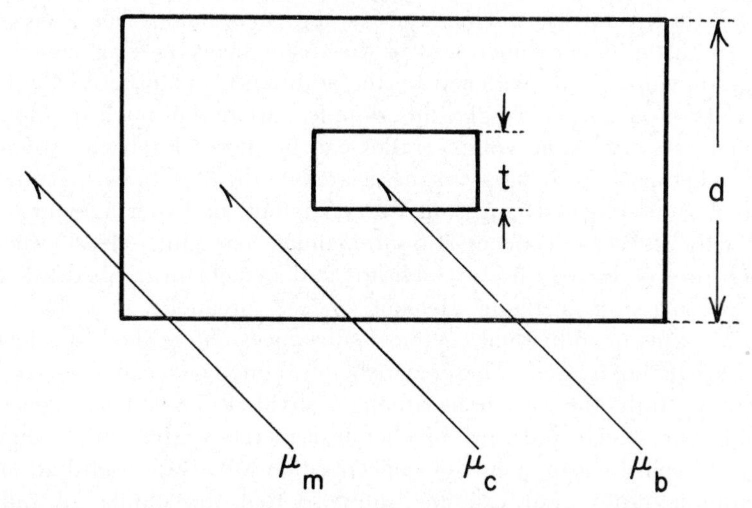

FIG. 28. Model of a body of refractive index μ_b, inside a cell of refractive index μ_c in a medium μ_m. The thickness of the body and cell are t and d, respectively.

be the thickness of the body and cell and Δ_1 and Δ_2 the optical path differences through the body and adjacent region, relative to the surround

$$\Delta_1 = \mu_b t + \mu_c(d - t) - \mu_m d \tag{34}$$

$$\Delta_2 = \mu_c d - \mu_m d \tag{35}$$

Eliminating μ_c and combining with Eq. 6 we obtain an expression for the dry mass M

$$M = \frac{A}{\chi}\left[\Delta_1 - \Delta_2\left(1 - \frac{t}{d}\right)\right] + \frac{A}{\chi}t(\mu_m - \mu_w) \tag{36}$$

The cell thickness may be obtained from measurements of o.p.d. in two media (cf. Eq. 18). Data on thickness of intracellular bodies can only be obtained if they are spheres or cylinders.

Difficulties from this cause are encountered in measurements on flattened cell nuclei in tissue culture, above and below which are thin layers of cytoplasm. Similar difficulties arise in measuring the mass of chromosomes in intact cells due to uncertainty in material above and below.

VI. ASSESSMENT OF RESULTS

A. Introduction

In the preceding sections the methods for measuring mass and concentration of dry substance, and water content of cells and other tissue elements have been described. It will be apparent that such data will be of use in several ways. Thus, for example, morphology has been concerned with the linear dimensions of the tissue elements. This geometric description can now be widened by the addition of quantitative data on mass and concentration. To give an example, variation of nuclear volumes and nuclear–cytoplasmic volume ratios can be more closely investigated to see whether the changes are associated with alterations in concentration due merely to changes in water content, or to changes in total mass with, perhaps, concentration remaining constant. Measurements of mass are, of course, useful in histo- and cytochemistry because the total dry mass of a tissue element or cell organelle is a basis to which amounts of individual chemical substances determined by specific methods can be related. These cytochemical methods can usefully be applied to study the variations among individuals in a heterogeneous assembly for which standard biochemical methods give only average values per cell. Measurements of concentration are also a useful adjunct to autoradiography. For example, suppose that the uptake of radioactive atoms is being measured at various sites in a cell or tissue. An increased uptake in one site could be due to either a greater specific activity of the substance or merely be due to the concentration of the substance being greater.

Such measurements of total mass and of chemical substances made at various stages during biological processes (cell division, locomotion, etc.) may be expected to assist in the understanding of the phenomena, that is, in explaining the sequences of events in terms of physics and chemistry. The results obtained by the various workers have been divided for convenience into three sections.

B. Growth, Synthesis, and Division

The interference microscope technique employs observations in visible light and hence it can be used, in principle, to follow changes in one individual living cell without damage to it. The changes in total dry mass may be used as a measure of the growth of the cell.

a. Pollen Grain Development.

An example of the use of the original Dyson (1950) microscope is the measurement of changes in dry mass during development of pollen grains in *Tradescantia bracteata* (Davies *et al.*, 1954b). Since it was not practicable to follow, in time, the development of an individual cell, measurements were made on living cells isolated in 5% sucrose from anthers at various stages of developments. In Fig. 29 the time intervals between well-defined stages of development have been arbitrarily taken as equal. Optical path differences were measured by a variation, not now used, of the method in Section IV, A, 1, b (i). Integration was carried out by dividing the image plane up into areas of approximately constant o.p.d. The total dry mass per cell was calculated from Eq. 13 and shown by the upper arrows in Fig. 29A. The lower arrows show the mass without the correction factor and hence the distance between arrows shows the magnitude of the factor. Cell volumes were calculated by assuming the shape to be ellipsoidal; hence average concentration could be calculated (Fig. 29B).

Due to the large variations in o.p.d. throughout the cell, results are only rough, but serve to indicate a synthesis of dry substance by a factor of ten, during development from microspore to mature pollen grain.

b. Growth of Schizosaccharomyces Pombe.

When the changes in mass in an individual cell can be measured, an absolute time scale can be used, and possible variations in growth among individuals in a population can be investigated. Such an elegant study has been reported briefly in Mitchison *et al.* (1956). *Schizo-*

saccharomyces pombe was grown in a weak gel, 17.5% gelatin and 2% wort broth, and the growth at 27.5° C. measured with the integrating attachment to the Baker-Smith interference microscope. The immersion medium served to reduce the optical path difference of the organism so

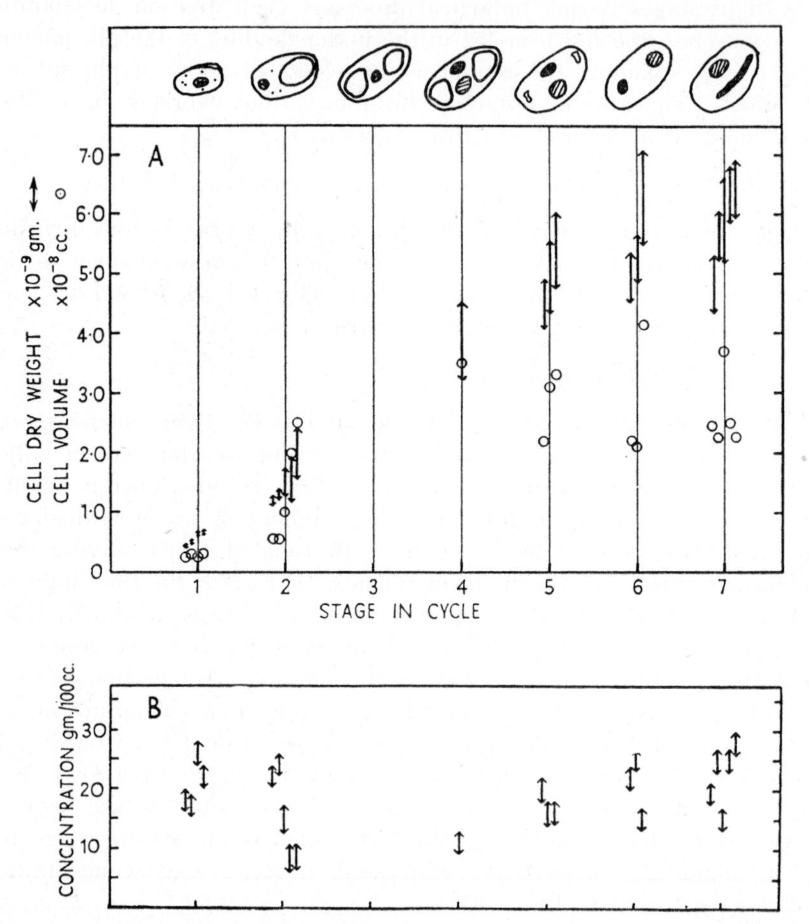

FIG. 29. Shows A, the increase in cell dry mass (arrows) cell columne (circles), and B, change in average concentration per cell, at various stages in pollen grain development in *Tradescantia bracteata.* The upper and lower arrows indicate the values with and without correction for immersion in 5% sucrose, cf. Eq. 13 (from Davies *et al.,* 1954b).

that it lay on the linear part of the interference pattern. This yeast is a rod-shaped organism so that volume and concentration of dry substance can be computed (cf. Eq. 13) as shown in Fig. 30. As a brief comment we note that the increase in total dry mass between divisions appears to be approximately linear, the mass also increasing during division.

c. Changes in Concentration in Germ Cells.

Measurements of the concentration of dry substance in nucleus and cytoplasm of living sea urchin eggs in sea water and the changes during fertilization and first division have been made (Mitchison and Swann, 1953) with the Baker-Smith interference microscope. Thickness could be measured since the cells and presumably the nuclei were spherical in shape. The concentration of dry substance in the cytoplasm of the

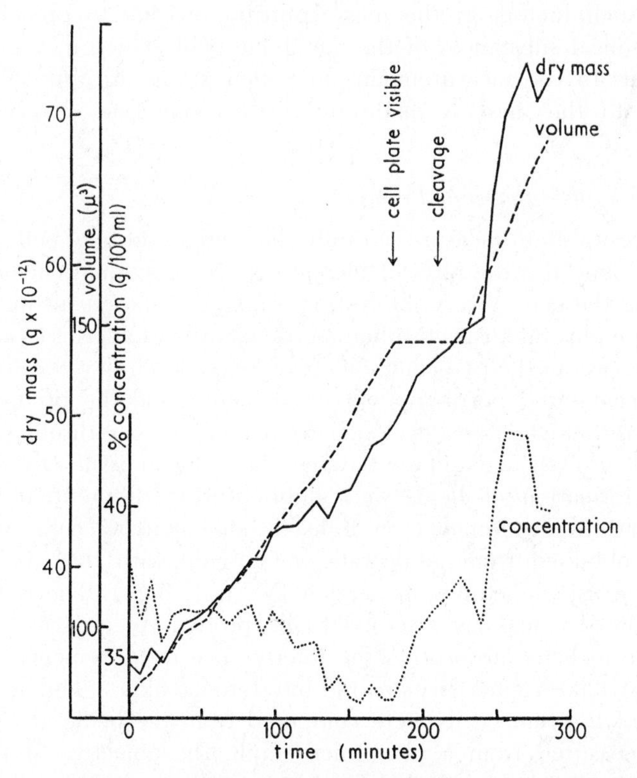

FIG. 30. Growth of *Schizosaccharomyces pombe.* One cell initially which then divides to form two daughter cells (from Mitchison *et al.*, 1956).

unfertilized egg was about 25 gm./100 cc. while that in the nucleus was about 16 gm./100 cc. Metaphase chromosomes also had a lower concentration than their surround. It was suggested that this somewhat surprising result may be restricted to egg cells since the nuclear material in tissue culture cells, amebae, and spit cells is known to have a higher concentration than the cytoplasm. The difference appears to be associated with the granules in the cytoplasm of egg cells.

An account of the changes in concentration during meiotic division in locust spermatocytes using the protein immersion method and the phase microscope has been given by Ross (1954).

The dry matter content of starfish oöcyte nucleoli changes from about 85% to 40% during development (Vincent and Huxley, 1954).

Measurements, above, have been made on living cells. In so far as the bulk of cell substance is protein the measurements are a measure of changes in protein content. In assessing the significance of changes in total mass in terms of growth, however, it is necessary to take into account such factors as the mass contribution due to precursors and other chemical substances of the metabolic pool, which may be varying in amount. In the measurements on pollen grains an appreciable contribution to the mass is made by the carbohydrate material of the cell wall.

d. Synthesis of Heme and Protein.

It is desirable to measure not only the increase in total cell substance with time, but also the rates of increase of the individual chemical components of the cell. A very interesting example is the recent interference microscope plus microspectrophotometric study by Lagerlöf *et al.* (1956) of the increase in the two components of hemoglobin in red cells, namely the iron-containing porphyrin derivative heme, and the protein globin. Mature erythrocytes are developed from immature erythroblasts by the process of growth and differentiation. The point at issue is the relative rates of formation of heme and globin during differentiation of the erythroblast. Measurements on unfixed bone marrow cells, in Ringer solution, obtained from adult rats were made with the Cooke-Dyson microscope using methods in Section IV, A, 1, b (i). The bulk of the cytoplasmic dry mass is assumed to be protein. In order to minimize errors due to heme absorption the interference measurements were carried out at 4900 Å (mercury arc and interference filter). The hemoglobin contents of the same cells were computed from the heme absorption at 4047 Å measured from a photomicrograph; the objective of the interference microscope was replaced with an ordinary apochromat.

In order to check the experimental procedure measurements were made on mature erythrocytes at different degrees of hemolysis. As more than 90% of the mass of the mature erythrocyte is hemoglobin a linear relationship between dry mass (by interference) and hemoglobin (by absorption) might be expected. The dotted line in Fig. 31 shows that this was the case. The results of measurements (Fig. 31) on erythroblasts at various stages of development in the formation of hemoglobin show that the bulk of the dry mass (globin) is formed prior to the formation of hemoglobin. This supports the contention that most of the protein is formed first and later converted into hemoglobin by the addition of heme. The possible (slight) extent to which protein is formed during hemoglobin production would seem to require additional measurements of the changes in area, and hence total mass, undergone by the cells.

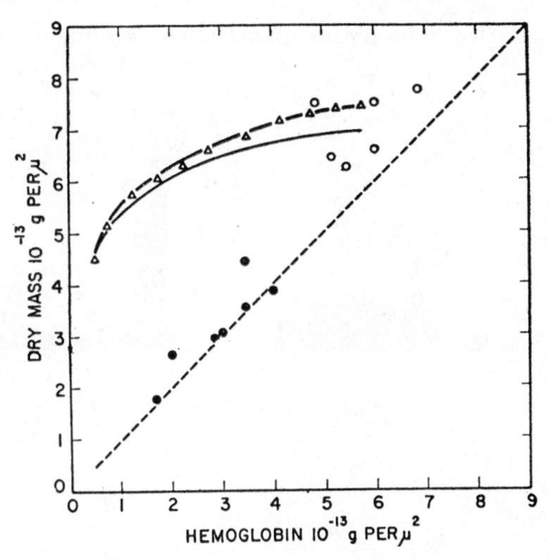

FIG. 31. The relationship between dry mass and hemoglobin per unit area in erythroblasts at various stages of development. The curve $\Delta - \Delta$ ($\chi = 0.18$) represents the successive mean values from 36 cells. The curve underneath is for $\chi = 0.19$. The unfilled circles are values from mature erythrocytes. The filled circles represent erythrocytes at different degrees of hemolysis (from Langerlöf *et al.*, 1956).

e. Synthesis of Protein and DNA in Somatic Nuclei.

The relative changes in amounts of total protein and DNA in nuclei of mouse fibroblasts in tissue culture and mouse ascites cells have been investigated by Richards and Davies (1958). Briefly, cells were fixed by freeze-substitution (see Appendix 3) and in the same nuclei total mass was measured by interferometry (Cooke photometer eyepiece plus photomicrography) and total DNA by scanning photometry (Deeley, 1955) of Feulgen-stained cells. It was assumed that the bulk of the dry mass of the fixed cell was protein and DNA (the nucleoli were excluded from the measurement). Loss of dry mass during fixation was found to range from about 10 to 20%: this is an upper limit since χ for the fixed tissue was taken as 0.18. Total protein content could be calculated by subtracting from the dry mass the DNA content computed from Feulgen stain plus the biochemical value for the diploid mouse nuclei.

The raw experimental data were used in two ways. First histograms were constructed, one relating to total protein, the other to total nuclear DNA (Fig. 32A, B). An attempt was made to deduce the average behavior of the cell nucleus, with regard to variation in DNA and protein content with time, from the shape of the histograms, using the formula given by Walker (1954). This could only be done for the mouse ascites cells for which the growth was known to be exponential.

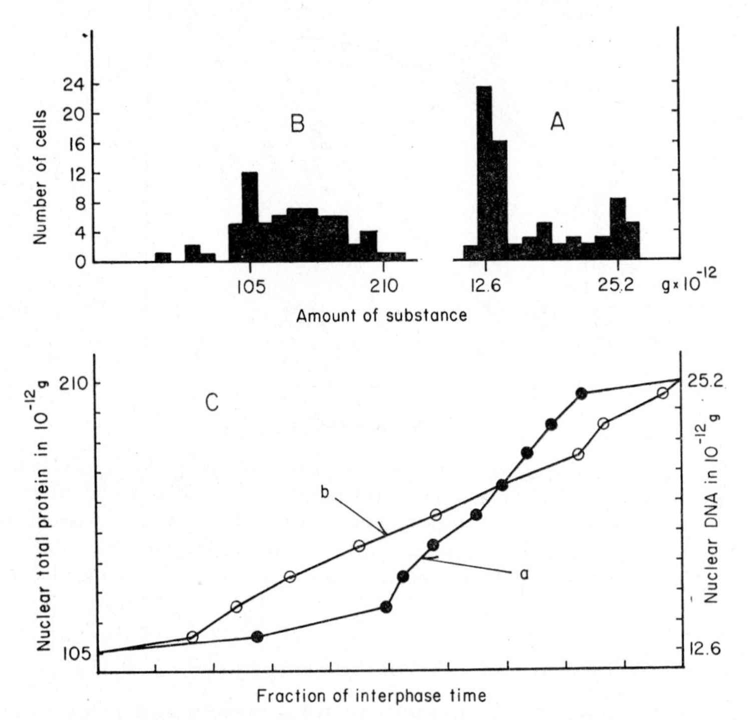

FIG. 32. Shows nuclear dry mass and DNA content in mouse ascites cells fixed by methanol freeze-substitution. A: Histogram for nuclear DNA. B: Histogram for nuclear total-protein. C: Calculated average synthesis curves for nuclear DNA (a) and total protein (b) (from Richards and Davies, 1958 modified).

It appears (Fig. 32C) that on average increase in protein content is approximately linear during interphase while the doubling of DNA occurs over a relatively short period towards the second half of interphase.

The average curves are somewhat misleading. For example, a fraction of the protein (say the histone associated with the DNA) may be synthesized during a short period of interphase and this would be lost due to the effects of averaging. The average data also suggest some close association in time between the synthesis of DNA and protein. However in the majority of cells this appears not to be so. This was demonstrated by using the raw data in a second way, that is, by plotting nuclear DNA content as a function of protein content (Fig. 33). There it will be seen that in some cells the DNA content has increased without synthesis of protein, while in other cell nuclei the reverse is true. That there appears to be no close relationship between synthesis of DNA and *total* protein

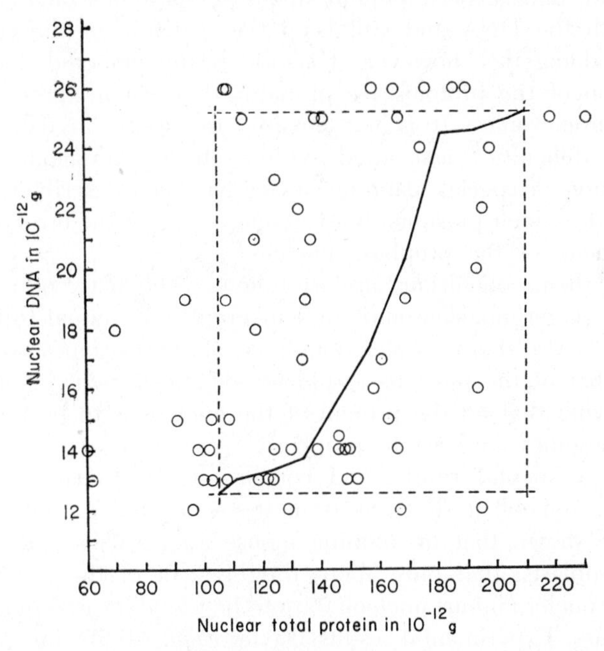

FIG. 33. Shows the variation of DNA with total protein content in individual nuclei; data from Fig. 32 replotted. The solid curve shows the relationship between DNA and total protein on average, deduced from Fig. 31C, a, b. The data for individual nuclei lie mainly off the average (from Richards and Davies, 1958).

may perhaps not be surprising, when account is taken of the large variations in nuclear protein content which may occur in some cells without alteration in DNA content.

f. Protein and DNA in Chromosomes and Nuclei.

A most interesting application of interferometry is due to Mellors and co-workers (summarized in Mellors, 1956) who have attempted to measure the mass of total protein in mouse sperm heads, chromosome sets, and interphase nuclei. The idea presumably is that such measurements can be used to make what might be termed a structural distinction between the proteins of the cell, that is, an evaluation of the mass of protein which is located in the same morphological volume as is the

DNA. If, for example, only a fraction of the proteins located in the interphase (pre-prophase) nucleus were to be found on the chromosomes then special interest would attach to them, since they, together with the DNA would be divided into two halves with precision during mitosis. And again, it is obviously of interest to discover what fraction of the mass of the chromosomal proteins are present in the mature sperm head, that is the minimum amount of protein which can be associated with the DNA and still keep the structure of the chromosome intact. In doing this, however, it should be remembered that chemical composition of the proteins are probably different in sperm heads and mitotic chromosomes. It is not possible to discuss in detail here the numerous difficulties associated with such measurements, and their interpretation. A special scanning device has been described by Mellors *et al.* (1954) which presumably overcomes some of the errors due to the inhomogeneity of the prophase nucleus.

Briefly their conclusions are as follows. The mass of a single set (haploid) of germinal chromosomes in prophase is equal to that of the sperm head. Also the mass of a double set of chromosomes is very nearly equal to that of the interphase diploid somatic nucleus of mouse liver, hence proving that all the protein of the nucleus is to be found within the chromosomes, at least in prophase.

The experimental results and conclusions of Richards and Davies (1958) are somewhat different from those of Mellors and co-workers. They have shown that in dividing mouse ascites cells and also in the mouse fibroblasts, that only about roughly half of the proteins of the interphase nuclei (minus nucleoli) are to be found located in the somatic chromosomes. Experimental results Davies *et al.* (1957) on sperm heads from mouse are also different from those reported by Mellors and Hlinka (1955). They, together with results just mentioned, indicate that the protein mass of a haploid set of somatic chromosomes is greater than that found associated with the same amount of DNA in the sperm heads of mouse. Further work is required to resolve the differences between the two groups of workers.

C. Further Quantitative Data

These investigations have not been so closely related to variations of the biological system in time as was the case in Section VI, B.

a. Concentration or Density, and Thickness.

(i) *Living cells.* Data on the concentration of dry substance in various living cells made by immersion in protein solutions and measurement of the null-point with phase microscopy, and due mainly to

Barer and co-workers, have recently been summarized by Barer (1956). For example, the concentration of dry substance in the cytoplasm (granule-free) of most tissue cells lies between about 10 and 25 gm./100 cc. At the upper extreme of concentration are some bacterial cells, fungal spores, cilia, and some sperm heads which have a concentration too high ($>$ 55 gm./100 cc.) to be measured by the technique. In sperm heads of bull the postnuclear cap gave a value of about 48 gm./100 cc., in ram about 50 gm./100 cc.

Barer *et al.* (1953) reported a range of concentrations, 28–34 gm./100 cc. for human red blood cells with a median value of 31–33 gm./100 cc. They state that these measurements on red cells must be regarded as comparative rather than absolute on account of possible changes in cell volume. The accepted (median) value in clinical hematology is somewhat higher, namely 35 ± 3 gm./100 cc.

An example of the method of measuring the thickness and concentration of living cells by measuring optical path differences in two media, a dilute salt solution and a protein solution has been given by Barer (1953, 1956). The concentration of dry substance in the cytoplasm of human oral epithelial (spit) cells was found to lie between 13 and 18 gm./100 cc.

(ii) *Fixed cells and tissue. Sperm heads.* Data (Davies *et al.*, 1954) on the heads of ram spermatozoa fixed in neutral formalin illustrate the observations made in Section II, C, 2, a. Measurements (method in Section IV, A, 1, b (i)) of optical path difference in the center of the sperm heads, oriented at right angles to the optical axis, were made in a series of solutions of sucrose and bovine plasma albumin (Fig. 34). Optical path difference plotted as a function of refractive index difference (medium minus water) gives a straight line for both sets of data. This is to be expected from Eq. 13 whether t has the value for the effective or geometrical thickness. Although spermatozoa are a homogeneous population, the dry mass of individual sperm heads does vary slightly and this may account for the fact that the lines do not intersect at zero abscissa as might be expected.

The refractive index at which the sperm head vanishes, obtained by extrapolation, in protein solution is 1.46 and in sucrose solutions it is 1.54. As an explanation of this difference it was suggested that sucrose molecules penetrate the sperm head, while the protein molecules do not. The geometrical thickness at the center of the sperm head, calculated from the data in protein is 0.45 μ. The so-called effective thickness calculated from the data in sucrose solution is 0.28 μ. Equation 18 was used. Hence the fraction F of the geometric volume occupied by particles is 62%. The average concentration obtained by dividing the mass per unit area by the geometric thickness is 72 ($\chi = 0.18$) to 87 ($\chi = 0.15$) gm./100 cc.

Bone tissue. The inorganic content of adult human femur has been compared by optical interference and X-ray methods, and good relative agreement was obtained (Davies and Engström, 1954). Quantitative data was also obtained by interferometry on the relative amounts of organic and inorganic substance throughout the bone, by measurements before and after decalcification. Here, however, we are only concerned with

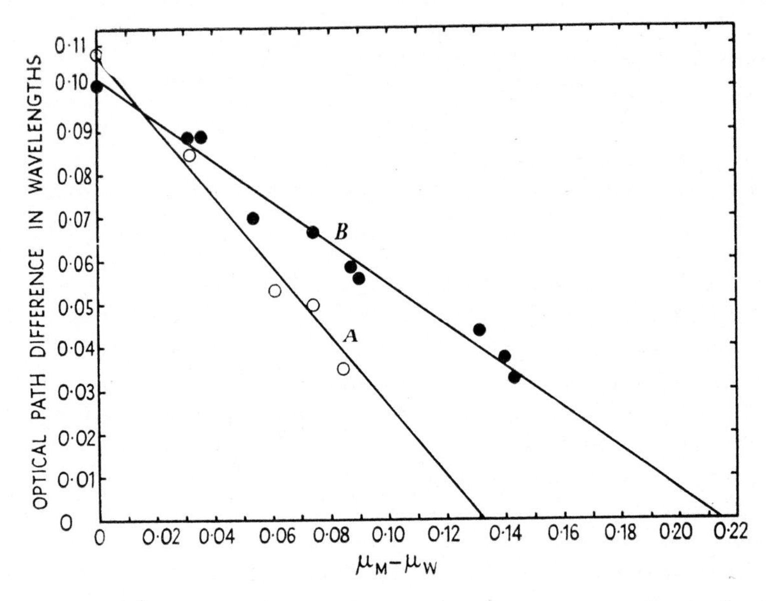

FIG. 34. Variation of o.p.d. at the center of ram sperm heads, fixed in 10% neutral formalin, as a function of $\mu_m - \mu_w$, where μ_m is the refractive index of the medium in which they are immersed and μ_w the refractive index of water. A, sperm immersed in solutions of bovine plasma albumin. B, sperm in aqueous solutions of sucrose (from Davies *et al.*, 1954b).

the data on the effective thickness at different sites in thin sections ($\sim 10\ \mu$) of alcohol-fixed adult bone. Measurements of o.p.d. were made by measuring band displacement from photomicrographs (e.g., Fig. 35), of sections immersed in liquid paraffin ($\mu = 1.483$) and water ($\mu = 1.333$). The effective thickness of the bone section was found to be greater in the so-called "old" bone than in adjacent sites in "young" bone. By "young" bone we refer to the less calcified regions surrounding the Haversian canal. At these adjacent points the geometrical section thickness could be assumed to be the same and hence this result indicated that a greater fraction of the volume was occupied by tissue substance in the old bone. The localized concentration or density was approximately the same in "old" and "young" bone, the slightly increased value in "old" bone being attributable to a slightly higher inorganic content. It is important to consider whether the density refers to the alcohol-dried bone tissue as it exists in liquid paraffin or in water. It can be shown that the effective thickness and density refer to the former condition. The average value of about 1.9 is in good agreement with the value obtained by conventional methods of 2.0 to 2.1. In this work χ-values of 0.17 and 0.10 were used for the protein and mineral substance, respectively.

b. Isolated Nuclei.

Investigations of the protein and DNA content of individual nuclei in intact cells, from different tissues of the same animal and, from animals under varying physiological conditions have been made by stain photometry; there have also been extensive biochemical studies on isolated nuclei (see for example reviews, Pollister, 1952; Swift, 1953; Alfert *et al.*, 1955; Allfrey *et al.*, 1955). In contrast to the constancy in mass of DNA per chromosome set in the cell nuclei from different tissues of one species, large variations are found in total protein content per nucleus. For various reasons interferometry provides a more reliable absolute measure of total protein content than ultraviolet light absorption or presently available stain photometry. Methods for measuring certain protein fractions of the nucleus by combining mass measurements with extraction procedures are being investigated. Consequently microscope interferometry is a new technique which has applications in studying the interesting relationships between the nucleic acids and protein in cells.

Hale and Kay (1956) have measured the dry mass (mainly protein plus nucleic acid) of calf thymus nuclei isolated in citric acid (CN) in sucrose-calcium chloride (SN) and in non-aqueous (NAN) media. They used the Cooke-Dyson interference microscope and the method in Sections IV, A, 1, b (i) and IV, B, a. The average values of the dry mass per nucleus in the three media obtained by interferometry (χ = 0.18), (CN, 19.8×10^{-12} gm.; SN, 21.1×10^{-12} gm.; NAN, 35×10^{-12} gm.) were in remarkably good agreement with those obtained by ordinary weighing procedures. The greater content of protein is nuclei isolated by a nonaqueous procedure is of course well established.

A point at issue about these results is how far the considerable spread of values about the mean is due to errors of measurement or to a true variation of nuclear protein content in the cells of this tissue. This spread is especially great in the nuclei prepared in nonaqueous media in which inaccuracy of integration must surely be high on account of the extreme inhomogeneity. More accurate measurements with an integrating photometer (e.g., Section IV, B, c) would help settle this point.

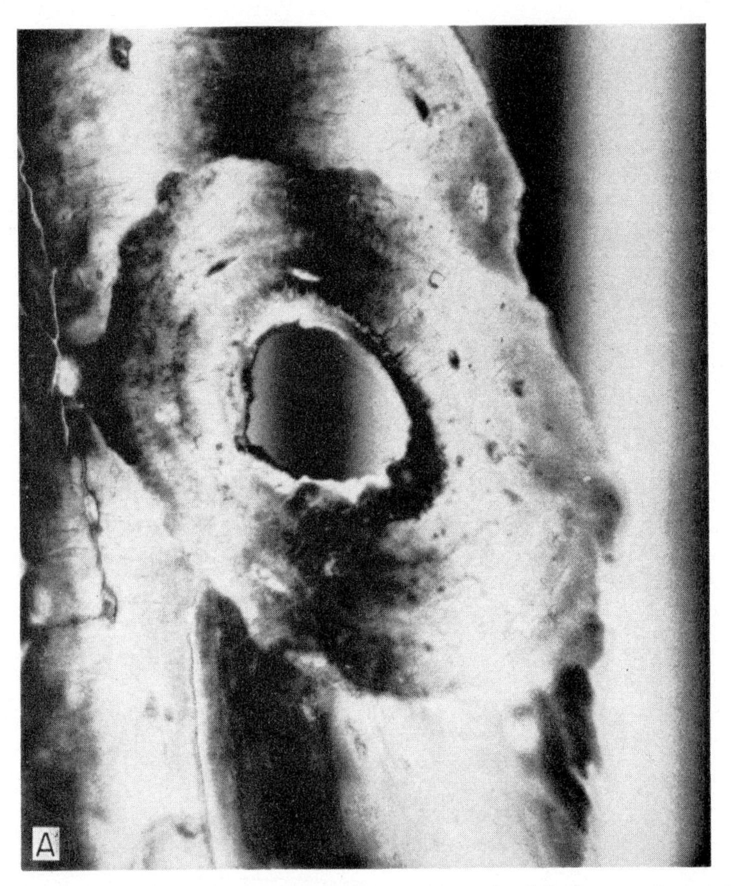

FIG. 35. A: photomicrograph of thin section of adult human femur. Taken with a Cooke-Dyson interference microscope, with objective of NA 0.85, condenser NA $\sim$ 0.5 and is at magnification $\times$420. B: schematic diagram of interference bands in image and surround. The lines (broken) just corresponding points (Davies and Engström, from C.T.S. Booklet).

Swartz (1956) has recently reported, in abstract, interference measurements on rat liver nuclei, frozen-dried and isolated in anhydrous glycerol. Presumably a high homogeneity of nuclear substance in nuclei prepared in this way can be obtained by using the freeze-drying technique conventional in cytology. The biochemist is often not concerned with avoiding ice crystal artifact in his isolated nuclei; and hence takes no special precaution to rapidly freeze the tissue.

Measurements of total dry mass (mainly protein plus nucleic acid) of cell nuclei from 7-w mouse liver rapidly ($<$ 10 minutes) isolated in Tyrode solution and fixed by methanol freeze-substitution were made

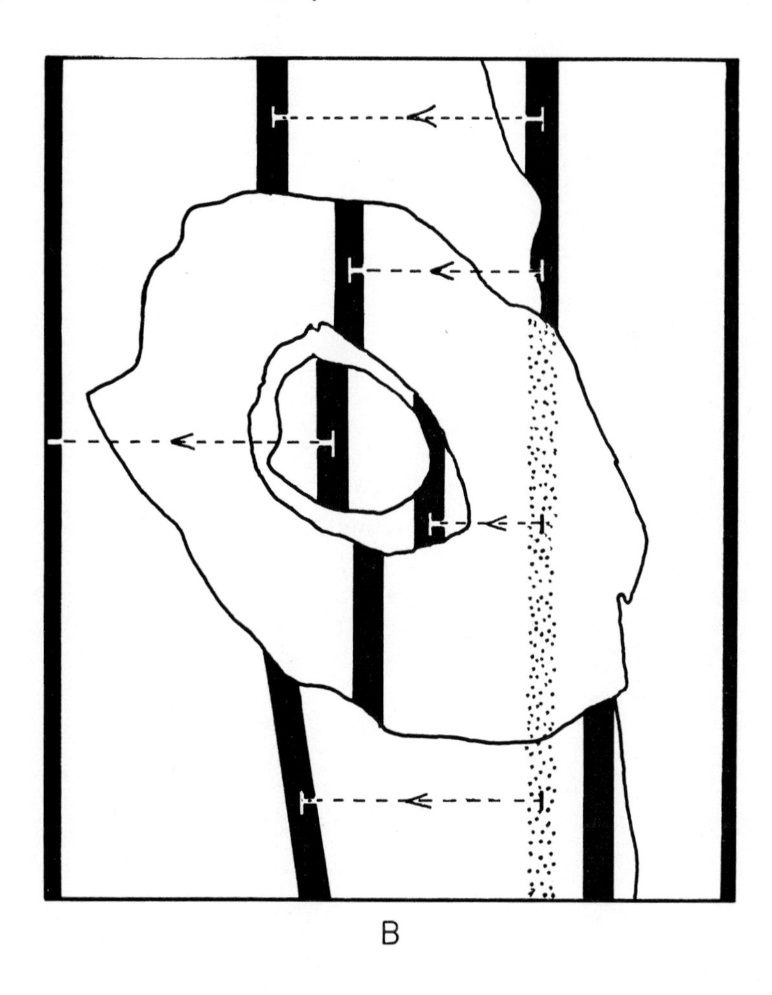

B

with the integrating photometer (Section IV, B, c (i)) (Davies *et al.*, 1957). The values for nuclear dry mass were found to lie in well defined classes with average values 0.5:1.0:2.1:4.2 (Fig. 36). Similar data (unpublished) was obtained on nuclei isolated in a sucrose-calcium chloride medium. These values for total mass may be compared with the ratio 1:2:4 found for DNA in the diploid, tetraploid, and octoploid parenchyma cells of liver. The apparent existence of four classes for total mass may be due to the presence of cells which contain the same amount of DNA as the diploid cell but different amounts of protein. The problem is being investigated.

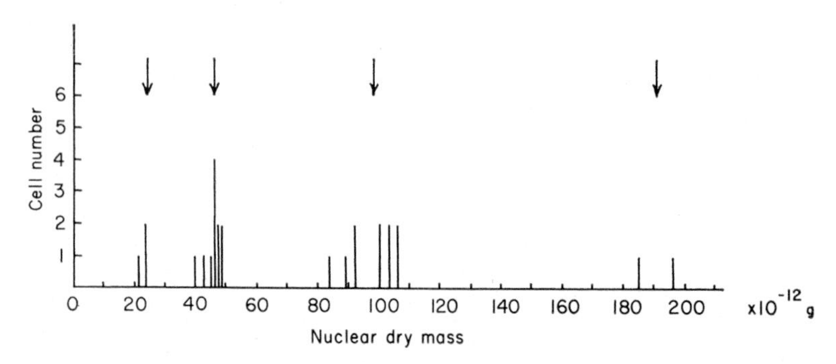

FIG. 36. Shows the distribution of dry mass ($\chi = 0.18$) in nuclei of mouse liver isolated rapidly at 0° C. in Tyrode solution and fixed by methanol freeze-substitution. The arrows indicate the arithmetic means at 23.3×10^{-12} gm., 45.6×10^{-12} gm., 97.5×10^{-12} gm., and 190.5×10^{-12} gm. (from Davies *et al.*, 1957).

c. Protein Distribution in Striated Muscle Fibrils.

Observations on the chemical composition and structure of muscle fibrils may be used as a basis for theories of muscle contraction. As a part of such studies, Huxley and Hanson (1957) have used the interference microscope to measure the distribution of protein in striated muscle fibers. Figure 37A is a photomicrograph in even-field interference contrast, of an isolated muscle fibril from the psoas muscle of rabbit mounted in isotonic potassium chloride. In these fibrils more than 96% of the total dry mass is protein as shown by biochemical tests. The optical path difference of these fibrils does not exceed about 0.1 wavelengths. Hence the light intensity distribution in the image plane closely follows the distribution of dry mass per unit area in so far as it is not limited by diffraction. A microphotometer record along the film negative (Fig. 37B) shows that the concentration (fiber thickness being approximately constant) is greater in the anisotropic (A) than in the isotropic (I) band. It is to be expected that in such small structures some blurring of the actual distribution of mass in the fibril will have occurred because of the phenomenon of light diffraction in the microscope. This will especially apply to the Z and H lines the dimensions of which approach the resolving power of the light microscope. This sets a limit to the accuracy with which the distribution of dry mass per unit area along the length of the fiber can be determined.

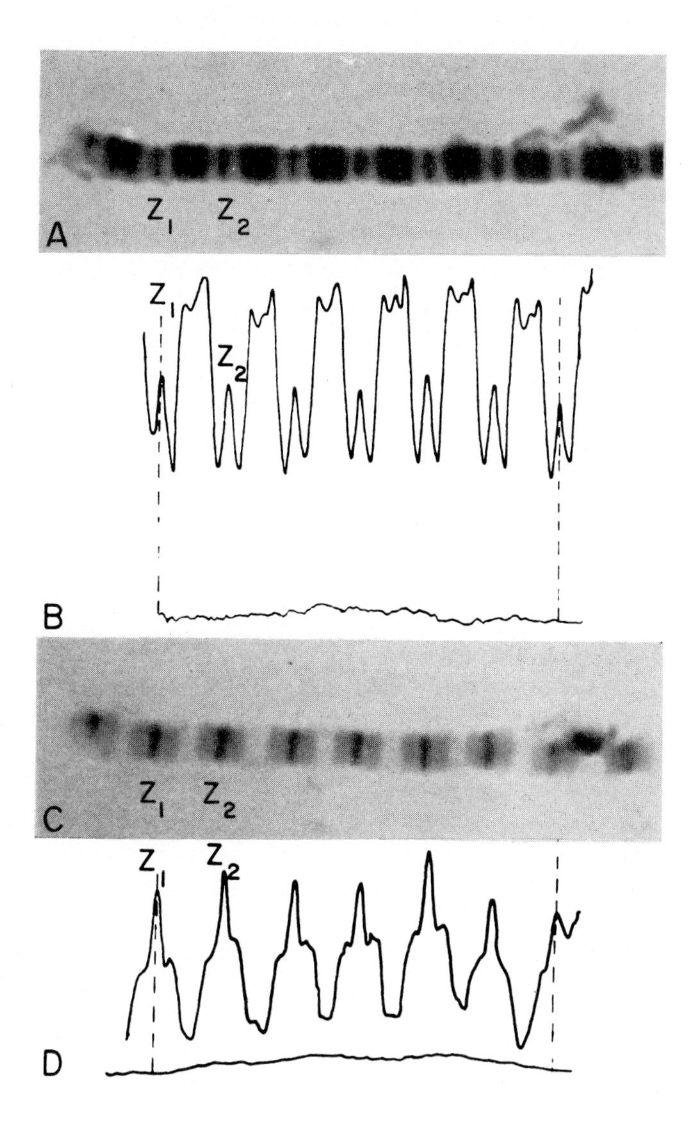

FIG. 37. A: Photomicrograph of a myofibril from psoas muscle of rabbit. Taken with the Cooke-Dyson interference microscope, even-field contrast, objective NA 1.3, condenser NA $\sim$ 0.5. B: Densitometer (Walker, 1955) tracing along the length of myofibril (A) using a narrow slit of length slightly greater than the width of the myofibril. The baseline is a tracing through the background taken with the same slit. C and D show photomicrograph and densitometer records on the same myofibril after extraction of myosin (from Huxley and Hanson, 1957). Magnification $\times$2400.

The localization of myosin in the fibril was investigated by specifically extracting this protein. This can be done by using potassium chloride solutions of high ionic strength. The photomicrograph (Fig. 31C) indicated, and the microphotometer record (Fig. 31D) upon analysis showed that the bulk of myosin is located in the anisotropic band. This experiment was carried out by irrigating the specimen, under the microscope. It was also found that the results on the total fractional decrease in mass (not depending on the accuracy of localization) obtained by optical interference were in good agreement with those obtained by parallel biochemical experiments in which protein nitrogen was determined by a micro-Kjeldhal.

Measurements of band density in myofibrils have also been attempted by Bennett (1955). Huxley and Niedergerke (1954) have made use of the halo-free property of the interference microscope to measure accurately the dimensions of the bands in muscle fibers.

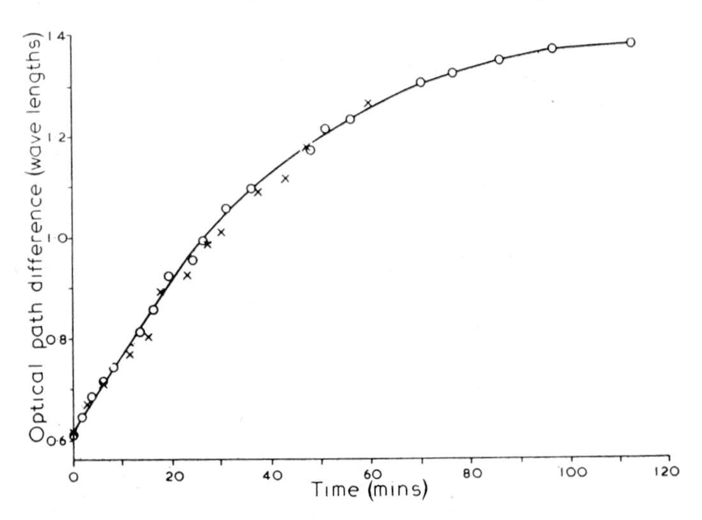

FIG. 38. Shows the increase in optical path difference, proportional to mass of calcium phosphate per unit area, in kidney brush border during first incubation (O), and second incubation (X) (from Davies *et al.*, 1954a).

d. Enzyme Activity.

Interferometry can, in principle, be used to measure the activity of enzymes which as a result of their activity cause a measurable change in o.p.d. due to deposition of a transparent end-product at the site of their activity. The method has been applied (Davies *et al.*, 1954a; Barter *et al.*, 1955) to measure alkaline phosphatase activity at sites in brush borders in fixed sections of kidney and duodenum. Briefly the results are as follows. Figure 38 shows the change in o.p.d. at a site in kidney brush-border due to the precipitation of calcium phosphate. The latter is formed as a result of combination of calcium ions with the phosphate ions, liberated from the substrate sodium glycerophosphate as a result of enzyme activity (Robison, 1923). The initial slope of this incubation curve may be used as a measure of enzyme activity. If Δ_1 is the initial increase in o.p.d. per minute and Δ_2 is the o.p.d. due to the tissue, then the enzyme activity in grams phosphate ion liberated per minute per unit mass of tissue is given by

$$\text{Activity} = \frac{\Delta_1 \chi_2}{\Delta_2 \chi_1} p \tag{37}$$

where χ_1, χ_2 are the values of χ for calcium phosphate and tissue, respectively, and p is the fraction of phosphate ion in calcium phosphate. Absolute values for enzyme activity lack precision because of uncertainty in the latter factors; the value of $\chi_2 p/\chi_1$ is near unity.

It was observed that the calcium phosphate could be removed from the tissue and that during a second incubation the rate of deposition remained unaltered (Fig. 38). Hence the effect of various factors affecting enzyme activity could be studied. For example, the variation in enzyme activity with pH is shown in Fig. 39. Errors due to possible diffusion of phosphate ions or end-product were also investigated and found to be small in the brush border.

e. Mammalian Sperm Heads.

Leuchtenberger *et al.* (1956) have reported measurements of total mass of bull sperm heads made with the Baker-Smith interference microscope. Optical path difference in the central region of head was measured in even-field contrast using a photomultiplier to detect the shift in the interference pattern. Area was presumably measured separately. The DNA and basic protein (arginine) contents were measured by photometry of Feulgen stain and Fast Green stain, respectively. Spermatozoa from infertile bull were found to contain appreciably less DNA than fertile bull, but to contain very nearly the same dry mass. This observation led to the conclusion that the protein content of the sperm head was greater in spermatozoa from infertile bull; this was supported by the slightly increased staining for arginine.

Mellors and Hlinka (1955) have measured the dry mass of unfixed nonmotile sperm heads of mouse, rat, and guinea pig, and compared these values with those obtained after fixation in a mixture of alcohol and ether. The difference in values is used as a measure of lipid. Some remarks on the formula used by them are given in Appendix 2. The average values for unfixed sperm heads are, mouse 16.0×10^{-12} gm. (measured in glycerol or water); rat, 11.0×10^{-12} gm. and for fixed sperm heads; mouse, 13×10^{-12} gm.; rat, 9.3×10^{-12} gm.; and guinea pig, 11.5×10^{-12} gm. Hence the fat-extractable material is 19% in mouse sperm heads and 15% in rat. These measurements refer to the sperm

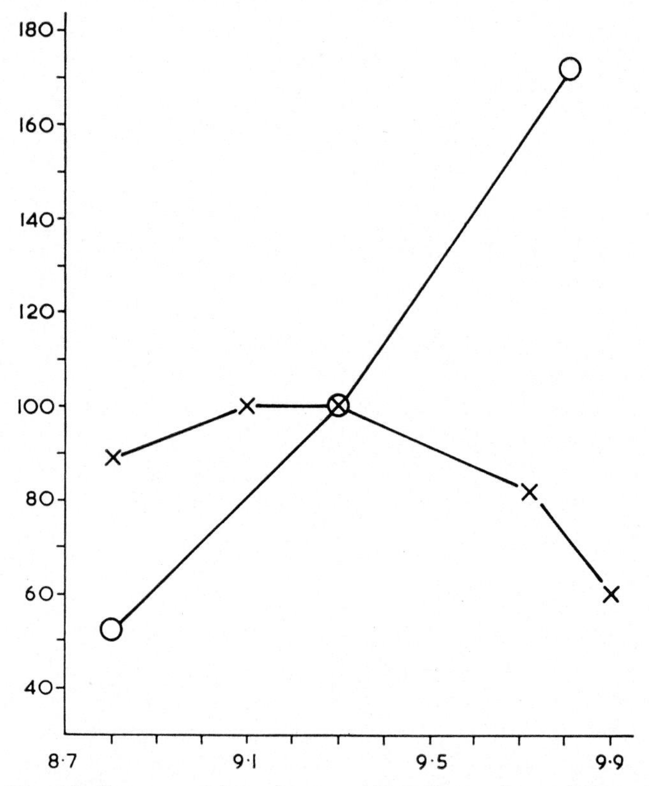

FIG. 39. (Ordinate: activity; abscissa: pH.) Shows the variations in enzyme activity (proportional to the initial slopes of the incubation curves, cf. Fig. 38) with pH, at sites in the brush borders of rat kidney cells (O), and rat duodenum (X), fixed by freeze-drying and ethanol; the slopes at pH 9.3 are taken as 100% (from Barter *et al.*, 1955).

nucleus, the area of which expressed as per cent of the sperm head is approximately 100, 90, and 50, respectively, in mouse, rat, and guinea pig. It would be interesting to know how these percentages were arrived at; whether, for example, they represent the relative area stained with Feulgen.

Davies *et al.* (1957) using the integrating attachment to the Cooke-Dyson microscope obtained a value of 7.9–8.6 for lipid-extracted mouse sperm heads. The discrepancy has not been resolved. These workers have also criticized the method for lipid estimation used by Mellors and Hlinka on the grounds that the alcohol-ether may extract substance other than lipid from the unfixed cell. In their (Davies *et al.*, 1957) estimation of lipid content of ram sperm heads fixation was carried out with 8% trichloroacetic acid first, and followed by extraction with ethanol-ether. The change was about 13% (Fig. 40). A similar value was found for

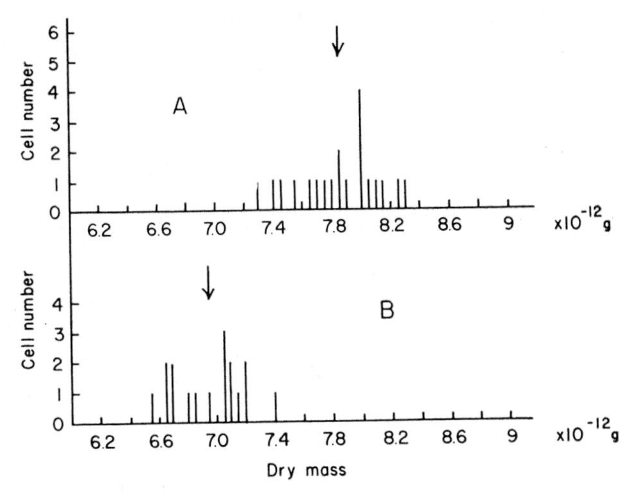

FIG. 40. Shows the dry mass ($\chi = 0.18$) in sperm heads of ram fixed in trichloroacetic acid (A) and after extraction in ethanol-ether (B). The arithmetic means, indicated by arrows, are at 8.3×10^{-12} gm. and 7.3×10^{-12} gm. (from Davies *et al.*, 1957).

the loss on ethanol-ether extraction of neutral formalin-fixed sperm heads. Data on dry mass, mainly protein and nucleic acid, in heads of spermatozoa from three mammals are shown in Table IV, and compared with

TABLE IV
RELATION OF DRY MASS TO DNA CONTENT IN LIPID-EXTRACTED SPERM HEADS

Object	Dry Mass $\times 10^{-12}$ gm.	DNA $\times 10^{-12}$ gm.
Bull sperm head	7.3–7.5	3.3[a] 2.8[b]
Ram sperm head	6.7–7.0	3.2[c]
Mouse sperm head	7.9–8.6	~ 3[a]

[a] Vendrely and Vendrely (1948, 1949).
[b] Mirsky and Ris (1949).
[c] Mann (1954).
From Davies *et al.* (1957).

the biochemical values for DNA. It appears that not only is the DNA content approximately constant in these mammalian species but also the total protein content.

D. Dry Mass and Concentration under Experimentally Changed Physiological Conditions

Easty *et al.* (1956) have studied the action of ribonuclease on various forms of ascites tumor cells using the Baker-Smith interference microscope and the method of setting to minimum light intensity. By using a special chamber they were apparently successful in slightly compressing these cells so as to make their shape more amenable for measurement, and keep them surviving for about 24 hours. A remarkable feature of their results is the large increase in mass (nearly double in an extreme case) of cytoplasm, nuclear sap, and nuclear inclusions, taking place in about 1 hour. No measurements of absolute mass are given. Since no measurements of area or refractive index of the medium are given, it is to be presumed that the former did not change and that the latter had no effect on the measurements.

Briefly, experiments have been made on the rate of hemolysis of red cells (Marsden, 1956, with Cooke-Dyson), on the comparative changes in concentration in colloid of thyroid gland in guinea pigs fed on cabbage and ascorbic acid (Hale, 1956 with Cooke-Dyson), and on the changes in water content and concentration of dry substance in cytoplasm of chick fibroblasts during alterations in the tonicity of the medium (Barer and Dick, 1955). The latter observations were made with the Baker-Smith microscope and half-shade eyepiece, and presumably by varying the protein and salt concentration of the medium. They illustrate clearly how sensitive these cells are to small changes in salt concentration in the external medium. Thus a change of salt concentration from 0.7 to 1% results in an alteration of concentration of dry substance from 11.7 to 18.1 gm./100 cc.

Acknowledgments

I wish to express my gratitude to Prof. J. T. Randall, F.R.S., and Dr. M. H. F. Wilkins who have been my mentors from the start, to Dr. E. M. Deeley for valuable collaboration in this field, and to Dr. A. R. Stokes for helpful discussion and suggestions. I have also had the benefit of discussion with Mr. F. H. Smith, and Mr. B. O. Payne.

VII. APPENDICES

Appendix 1: Physical Optics of Light Interference

A brief introductory account of the basic physical concepts necessary for an understanding of interferometry will be given. For a fuller explanation one of the standard text books should be consulted, e.g., Jenkins and White (1957), Williams (1948), Ditchburn (1952).

Two models have been used to explain the properties of light. According to one, with which we are not concerned here, light is regarded as a collection of particles or photons. The phenomenon of interference is, however, explained by considering light to consist of waves.

We consider a source of light to be emitting waves. The basic hypothesis is that, at any point in the surrounding space, there is some physical property, which we need not for the moment specify which has a measurable value. This property undergoes a periodic fluctuation with a period T, say. A disturbance at one point at a given time is considered to produce a disturbance of similar magnitude at a neighbor-

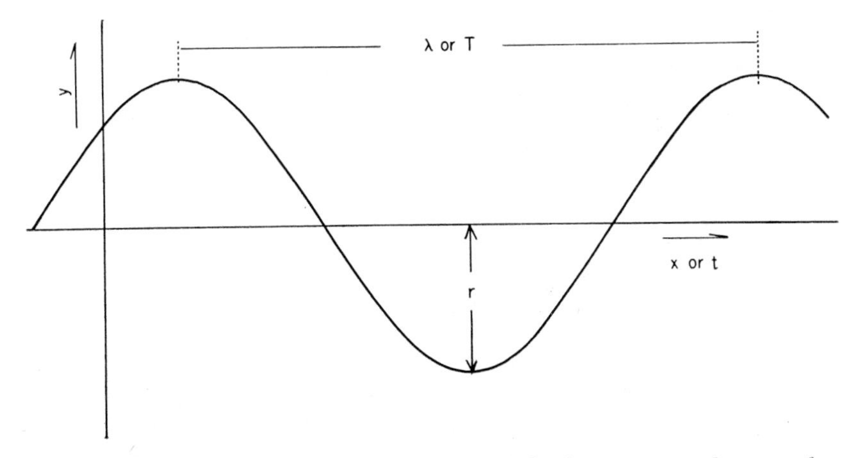

Fig. 41. Shows a simple harmonic wave of displacement y, either as a function of distance x at an instant of time, or as a function of time at a fixed distance. r is the amplitude. λ is the wavelength. T is the periodic time.

ing point, distance λ away, at a slightly later time T. It follows that the disturbance is propagated with a velocity v given by λ/T. Figure 41 shows a simple harmonic wave with the form of a sine or cosine curve. The quantity λ is called the *wavelength,* and the maximum value of the disturbance is called the *amplitude* r. It may be shown that the energy associated with the wave (*relative energy* or *intensity*) is proportional to the square of the amplitude. These concepts, together with one other, are sufficient to explain interference phenomena.

For a detailed quantitative description of such phenomena as reflection, refraction, and diffraction of light we require the electromagnetic theory due to Maxwell. This theory identifies the fluctuating physical property as electric and magnetic fields which have a direction at right angles to each other and to the direction of propagation.

The curve of a simple harmonic wave (Fig. 41) relates displacement with either distance (x) from the source at an instant of time (t), or at a constant distance it shows the variation of the displacement in time. This so-called plane wave is described by the expression

$$y = r \cos p \tag{38}$$

where p is known as the *phase* and depends on x and t. According to the property of a cosine function y assumes the same value whenever p increases by 2π. Since y repeats itself whenever t increases by T or x by λ we can specify p as

$$p = \frac{2\pi}{T} t + \frac{2\pi}{\lambda} x \tag{39}$$

Hence combining Eqs. 38 and 39 we obtain

$$y = r \cos 2\pi \left(\frac{t}{T} + \frac{x}{\lambda} \right) \tag{40}$$

According to Eq. 40 if the path increases by dx then the phase p increases by $\frac{2\pi}{\lambda} \cdot dx$, that is, a path increase of λ increases the phase by 2π radians (360°).

The interaction between two light waves is governed by the *principles of superposition.* This hypothesis states that the disturbance (Y) at a given place and time due to the passage of a number of waves ($y_1, y_2 \ldots$) is equal to the algebraic sum of the disturbances produced by the individual waves, that is

$$Y = y_1 + y_2 + \ldots \tag{41}$$

To make the point clear we consider the interaction at P due to the light waves, of the same wavelength, emitted by two point sources A and B (Fig. 42). In the classical experiment of Thomas Young (1773–1829), who first showed how interference may be explained in terms of the principle of superposition, the sources A and B were small apertures in a card illuminated by a distant point source. Light waves spread out from A and B due to diffraction. The effect at P is, according to Eq. 41

$$Y = r_1 \cos 2\pi \left(\frac{t}{T} + \frac{AB}{\lambda} \right) + r_2 \cos 2\pi \left(\frac{t}{T} + \frac{BP}{\lambda} \right) \tag{42}$$

where r_1 and r_2 are the amplitudes of the two beams. The *phase difference* (ϕ) between the two beams is $\frac{2\pi}{\lambda}(BP - AP)$, that is, $\frac{2\pi}{\lambda} \times$ *path difference* (Δ).

It is the amplitude of the new vibration or rather amplitude squared (proportional to relative energy or intensity) which is of interest. This can be obtained by solving Eq. 42 or more conveniently by a vector-addition method in which the vectors OQ, OR represent the amplitudes r_1 and r_2 and the angle ROQ represents the phase difference ϕ (Fig. 43).

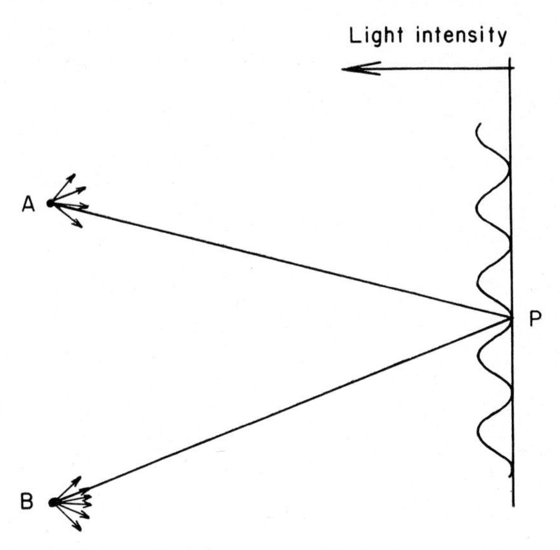

FIG. 42. A and B are coherent sources of light the waves from which interfere, for example, in the plane containing point P.

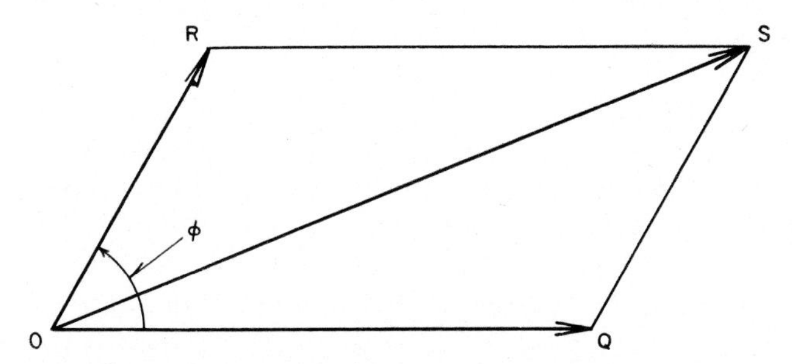

FIG. 43. Vector representation of interference. The length of the vectors OQ and OR represents the amplitude of the light vibration, and the angle ϕ between them is equal to the phase difference. Vector OS represents the resultant vibration.

According to the method of vector-addition the resultant amplitude OS is given by

$$OS^2 = Y^2 = r_1^2 + r_2^2 + 2r_1r_2 \cos \phi \qquad (43)$$

Equation 43 is the basic equation for interference between two beams and is shown graphically in Fig. 44 for the case of r_1 and r_2 equal (r) when

$$Y^2 = 2r^2(1 + \cos \phi) \tag{44}$$
$$= 4r^2 \cos^2 \phi/2 \tag{45}$$

Considering the light intensity (I) in the dotted plane (Fig. 42), as P moves the path difference is continually changing, with a corresponding variation in ϕ, and hence there appears a series of maxima and minima of light intensity ($I \propto Y^2$) known as an interference pattern.

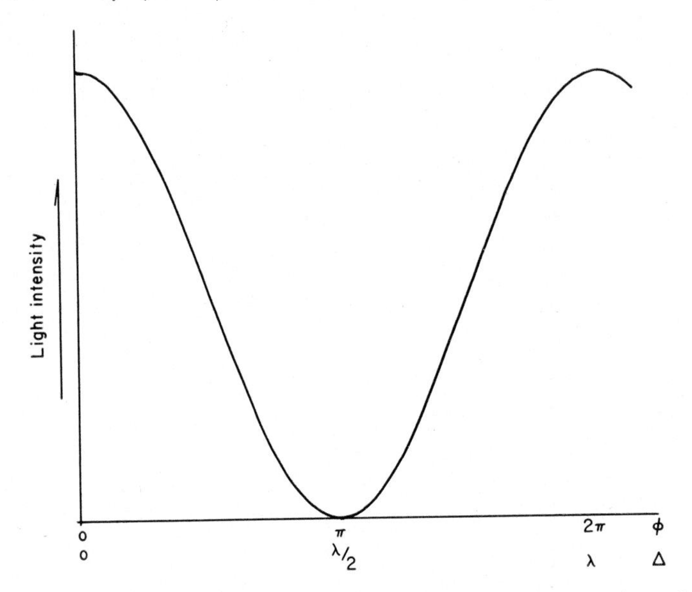

FIG. 44. Shows the variation in light intensity I when the phase difference ϕ between two interfering beams is varied; a plot of Eq. 44. Represents also the variation in light intensity when a plane-polarized light ray traverses a rotating polaroid. The angle ψ between the planes of vibration is equal to $\phi/2$ in this figure.

An essential feature, not brought out yet, is that for an interference pattern to be produced the wave trains from the two sources must preserve a constant phase relationship, that is, they must be *coherent.* If the phase difference fluctuates in time, the patterns shift, resulting in zero contrast. It is a matter of common experience that two light sources will not give rise to an interference pattern and this is due to the random way in which wave trains of finite length are emitted. The two point sources, A and B, satisfy the requirement of coherence since they are derived from a single point source. When the two sources are incoherent the total intensity (I) is obtained by simply adding the individual intensities, that is $r_1^2 + r_2^2$. In practice the coherence of two interfering sources may lie between the two extremes just mentioned and the visibility of the interference pattern, or the extent to which the minima are perfectly black, may be used as a measure of the coherence.

Suppose that a plate of thickness dx is placed in the path AP. During the time taken to go the distance dx in the medium the light waves would have travelled a distance $v/v_m\, dx$ *in vacuo* where v and v_m are the velocities *in vacuo* and in the medium constituting the plate, respectively. The distance $v/v_m\, dx$ is usually referred to as the *optical path* to distinguish it from the geometrical path (dx). The ratio v/v_m defines the *refractive index* (μ) of the plate. Hence the increase in optical path due to the introduction of the plate is ($\mu\, dx - dx$), that is $(\mu - 1)\, dx$ and this is usually referred to as the *optical path difference* (Δ). The corresponding phase difference is $2\pi/\lambda \times \Delta$.

A note on polarized light. Interference between plane-polarized light beams is governed by principles expressed by the Fresnel-Arago laws. (1) Two rays polarized at right angles do not interfere. (2) Two rays polarized at right angles (obtained from the same beam of polarized light) will interfere in the same manner as ordinary light only when brought into the same plane.

When a ray of plane-polarized light of unit amplitude is incident on a sheet of polaroid, it is resolved into two components of amplitudes $1.\cos \Psi$ and $1.\sin \Psi$ vibrating at right angles to one another, one of which is absorbed. The transmitted intensity is therefore proportional to, say, $\cos^2 \Psi$. Hence the light intensity variations which occur when the polaroid is rotated are similar to those which occur when the phase difference between two coherent interfering beams of equal amplitude is varied; in Fig. 44 the abscissa ϕ is replaced by $2\,\Psi$.

Appendix 2: Remarks on Refraction

In this appendix an approximate theoretical relationship is derived between specific refraction increment of a protein solution and the refractive index of the dry protein. Also the relationship recently used by Mellors and Hlinka (1955) for deriving the mass of fixed cells from interference data is discussed.

It is found experimentally that both the density ρ and refractive index μ of a substance depend on temperature and pressure. Attempts have therefore been made to determine a function of ρ and μ which is independent of both pressure and temperature. The function will be constant for one particular substance and is known as the specific refraction (R). An empirical formula proposed by Gladstone and Dale (1863) for the specific refraction is

$$R_{GD} = \frac{\mu - 1}{\rho} \tag{46}$$

Another formula which is derived in electromagnetic theory is due to Lorenz and Lorentz according to which the specific refraction R_{LL} is

$$R_{LL} = \frac{\mu^2 - 1}{\mu^2 + 2}\frac{1}{\rho} \tag{47}$$

A discussion of these and other formulae is given by Böttcher (1952).

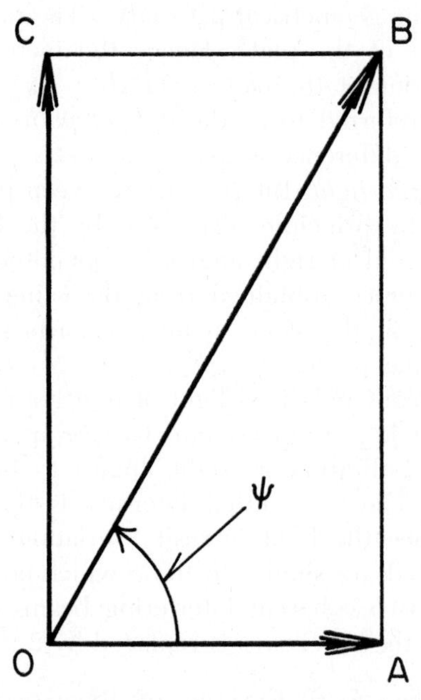

FIG. 45. A plane-polarized light ray represented by the vector *OB*, when incident upon polarized is resolved into two components represented by *OA* (*OB* cos ψ) and *OC* (*OB* sin ψ), the latter being absorbed.

The method of deriving the Lorentz-Lorenz relationship suggests that the specific refraction of a mixture can be calculated from that of the specific refractions of the components, that is it is an additive property, provided that, as a result of mixing, the properties (polarizability) of the individual components are not altered. A similar formula for mixtures can be derived using the Gladstone-Dale formula, and this was applied by Adair and Robinson (1930) to relate specific refraction increments of protein solutions to refractive index of dry protein. A similar relationship has been given by Oster (private communication;

see also Oster, 1949), and more recently by Mellors and Hlinka (1955), and is derived as follows.

Let R, R_1, and R_2 be the specific refractions of solution, solvent, and dry protein, respectively; let μ, μ_1, and μ_2 be the corresponding refractive indices and let v, v_1, and v_2 be the corresponding volumes. Assuming, according to the Gladstone-Dale rule for mixtures, that the specific refraction of the solution is obtained by adding the specific refractions of the components weighted with respect to the fractional mass present, we obtain

$$(\mu - 1)v = (\mu_1 - 1)v_1 + (\mu_2 - 1)v_2 \tag{48}$$

In deriving Eq. 48 and in using it below we assume that the density of dry protein is identical with the reciprocal of the partial specific volume in solution, that is, no volume change occurs on mixing.

Consider 1 cc. of the solution, that is $v = 1$. If the concentration of solid is c_2 gm. cc.$^{-1}$ and $\bar{v}_2$ is the partial specific volume of the solute, then

$$v_2 = c_2\bar{v}_2$$

and

$$v_1 = 1 - c_2\bar{v}_2$$

Inserting in Eq. 48, and rearranging we obtain

$$\frac{\mu - \mu_1}{c_2} = \bar{v}_2(\mu_2 - \mu_1) \tag{49}$$

That is, the specific refraction increment ($\times 100$) is equal to the product of the difference in refractive indices of water-free solute and solvent times the partial specific volume of the dissolved substance. The analysis given by Mellors and Hlinka (1955) is equivalent to substituting Eq. 49 in Eq. 1 when we obtain for the dry mass M the expression

$$M = \frac{\Delta_1 A}{\bar{v}_2(\mu_2 - \mu_1)} \tag{50}$$

where Δ_1 is the o.p.d. measured in a medium of refractive index μ_1. It should be noted that this equation is identical with Eq. 24 if in the latter the expression $(\mu_p - \mu_w)/\rho$ is substituted for χ and $1/\rho$ is equated to $\bar{v}_2$. These equations (24 and 50) serve the same purpose and are in fact identical, the uncertainty in both being due to that in χ for fixed protein.

In applying Eq. 50 to the derivation of dry mass of fixed cells, Mellors and Hlinka (1955) have chosen to give μ_2 the value 1.59 and $\bar{v}_2$ the value 0.75. Possible errors in assigning values to v_2 and μ_2 produce uncertainty in the calculated value for mass in the same way that they do when we attempt to evaluate χ. As an immersion medium for the fixed cell these workers chose glycerol. According to the data in Section II, F there is apparently an increase in the total refraction in glycerol compared with that in water. This is not taken into account in Eq. 50. An examination of the data suggests that the error due to increased refraction in glycerol is approximately balanced by the choice of the value 1.59 for the refractive index, which is higher than the observed value 1.56 for which fixed and dried cells disappear in this medium. As a result, in this particular instance, it may be expected that the mass calculated from data in water will be approximately the same as that derived from measurements in glycerol. In fact Mellors and Hlinka (1955) and Swartz (1956) report good agreement between measurements of dry mass of similar cells in water and glycerol. It appears, if our experimental data (Section II, F) is found to apply to different cell types, that this agreement is fortuitous and that in measurements in media of different refractive index that the values assumed for $\bar{v}_2$ and μ_2 will lead to error. This error will increase as the refractive index of the medium approaches that of the fixed cell.

Appendix 3: Preparation of Specimen

The preparation of the specimen depends on its nature and on the kind of problem to be investigated, and to some extent it is an art which will be well known to the biologist who wishes to apply interferometry to the material with which he is familiar. The following comments which relate to the special requirements of quantitative measurements of optical path difference by interferometry may prove helpful.

(1) *Dimensions of specimen and mounting.* Unlike phase contrast microscopy, interference microscopy places limits on the area of specimen and optical homogeneity of the mounting medium. This is due to the deleterious effects of optical inhomogeneity in the comparison beam. When isolated cells are being examined the number per unit area of slide can usually be so adjusted as to produce negligible effect in both the Cooke-Dyson and Baker-Smith microscopes. When sections are used the maximum viewing distance into the section can be calculated from the shear-distance in the Baker-Smith microscope. In the Cooke-Dyson microscope, provided the diameter of section does not exceed about 0.3 mm. no part will lie in the comparison beam. In sections larger than this the reduction in contrast may be small provided that the section is thin and does not scatter too much light energy.

In preparing specimens for the Cooke-Dyson microscope care is required to have an optically homogeneous mounting medium free from bubbles. It is not usually possible to observe living cells in liquid paraffin which is immiscible with water. In hanging drop tissue cultures which should be grown in fluid medium it may be desirable to excise the explant and to work with a chamber filled with physiological fluid, so as to avoid any variable thickness of air interface. Troubles from these causes are often less serious in the Baker-Smith microscope, on account of the smaller distance separating the two interfering beams. Bubbles in the immersion medium can easily be avoided by using grease-free slides and coverslips and by slowly contacting the coverslip on to the slide.

Simple chambers can easily be constructed to permit irrigation of the cell or tissue during observation. The simplest type is made by cutting strips of coverslip and cementing them on to a slide, the coverslip bearing the specimen being waxed on. If the ends are left open, evaporation of liquid causes curvature of the coverslip and alteration in contrast, so that any loss of liquid should be made good. It has been found that in making very accurate measurements of o.p.d. ($1/100\ \lambda$–$1/1000\ \lambda$) using the Davies-Deeley integrator that perfection of coverslips is important. Localized imperfections due to striae in the glass, etc. which will be approximately in focus with the object show up as phase changes and introduce error.

It may be desirable to squash living cells (see, e.g., Easty *et al.*, 1956) so as to make their geometry approximate to that of a parallel sided plate; this is rarely possible. Fixed cells may be squashed or crushed so as to reduce their optical path difference. When mounting fixed tissue use fresh solutions. Media contaminated with bacteria may contain proteolytic enzymes which cause gradual loss of mass.

(2) *Fixation.* Possible changes as a result of fixation include loss of mass due to removal of cell substance, increase of mass due to combination of the fixative with the cell, and alterations in total refraction due to changes in the physical properties of the protoplasm. For example, fixation in osmium tetroxide causes an apparent increase in dry mass of ram sperm heads of 10 to 20% (Davies and Deeley, unpublished). Fixatives containing heavy metals should be used with care. The loss of substance during fixation, for example some nonprotein when protein content is being assayed, is an advantage. The uncertainty in χ which occurs in fixed tissue has already been noted.

Fixatives which cause clumping of material should in general be avoided since they introduce difficulties into visual assay of optical path differences and may introduce systematic error as described in Section V. However, when the over-all optical path difference is less than about 0.1 wavelengths an inhomogeneous distribution will not introduce errors into measurements of total mass by, e.g., nonscanning integration. The protoplasm of living cells often appears homogeneous in the light microscope and negligible inhomogeneity is created when fixing in neutral formalin and after careful freeze-drying or freeze-substitution. The latter technique is proving particularly valuable in quantitative cytochemistry because of this fact and because it is simple as compared with freeze-drying.

When using freezing methods the volume of tissue should be kept at a minimum so as to avoid ice crystal formation during freezing. Single cell suspensions should be spread (or smeared) on a coverslip and then frozen immediately to prevent air drying. Precautions in freeze-substituting in Davies (1954) may be noted. During the smearing rounded cells are often flattened to the coverslip by surface tension and when fixed all parts of the cell appear may be approximately in one plane and hence in focus in the microscope. Material prepared in this way usually adheres to the coverslip and does not easily wash off during subsequent treatment. The amount of soluble material (protein) in the medium in which these cells are mounted prior to fixation should be kept to a minimum. Otherwise fixed protein threads in the background to the specimen may produce error especially in nonscanning integration.

(3) *Single cells and sections of fixed material.* For many reasons measurements on isolated cells or cell parts are easier to make and often give more "useful" information than measurements on sections of fixed tissue. With preparations of single cells upset to the comparison area can easily be avoided. A single layer of cells is also easier to fix in a homogeneous manner by freeze-substitution than a small piece of tissue.

It is often desirable to measure total mass per cell or cell nucleus. [We note that it is the constancy in total amount (*not concentration*) of DNA per nucleus which is partly responsible for causing us to attach special significance to DNA as genetic material. Measurements of concentration of DNA have another significance and we may suggest they are bound up with its "reactivity"; in sperm heads the concentration of DNA is relatively high, and it is presumably inactive, while its high concentration in heterochromatin may be associated with the inactivity of genes at these sites.] This is not easy to do in tissue sections since all boundaries are often indistinct and parts of a particular cell or nucleus may be missing. Presence of material over and underlying nuclei is not easy to allow for as it requires knowledge of thickness of section and nucleus (Eq. 36). Of course similar difficulties due to cytoplasm occur in single cells although in well-flattened cells in tissue culture the amount above and below the nucleus may be relatively small.

In order to avoid uncertainty from over and underlying substance methods of isolating cells from tissues (Moscona, 1952; Rinaldini, 1958) and of isolating cell parts, e.g., nuclei from cells are valuable. These methods introduce their own uncertainties due to possible loss and transfer of material.

Sections of fixed tissue are valuable for giving a measure of the distribution of mass throughout a tissue. They can also be useful when the ratio of a particular chemical substance to dry mass is being determined.

APPENDIX 4: SETTING UP THE MICROSCOPE

The detailed instructions on adjusting the interference microscopes are contained in the manufacturers handbooks. The following notes on illuminating the microscope may be helpful.

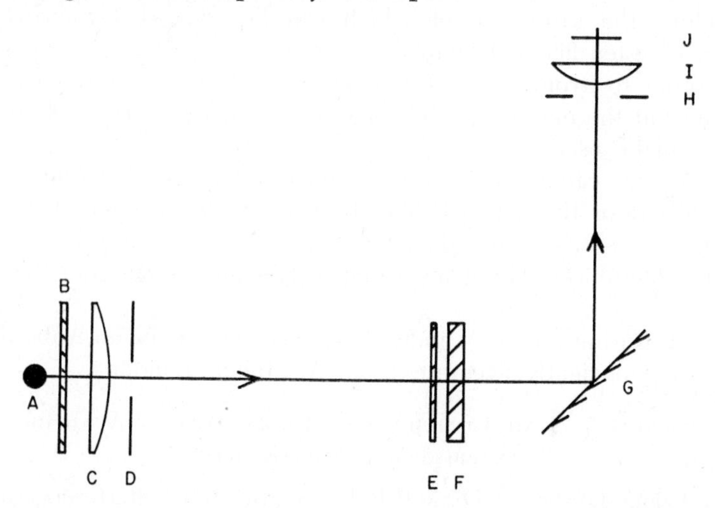

FIG. 46. In Köhler illumination of the microscope the field may be evenly illuminated despite the nonuniformity of light source. A, mercury arc. B, Chance ON 20 heat-filter; C, lens; D, iris diaphragm, *the field stop;* E, F, filters for isolating the spectral line λ 546 mμ; G, microscope mirror; H, iris diaphragm, *the aperture stop;* I, microscope condenser; J, specimen.

The microscope is illuminated according to Köhler, that is, the lens *C* focuses the source *A* onto the aperture stop, the condenser iris diaphragm, at *H*. The field stop is an iris diaphragm at *D*. The distance *DGH* will depend on the dimensions of the light source and the focal length of the lens *C*, the criterion being that the image of the light source should fill the aperture at *H*. This source *A* is a high-intensity mercury arc (see Fig. 46), for example a 250-watt compact source (Mazda, England). These arcs have a tendency to flicker and for some visual photometry a more stable but lower wattage arc may be used. Lens *C* should be of good quality and approximately plano-convex with the surface of least curvature facing the source—this reduces spherical aberration in the lens. A Watson-Conrady aplantic doublet, glycerine cemented (Watson, England) has been used, but any good quality short focus ($\sim 2''$) lens will do. The lens may be subject to considerable heating from the mercury arc and a heat filter may be inserted at *B* (e.g., Chance ON 20). *F* is a filter which isolates the green line (5461 Å) from the mercury arc, for example Wraten filter No. 77a or No. 77 and No. 58.

The unfiltered light of the mercury arc may be used as the "white-light" when determining the whole number part of the optical path difference. In a heterogeneous specimen it may be difficult to recognize the colors, the saturation of which can be somewhat improved by using a tungsten filament lamp.

In order to avoid a double image of the field stop formed by two reflections at the microscope substage mirror, a front-surface aluminized mirror can be used.

A 35 mm. camera is useful for recording images. As film Kodak Microfile Pan or Ilford Pan F have been used. The Microfile has a very high gamma ($\sim$ 3–4) in green light and is useful for recording faint intensity differences. The gamma of Pan F is much lower but the speed is greater.

For measuring refractive indices of immersion media an Abbe refractometer is sufficiently accurate (e.g., A. Hilger, London).

Appendix 5: Some Comments on Commercially Available Interference Microscopes

The Cooke-Dyson (C-D) and Baker-Smith (B-S) interference microscopes are both suitable for quantitative microinterferometry on biological material. Although no extensive series of measurements with the two instruments on the same specimen have been reported those which have been made on similar material indicate good agreement. Thus measurements of the dry mass of fixed sperm heads from bull made independently by Davies *et al.* (1957) and Leuchtenberger *et al.* (1956) with C-D and B-S microscopes gave average values of 7.1×10^{-12} and 7.4×10^{-12} gm. respectively. The small differences reported by Hale (1954) in measurements on blood corpuscles with the two microscopes are apparently due to the use of the double-focus Baker objectives and were absent when the shearing system was used (Hale, private communication). In an abstract, Casselman and Wilbur (1956) report differences but do not state which Baker objective was used.

Although both microscopes appear to be suitable for tackling most problems encountered in biological practice, there are certain special cases where this is not so. The comments, mainly factual, in Table V may be of help in making a choice. It is understood that the interfer-

TABLE V

COMMENTS ON THE BAKER-SMITH AND COOKE-DYSON INTERFERING MICROSCOPES

Baker-Smith	Cooke-Dyson
1. High, medium and low power objectives available: NA 1.3, ×100; 0.7, ×40; 0.25, ×10. Requires a different condenser for each objective but change-over is simple. Condenser numerical aperture cannot be increased above about 0.5. No immersion media required except for high power (water immersion). Water immersion must be an advantage when material is located in water much below coverslip (less objective aberrations). Both microscopes need more care in setting up and greater insight into theory than is normally required.	Only high and medium power objectives available: NA 1.3, ×90; 0.85, ×40. Same condenser used throughout. Condenser aperture can be increased to about 0.5–0.7 without appreciable contrast reduction. Immersion medium required at 3 sites between condenser and lower wedge, wedge and specimen, specimen and objective. (First can be conveniently left for a long period.) The necessity for immersion is considered by some to decrease ease of use. Advantages of homogeneous immersion include less critical choice of coverslip thickness and optical quality. Others consider that the C-D is easier to adjust to perfection than B-S (this refers to fluffing out bands in back focal plane of objective).
2. Medium light transmission. Compared with an ordinary microscope of same condenser NA, the light transmission varies from ∼ 25% (the transmission of parallel polaroids) downwards, depending on setting of analyzer.	Light transmission lower than B-S and not suitable for photographing moving objects in monochromatic light under high power. Compared with an ordinary microscope of condenser NA 0.6 the light transmission varies from ∼ 0.5% downwards, depending on setting of phase screw.
3. Special attachment (half-shade eyepiece) for accurate measurements of frational part of o.p.d.	Special attachment (eyepiece photometer) for accurate measurement of fractional part of o.p.d. Accuracy similar to half-shade eyepiece.
4. Special attachment (Wright eyepiece) required for producing a banded field. Each different band separation requires a different quartz wedge.	No special attachment for producing banded field which is obtained by rotating objective and upper wedge. Band separation can be continuously varied above a minimum distance which is fixed; bands are also rotated.
5. The high power objective has a relatively small amount of shear and with cells much greater than about 20 μ there is overlap. This is a distinct disadvantage. Ghost image is well located and a limited area of biological section of any area can be viewed with comparison beam clear of specimen.	No special difficulty due to comparison beam at high power. Sections must be cut to small size if comparison beam is to be kept clear.

TABLE V (Continued)

Baker-Smith	Cooke-Dyson
6. The conversion to ordinary illumination (no interference contrast) is very simple.	The microscope can be converted to ordinary illumination, but this involves widely separating the two interfering fields and the area of field will be small.

ence microscope described briefly by Johannson and Afzelius (1956) is being manufactured (Jungnerbologet, Stockholm 14) and it will be interesting to see if it has any special advantages. It appears that this microscope may have only a small amount of shear and if this proves to be insurmountable the applications to biological material will be restricted. Also the use of a slit aperture in the condenser may be objectionable.

REFERENCES

Adair, G. S., and Robinson, M. E. (1930). *Biochem. J.* **24,** 993.
Adair, G. S., Ogston, A. L., and Johnston, J. P. (1946). *Biochem. J.* **40,** 867.
Alfert, M., Bern, H. A., and Kahn, R. H. (1955). *Acta. Anat.* **23,** 185.
Allfrey, V. G., Mirsky, A. E., and Stern, H. (1955). *Advances in Enzymol.* **16,** 411.
Ambrose, E. J. Private communication.
Armstrong, S. H., Jr., Budka, M. T. E., Morrison, K. C., and Hasson, M. (1947). *J. Am. Chem. Soc.* **69,** 1747.
Baker, C. (1955). "The Baker Interference Microscope" 2nd ed. Mainsail Press, London.
Barer, R. (1952). *Nature* **169,** 366.
Barer, R. (1953). *Nature* **172,** 1097.
Barer, R. (1955). *In* "Analytical Cytology" (R. C. Mellors, ed.), Chapter 3. McGraw-Hill, New York.
Barer, R. (1956). *In* "Physical Techniques in Biological Research" (G. Oster and A. W. Pollister, eds.), Vol. 3, p. 30. Academic Press, New York.
Barer, R., and Dick, D. A. T. (1955). *J. Physiol. (London)* **128,** 25 p.
Barer, R., and Joseph, S. (1954). *Quart. J. Micrscop. Sci.* **95,** 399.
Barer, R., and Joseph S. (1955a). *Quart. J. Microscop. Sci.* **96,** 1.
Barer, R., and Joseph, S. (1955b). *Quart. J. Microscop. Sci.* **96,** 423.
Barer, R., and Ross, K. F. A. (1952). *J. Physiol. (London)* **118,** 38 p.
Barer, R., and Tkaczyk, S. (1954). *Nature* **173,** 84.
Barer, R., Howie, J. B., Ross, K. F. A., and Tkaczyk, S. (1953). *J. Physiol. (London)* **120,** 67.
Barter, R., Danielli, J. F., and Davies, H. G. (1955). *Proc. Roy. Soc.* **B144,** 412.
Bawden, F. C., and Pirie, N. W. (1937). *Proc. Roy. Soc.* **B123,** 274.
Bennett, H. S. (1955). *Anat. Rec.* **121,** 263.
Bernal, J., and Fankuchen, I. (1941). *J. Gen. Physiol.* **25,** 111.
Böttcher, C. J. F. (1952). "Theory of Electric Polarisation." Von Nostrand, Princeton, New Jersey.

Brown, G. L., McEwen, M., and Pratt, M. (1955). *Nature* **176,** 161.
Caspersson, T. (1936). *Skand. Arch. Physic.* **73,** Suppl. 8.
Caspersson, T. (1950). "Cell Growth and Cell Function." Norton, New York.
Caspersson, T., Carlson, L., and Svensson, G. (1954). *Exptl. Cell Research* **7,** 601.
Caspersson, T., Lomakka, G., Svensson, G., and Säfström, R. (1955). *Exptl. Cell Research* **3,** 40.
Casselman, B., and Wilbur, M. (1956). *Proc. Histochem. Soc.* **4,** 437.
Chick, H., and Martin, J. (1913). *Biochem. J.* **7,** 92.
Cohn, E. J., and Edsall, J. T. (1943). "Proteins, Amino Acids and Peptides." Reinhold, New York.
Crossmon, G. C. (1949). *Stain Technol.* **24,** 244.
Davies, H. G. (1954). *Quart. J. Microscop. Sci.* **95,** 433.
Davies, H. G., and Deeley, E. M. (1956). *Exptl. Cell Research* **11,** 169.
Davies, H. G., and Engström, A. (1954). *Exptl. Cell Research* **7,** 243.
Davies, H. G., and Walker, P. M. B. (1953). *Progr. in Biophys. and Biophys. Chem.* **3,** 195.
Davies, H. G., and Wilkins, M. H. F. Unpublished.
Davies, H. G., and Wilkins, M. H. F. (1951). Report to the Cytochemistry Commission of the Society for Cell Biology. Physical aspects of Cytochemical Methods, Stockholm.
Davies, H. G., and Wilkins, M. H. F. (1952). *Nature* **169,** 541.
Davies, H. G., Barter, R., and Danielli, J. F. (1954a). *Nature* **173,** 1234.
Davies, H. G., Deeley, E. M., and Denby, E. F. (1957). *Exptl. Cell Research Suppl.* **4,** 136.
Davies, H. G., Engström, A., and Lindström, B. (1953). *Nature* **172,** 1041.
Davies, H. G., Wilkins, M. H. F., Chayen, J., and La Cour, L. F. (1954b). *Quart. J. Microscop. Sci.* **95,** 271.
Deeley, E. M. (1955). *J. Sci. Instr.* **32,** 263.
Ditchburn, R. W. (1952). "Light." Interscience, New York.
Dyson, J. (1950). *Proc. Roy. Soc.* **A204,** 170.
Dyson, J. (1953). *Nature* **171,** 743.
Dyson, J. Private communication.
Easty, D. M., Ledoux, L., and Ambrose, E. J. (1956). *Biochim. et Biophys. Acta* **20,** 528.
Engström, A., and Lindström, B. (1949). *Nature* **163,** 563.
Engström, A., and Lindström, B. (1950). *Biochim. et Biophys. Acta* **4,** 351.
Faust, R. C., and Marrinan, H. J. (1955). *Brit. J. Applied Phys.* **6,** 351.
Folch, J., and Lees, M. (1951). *J. Biol. Chem.* **191,** 807.
Gladstone, J. H., Dale, J. (1863). *Phil. Trans. Roy. Soc.* **153,** 317.
Hale, A. J. (1954). *J. Physiol. (London)* **125,** 50 p.
Hale, A. J. (1956). *Exptl. Cell Research* **10,** 132.
Hale, A. J., and Kay, E. R. M. (1956). *J. Biophys. and Biochem. Cytol.* **2,** 147.
"Handbook of Chemistry and Physics." (C. D. Hodgman, ed.), 30th ed. Chemical Rubber Publishing, Cleveland, Ohio, 1946.
Haurowitz, F. (1950). "Chemistry and Biology of Proteins." Academic Press, New York.
Huxley, A. F. (1952). *J. Physiol.* **117,** 52, p.
Huxley, A. F. (1954). *J. Physiol.* **125,** 11, p.
Huxley, A. F., and Niedergerke, R. (1954). *Nature* **173,** 971.
Huxley, H. E., and Hanson, J. (1957). *Biochim. et Biophys. Acta* **23,** 229.

Jenkins, F. A., and White, H. E. (1957). "Fundamentals of Optics," 3rd ed. McGraw-Hill, New York.
Johansson, L. P., and Afzelius, B. M. (1956). *Nature* **178,** 137.
Klug, A. Private communication.
Lagerlöf, B., Thorell, B., and Åkerman, L. (1956). *Exptl. Cell Research* **10,** 752.
Lebedeff, A. A. (1930). *Rev. opt.* **9,** 335.
Leuchtenberger, C., Murmanis, I., Murmanis, L., Ito, S., and Weir, D. R. (1956). *Chromosoma* **8,** 73.
Lomakka, G. (1954). *Exptl. Cell Research* **7,** 603.
Lomakka, G. (1955). *Exptl. Cell Research* **9,** 434.
McFarlane, A. S. (1935). *Biochem. J.* **29,** 407.
Mann, T. (1949). *Advances in Enzymol.* **9,** 329.
Mann, T. (1954). "The Biochemistry of Semen." Wiley, New York.
Markham, R. Private communication.
Marsden, N. V. B. (1956). *Exptl. Cell Research* **10,** 755.
Mellors, R. C. (1956). *Ann. N. Y. Acad. Sci.* **63,** 1177.
Mellors, R. C., Stoholski, A., and Beyer, R. (1954). *Cancer* **7,** 873.
Mellors, R. C., and Hlinka, J. (1955). *Exptl. Cell Research* **9,** 128.
Mirsky, A. E., and Ris, H. (1949). *Nature* **183,** 666.
Mitchison, J. M., and Swann, M. M. (1953). *Quart. J. Microscop. Sci.* **94,** 381.
Mitchison, J. M., Passano, L. M., and Smith, F. H. (1956). *Quart. J. Microscop. Sci.* **97,** 287.
Moscona, A. (1952). *Exptl. Cell Research.* **3,** 535.
Northrop, T. G., and Sinsheimer, R. L. (1954). *J. Chem. Phys.* **22,** 703.
Northrop, T. G., Nutter, R. L., and Sinsheimer, R. L. (1953). *J. Am. Chem Soc.* **75,** 5134.
Oettlé, A. G. (1950). *J. Roy. Microscop. Soc.* **70,** 232.
Oster, G. (1949). *Bull. soc. chim. France* **D353,** p.
Oster, G. Private communication.
Perlman, G. E., and Longsworth, L. G. (1948). *J. Am. Chem. Soc.* **70,** 2719.
Pollister, A. W. (1952). *Exptl. Cell Research* Suppl. 2, 59.
Pringsheim, E. (1898). *Verhandl. deut. physik. Ges.* **17,** 152.
Putzuys, P., and Brosteaux, J. (1936). *Bull. soc. chim. biol.* **18,** 1681.
Reichmann, M. E., Rice, S. A., Thomas, C. A., and Doty, P. (1954). *J. Am. Chem. Soc.* **76,** 3047.
Reiss, E. (1930). *Arch. exptl. Pathol. Pharmakol. Naunyn-Schmiedelberg's* **51,** 18.
Richards, B. M., and Davies, H. G. (1958). In preparation.
Rinaldini, L. M. (1958). *Intern. Rev. Cytol.* In press.
Robison, R. (1923). *Biochem. J.* **17,** 286.
Ross, K. F. A. (1953). *Quart. J. Microscop. Sci.* **94,** 125.
Ross, K. F. A. (1954). *Quart. J. Microscop. Sci.* **95,** 425.
Schneider, W. C. (1945). *J. Biol. Chem.* **161,** 293.
Simpson, W. L. (1941). *Anat. Rec.* **80,** 173.
Sirks, J. L. (1893). *Handel. Ned. Natuur- en Geneesk. Congr. Groningen,* p. 92.
Smith, F. H. (1950). *British Patent Spec.* 639041.
Smith, F. H. (1954). *Nature* **173,** 362.
Smith, F. H. (1955). *Research (London)* **8,** 385.
Stokes, A. R. Unpublished.
Swartz, F. (1956). *J. Histochem. and Cytochem.* **4,** 436.
Swift, H. (1953). *Intern. Rev. Cytol.* **2,** 1.

Swift, H., and Rasch, E. (1956). "Physical Techniques in Biological Research" (G. Oster and A. W. Pollister, eds.), Vol. III. Academic Press, New York.
Tennent, H. G., and Villbrandt, C. F. (1943). *J. Am. Chem. Soc.* **65,** 424.
Thorell, B. (1947). "Studies on the Formation of Cellular Substances during Blood Cell Formation." Kimpton, London.
Vendrely, R., and Vendrely, C. (1948). *Experientia* **4,** 434.
Vendrely, R., and Vendrely, C. (1949). *Experientia* **5,** 327.
Vincent, W. S., and Huxley, A. H. (1954). *Biol. Bull.* **107,** 290.
Walker, P. M. B. (1954). *J. Exptl. Biol.* **31,** 8.
Walker, P. M. B. (1955). *Exptl. Cell Research* **8,** 567.
Wilkins, M. H. F. (1950). *Discussions Faraday Soc.* **No. 9,** 363.
Williams, W. E. (1948). "Applications of Interferometry." Wiley, New York.
Zernike, F. (1935). *Physik. Z.* **36,** 848.

ULTRAVIOLET MICROSPECTROPHOTOMETRY

P. M. B. Walker

Biophysics Research Unit, King's College, London, W. C. 2, England

I. INTRODUCTION

All components of cells and tissues absorb in some region of the spectrum, but in most instances the cytochemist is unable to derive much useful information from this. For example, infrared radiation has too long a wavelength to resolve the structures of most cells, while at wavelengths shorter than 230 mμ all proteins have so high an absorption that identification is very difficult. However, in the intermediate spectral range between 230 and 700 mμ it is possible both to identify certain important substances and to measure the quantities present in a single cell or nucleus.

This subject has already been extensively reviewed (Caspersson, 1950; Loofbourow, 1950; Scott and Sinsheimer, 1950; Blout, 1953; Davies and Walker, 1953; Nurnberger, 1955) and has also been considered at some length by this author (Walker, 1956). It is therefore necessary to define the aims and limitations of this chapter which will reduce overlapping, while giving sufficient information to enable the reader to estimate the difficulties and advantages of the method. It is also hoped that a basis for a critical assessment of the results obtained by microspectrometric techniques will be provided.

With the editor's permission the arrangement common to the other chapters will be abandoned and the following layout substituted. After a brief introduction on the theory of the method, there will be a fuller discussion of two topics: the relative merits of photographic and photoelectric recording and of the various types of objective suitable for use in the ultraviolet. A decision on these subjects will govern the user's choice of equipment. There will be a minimum of description of apparatus in the text, but annotated drawings of a number of complete instruments and of certain important details will be found in the Appendix. There will, however, be a discussion on instrumental errors and suitable test procedures for detecting them.

Unfortunately, no review of ultraviolet absorption techniques is complete without a section on errors in measurement, that is, on those errors introduced by the biological specimen. The emphasis here will be on practical methods of detecting and mitigating them rather than on theoretical discussion. There will also be a discussion of the factors within the cell which may influence the absorption of cellular constituents. Finally, there will be some comment on recent results.

II. THEORY OF THE METHOD

We have adopted here the notation, now generally accepted by spectroscopists, which substitutes the term absorbance (A) for extinction or optical density, and absorptivity (k) for extinction coefficient.

The term density is however retained for photographic processes, where it is convenient to differentiate between the absorbance of the object and the density of the film. The absorption law now reads

$$A_\lambda = \log_{10} \frac{I_0}{I} = kcl \tag{1}$$

where λ is the wavelength at which measurement is made; I_0 and I the incident and transmitted energies and c the concentration in grams per liter of the substance of absorptivity k; l is the length or thickness of the object in centimeters. It is often difficult to measure the thickness of cells, but this is unnecessary provided the result can be expressed in terms of mass per unit area, when

$$m = \frac{Aa}{k} \tag{2}$$

if m is the mass in grams and a the area in 10^3 cm.^{2}. It follows from this relation that the smaller the area measured, the greater the sensitivity, and indeed it is possible to detect about 5×10^{-14} gm. of nucleic acid within a projected area of 1 μ^2. The absorptivity (k) of polymerized nucleic acid is about 20 at 260 mμ, and A would be in this instance 0.1, which is probably the smallest value which can be measured with reasonable accuracy ($\pm$ 10%). Other cellular components have different values for their maximum absorptivity, and it is possible therefore to compile a table showing their relative importance as defined by the smallest mass per μ^2 which may be measured (Table I). As a very rough guide, if only the amount shown in the last column of the table is thought to be present within the cell, there will be little chance of even detecting it; ten times this value could be measured only if it were strictly localized, while a hundred times should be quite easy to measure.

The comparatively small number of cellular substances which can be investigated by this method may be considered as a limitation of the technique. On the other hand, if there were many more or indeed if all those listed were found simultaneously, analysis would be almost impossible because of the broad peaks and general similarity of some of their absorption curves (Fig. 1).

It follows from the preceding paragraphs that the basic requirements of this technique are: (1) a monochromator and source capable of providing energy of sufficient spectral purity that cellular constituents may be identified and their absorbance measured at a specific wavelength, (2) a microscope to define a small part of a cell and to pass energy through it, and (3) a detector capable of measuring the energy transmitted by the defined part of the cell and by the background (I and I_0).

TABLE I
SOME OF THE MORE IMPORTANT ULTRAVIOLET-ABSORBING SUBSTANCES OCCURRING IN CELLS

Substance	$\varepsilon_{mol} \times 10^{-3}$	Mol wt. = M	$k = \varepsilon/M$	λ_{max} (mμ)	Mass (10^{-14} g.) per μ^2 to give 0.1 A	References for ε_{mol} or k
DNA, RNA[a]	—	—	20	260	5	
Tyrosine in						Beaven and Holiday (1952)
N/10 NaOH	2.33	181	12.9	293.5	7.6	
N/10 HCl	1.34	181	7.4	274.5	13.5	
Tryptophan	5.0	204	21.8	280.5	4.6	Beaven and Holiday (1952)
"Standard[b] protein"	—	—	0.66	280	151	Caspersson (1940)
Phenylalanine	0.2	165	1.21	258	83	Beaven and Holiday (1952)
Ascorbic acid[c]	99	176	56.5	265	1.8	Chayen (unpublished)
Catechol[c]	23	110	21.4	276	4.8	Chayen (unpublished)
Folic acid[c] (Light)	23.3	441	52.8	255	1.9	Chayen (unpublished)
Adenosine[d]	14.9	267	56.5	259.5	1.8	Beaven *et al.* (1955)
Cytidine[d]	9.1	243	37.5	271	2.7	Beaven *et al.* (1955)
Guanosine[d]	13.65	283	48.2	252.5	2.1	Beaven *et al.* (1955)
Thymidine[d]	7.38	242	30.5	267	3.3	Beaven *et al.* (1955)
Uridine[d]	8.5	244	34.8	262	2.9	Beaven *et al.* (1955)

[a] Caspersson gives a value of k for RNA as 22.
[b] Standard protein refers to 5% tyrosine and 1% tryptophan.
[c] Not specially purified.
[d] All the nucleoside values are for neutral pH, see reference for full series of curves. The corresponding nucleotides have almost identical absorption spectra.

The first and third requirements are common to both macroscale spectrophotometry and to microspectrophotometry, and the complete supersession of the old laborious photographic method of spectrometry by Beckman, Uvispeck, and similar instruments has led many workers in microspectrophotometry to concentrate their attention on photoelectric instruments. This is probably justified if the research worker has the time and facilities to construct double-beam recording instruments, but in many instances the merits of photographic recording should be seriously considered.

III. THE TECHNIQUES OF MEASUREMENT

The stages of the photographic method can be summarized as follows. The negative with the images of the specimen and of the calibration step wedge (see Fig. 2) are exposed simultaneously at the desired wavelength. After even development of the film to ensure both specimen and wedge images receive the same treatment, the densities of the wedge steps on the film are measured and plotted against the absorbances of the original wedge. This should give a straight line relation over the middle part of its range, which may be then used to convert the densities measured on the specimen negative to absorbance units.

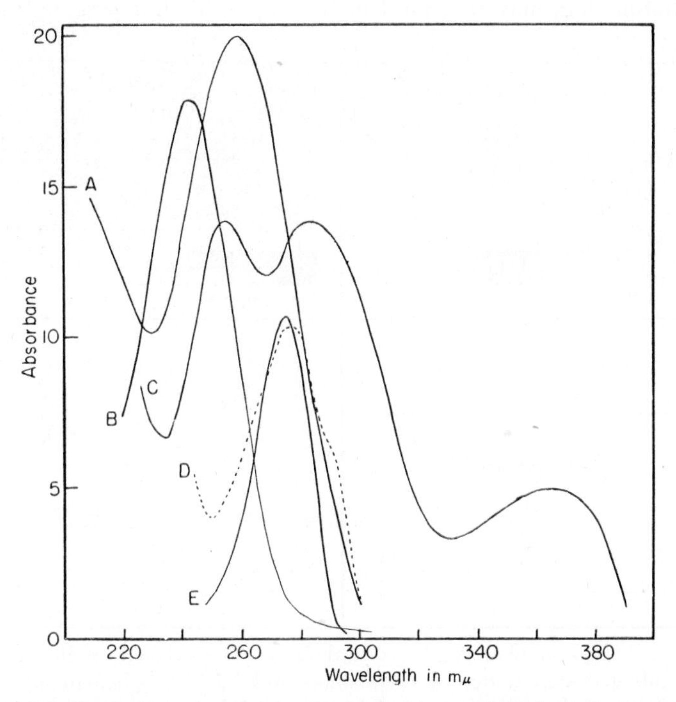

FIG. 1. Absorption curves of some substances of biological importance. A, deoxyribose nucleic acid (Caspersson, 1936); B, ascorbic acid in metaphosphoric acid (Chayen, unpublished). Note: the peak of the absorption curve is shifted to 265 mμ in water at pH 7.0 (Baird *et al.*, 1934). C, Folic acid in NaOH (Chayen, unpublished). D, Dotted curve, serum globulin in H_2O, pH 6.0 (Beaven and Holiday, 1953); E, pyrocatechol in 10% acetic acid (Chayen, unpublished).

In contrast, all that is required in the photoelectric method is to make measurements of I_0 and I, while maintaining all other conditions, such as source energy and photomultiplier voltage, constant. The advantages of this method are the elimination of the intermediate photographic steps, the greater accuracy, and the ability to read I and I_0 from the central portion of the specimen without irradiating the rest of the specimen. The advantages of the photographic method are (1) the simultaneous recording of I and I_0 with reasonable accuracy provided the field can be evenly illuminated and (2) the furnishing of a permanent record on which the focus may be checked and the area of the specimen measured.

The various steps in the photographic procedure are illustrated in Fig. 2, from which it will be seen that, in addition to the basic microscope, monochromator, and camera, some method of measuring the densities of photographic film is required. Some simple designs for microphotometers may be found in the appendix, but if many measure-

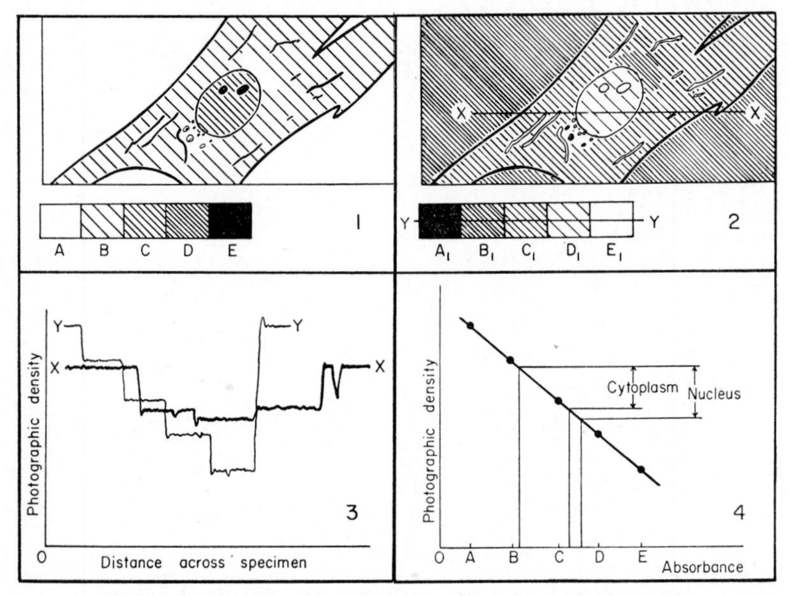

FIG. 2. Steps in photographic recording. 1: The original specimen, together with a calibrated step wedge of absorbances A–E. 2: The photographic negative with the image of the specimen and of the step wedge of densities A_1–E_1. 3: Microdensitometer traces across the specimen (XX) and the step wedge (YY). 4: The relation between the densities and the known absorbances of each step of the wedge are now plotted and from this relation the absorbances of the nucleus and cytoplasm can be readily obtained.

ments are required it may be necessary to use a more complicated recording instrument. On p. 174 is discussed the testing of microspectrophotometers, which may require a quartz-window photomultiplier and galvanometer even for the photographic method. The availability of these ancillary instruments should be considered when a decision has to be made between photographic and photoelectric recording.

Since photocells, unlike photographic emulsions, do not integrate the incident energy, they are much more sensitive to short-term fluctuations in the source, and high-energy lamps will also be required if measurement is to be reasonably rapid. Thus the easily stabilized hydrogen arc cannot provide sufficient energy at the shorter wavelengths unless very large measuring apertures are used. Even the special high-energy hydrogen arc developed by Wyckoff (1952) could only be used with a 5 μ diameter measuring aperture in the M.I.T. recording microspectrophotometer because, in addition to decline of the source energy, the optics have a reduced transmission and photomultiplier sensitivity falls off at shorter wavelengths. Higher energy sources like the mercury or xenon arcs are more difficult to stabilize, and the fullest advantages of photoelectric recording are not therefore apparent except with the more complicated double-beam recording instruments, in which the effects of source fluctuation are reduced. We have illustrated in the Appendix, however, certain simpler designs of photoelectric instruments, which will help the reader to make his final choice of recording method.

A. Objectives for Use in the Ultraviolet

There are three main classes of objective that can be used in an ultraviolet microscope.

1. *Monochromats*

These are made entirely of quartz and are designed to give their best performance at one selected wavelength. They may be obtained with a high numerical aperture and provide a bright and well-detailed picture (e.g., Ludford and Smiles, 1950). However, their image quality deteriorates if used at wavelengths other than that for which they were designed, or if the radiation used is not strictly monochromatic. Focusing is difficult unless a Köhler eyepiece or a matched visible objective and a very good fine adjustment are used (Barnard and Welch, 1936; Taylor, 1953). They also have a low ultraviolet transmission and considerable glare may be experienced due to the large number of quartz-air interfaces. A thorough experimental investigation of the properties of these objectives has been made by King and Roe (1953).

2. *Refracting Achromats*

Owing to the small number and variety of optical materials which transmit in the 240–300 mμ region of the spectrum, it is very difficult to design an achromatic refracting objective for this range of wavelengths. A few objectives of this type have been designed (Johnson, 1939; Foster and Thiel, 1948), which are of limited usefulness. In addition we have used (Walker and Davies, 1950) a 1.2 numerical aperture (NA) quartz lithium fluoride objective designed by Bracey and it is reported that a 0.9 NA design by Zeiss is in preparation, but little information is available about the latter. The image quality of the

1.2 NA objective is inferior to that of a quartz monochromat of the same numerical aperture, and it suffers from the same disadvantages of low ultraviolet transmission and glare. It also requires slight refocusing ($\sim 1\ \mu$) between 248 and 312 mμ. Its chief merits are that it does not require strictly monochromatic illumination and therefore can be used with the diffuse or doublet lines of the mercury arc, and the difference in focus between ultraviolet and visible regions can be easily corrected by inserting a simple lens (Davies and Wilkins, 1950).

3. *Reflecting Objectives*

A simple achromatic objective, which may be used over a very wide spectral region, can be made with two mirrors (Fig. 3), but such

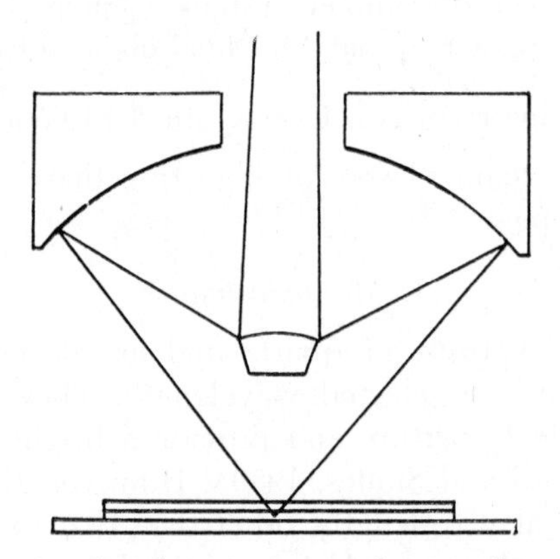

FIG. 3. Simple two-mirror reflecting objective.

objectives have important limitations. It will be seen that light may only be collected from the object in a hollow cone, since the central region is obscured by the small mirror. This causes energy to be transferred from the central portion of the Airey disc diffraction pattern into the first ring to a degree dependent on the obscuration ratio of the objective (Steel, 1953). If this is large the deterioration of the image quality will be noticeable, and accurate measurement of the absorbance of small particles ($< 1\ \mu$) will not be possible. It is also difficult to obtain numerical apertures above about 0.5 and at the same time avoid aberrations reducing the useful field size, although several at least partially successful solutions have been achieved.

a. Catadioptric Systems.

These objectives, which have been most fully investigated by Grey (1950) have both reflecting and refracting components. High numerical apertures and good quality of image are attainable provided the manufacturers can maintain the high accuracy of assembly that the designs require. The higher numerical aperture examples approach in complexity the plain refracting types and have therefore a low transmission. This objective is manufactured by Messrs. Bausch and Lomb.

b. Aspheric Mirror Objectives.

An objective with a large well-corrected field area and low obscuration ratio has been designed by Burch (1947). This has an aspheric large mirror in the conventional two-mirror design, which requires considerable technical skill in hand figuring and in testing. Consequently very few have been made. Their main practical disadvantage is their large size, which requires a large and expensive microscope to carry it. The basic numerical aperture of the Burch objectives is 0.65, but this may be considerably increased by adding an immersion front component.

c. Objectives with Semireflecting Surfaces.

The effective obscuration ratio may be considerably reduced if the mirrors are made semireflecting. Such an objective of 0.85 NA has been designed by Blaisse *et al.* (1952) and is being manufactured by Leitz. This should be a most useful objective provided that the inevitable loss in transmission is not too serious.

d. Non-concentric Mirror Objectives.

By departing from the monocentric arrangement of the mirrors, it is possible to design (Norris *et al.*, 1951) a two-mirror objective with a NA of 0.6, but with a somewhat restricted field and high obscuration ratio compared for example with the Burch objective. A most useful improvement on this design is the so-called solid objective of NA 0.9 (Fig. 4) in which the reflecting surfaces are formed on quartz elements, in front of which is placed the immersion component (Wilkins, 1953). This simple and easily used objective has a high transmission and is manufactured by R. & J. Beck.

Thus the investigator is faced with a rather bewildering choice, but it is possible to narrow this if the objectives are evaluated on their suitability for making measurements and not just for taking photographs. From this point of view monochromats are unsuitable unless measurement is confined (as it should not be) to one or a limited number of wavelengths, although Caspersson has in part overcome these difficulties by using a rather complicated procedure. The development of the Zeiss 0.9 NA achromat should be interesting and helpful, since the 1.2 NA achromatic objective is not generally available.

The relative merits of the different types of reflecting objectives require further comment in view of the discussion on field size and central obstruction. Much of this appears to this author to be founded on misconception regarding the kinds of biological material and problems for which the objectives were required. One of the chief merits of the photoelectric method is that it is unnecessary to make measurements other than at the center of the field since the specimen is moved between recording I and I_0. Even in the photographic method, it is easy to center that part of the specimen from which the highest accuracy is required and aberrations will not effect measurement in the clear field. A large field is only required when photographic recording is used to obtain the total amount of absorbing substance in, say, nucleus or cytoplasm. The error introduced by objective aberrations will then depend on the degree of heterogeneity and the "average aberration" over the area of the structure.

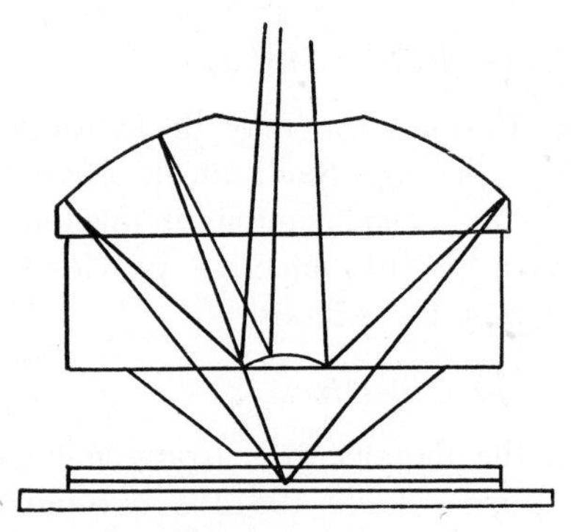

FIG. 4. Arrangements of components of the 0.9 NA solid quartz objective designed by Wilkins and manufactured by R. and J. Beck, London.

The general problem of specimen heterogeneity is discussed in a later section on specimen errors, but a short further comment on objective aberrations is required here. Even with a perfect objective there is a lower limit to the size of particle which may be measured to a given degree of accuracy, since the smaller the particle the greater the fraction of energy that is thrown into the Airey diffraction pattern surrounding it. This lower limit has been calculated in several ways (Caspersson, 1936 discussed in Davies and Walker, 1953) to be about 4 λ for an objective of NA 0.9, but see Wilkins (1950) for a lower value. Central obstruction, spherical aberration, and coma will all increase this minimum area. For example, Steel (1953) has calculated that with an 0.9 NA objective with 0.45 linear central obstruction, the error in measuring the central part of an absorbing particle (at 260 mμ) approximates to that of a perfect objective for particles greater than about 0.2 μ in diameter and that the most serious difference occurs for particles half this diameter. Therefore, if small biological objects such as mitochondria must be measured, the more complicated objectives with their lower central obstruction should be used, provided two conditions hold. The transmission at short wavelengths should not be so low as to prohibit measurements with the requisite aperture with available light sources, and glare from a variety of causes should not introduce errors more serious than those resulting from central obstruction.

All dry objectives for use in the ultraviolet have about half the tolerance for coverslip thickness compared with similar visible objectives (Grey, 1950) and since homogeneous immersion is not often practicable in the ultraviolet, coverslips for immersion objectives must also be carefully checked. A practical disadvantage of reflecting objectives other than the Beck 0.9 solid, the Leitz, and the Bausch and Lomb objectives mentioned, is that their reflecting surfaces are open to the atmosphere and are subject to serious deterioration in their reflectance unless they are frequently realuminized or carefully cleaned.

Reflecting objectives also make satisfactory condensers, with the advantage over their refracting counterparts that the aperture which requires illumination is usually small ($\sim$ 3 mm. diameter). In all the designs of instrument which are illustrated in the Appendix a reflecting condenser should be used although the objective may often be varied to suit a particular requirement.

B. Apparatus

In order not to overburden this chapter with discussions of apparatus, we have relegated to the Appendix drawings and short descriptions of the following instruments or parts thereof.

(1) Basic instruments.
  - (a) Photographic equipment with a monochromator designed by Seeds and Wilkins.
  - (b) Photoelectric equipment with a Czerny-Turner monochromator.
  - (c) Double beam variants for photographic and photoelectric recording.

(2) Instrument with a spectrograph following the microscope.

(3) A "minimum" ultraviolet microscope.

(4) Microdensitometers.
  - (a) Enlarger type.
  - (b) Double-beam type.

(5) Monochromator for photographic recording.

(6) Optical matching of a commercial monochromator to a reflecting condenser.
(7) Slit designs.
(8) Micrometer moving stage.
(9) Methods of using phase contrast illumination with reflecting microscopes.
(10) Photomultiplier circuits.
(11) Lamps and lamp stabilization.

C. The Testing of Apparatus

1. *Monochromator Stray Light*

The perfect monochromator should reject all wavelengths other than that desired, allowing perhaps one part in a million of the energy entering the system to leave the exit slit. All this unwanted energy must be absorbed by the walls and baffles of the monochromator in such a way that stray reflections do not contribute to the energy falling on the specimen. This is difficult to achieve and in practice the useful energy passing the exit slit can be divided into two components.

There is first a distribution of energy with a peak at the nominal wavelength at which the monochromator is set, and second, a background of energy spread throughout the spectrum. The former will determine the spectral resolution of the instrument and will cause a loss of fine spectral detail if it is too large. It is normally not a serious fault in the microspectrophotometry of biological cells, since the absorption curves of substances in cells tend to have broad peaks. Indeed rather wide band passes are often necessary at the shorter wavelengths if there is to be sufficient energy for continuous recording. The general background or stray light is, on the other hand, far more serious since it leads to an underestimation of the absorbance, which will be very large if the true absorbance is high. This is shown by the relation given by Holiday and Beaven (1950),

$$10^{A_1} = \frac{x + 1}{10^{-A} + x} \tag{3}$$

where x is the proportion of effective stray light and A and A_1 the true and apparent absorbances. Only a total of 0.33% stray light is required to give an error of 1% in determining an absorbance of one as is given in Table II. A much fuller table of corrections for stray light has been given by Opler (1950).

The best way of eliminating stray light is to use two monochromators in tandem so that only a narrow band-width of radiation with an already

TABLE II
ERROR INTRODUCED IN MEASURING ABSORBANCE BY 0.33% STRAY LIGHT IN THE MONOCHROMATOR [a]

True Absorbance	Apparent Absorbance	Percentage Error
2.0	1.877	− 6.13
1.0	0.987	− 1.28
0.5	0.497	− 0.60
0.3	0.2987	− 0.43
0.2	0.1993	− 0.35

[a] From Holiday and Beaven (1950).

much reduced background is able to enter the second monochromator, but the number of optical surfaces causes a low overall transmission. Monochromators, such as those illustrated in the Appendix, have relatively few optical components, and are designed to work at low apertures (*f*/80). In these instruments suitably placed baffles and lining with velvet antireflection paper can reduce the stray light to an acceptably low value provided that the height of the input slit is reduced to the minimum required to give an evenly illuminated output slit. If dust and fumes can be excluded from the laboratory, the monochromator may be placed as a temporary expedient on the bench top and reflections from inside a box eliminated.

There are several methods of testing for stray light (see Photoelectric Spectrometry Group Bulletin, 1950) that is, of determining x in Eq. 3. The simplest method is to measure the apparent absorbance at a short wavelength of a low-frequency cut-off filter or solution, which transmits all wavelengths longer than that measured. Suitable filters are the Corning 790, cutting off at 235 mμ, and 0.5 cm. of carbon tetrachloride, which cuts off at 265 mμ. This method is based on the reasonable assumption that at these short wavelengths, where detector sensitivity and source intensity are declining and the transmission of the optical components much reduced, the only effective stray light will be that which originates from longer wavelengths where these defects are absent.

Another method is to analyze the emergent radiation with a second monochromator employing a detector similar to that normally used, but this method may not be practicable in the average laboratory. A third method is to test the absorption law with a solution which has an absorption peak at the shorter wavelengths. If the solution fills all the microscope field which is illuminated then a departure from linearity as the concentration increases will indicate the presence of stray light.

3. *Microscope Glare*

This is another important source of instrumental inaccuracy. It has analogous origins to the fault just discussed, namely unwanted reflections within the objective or from mechanical parts of the microscope. It is discovered by measuring the apparent absorbances of opaque objects like mercury droplets or iron spheres greater than, say 1 μ in diameter (smaller objects will show in addition effects due to objective aberrations). Knowing this measurement it is possible to derive from a relation given in Davies and Walker, 1953 (see also, King and Roe, 1953), the correction required for all absorbance measurements which have been made under the same conditions. For example an apparent absorbance of 1.24 for an opaque object means that objects of measured absorbance of 1.2 and 0.8 have a true absorbance of 1.8 and 0.92, respectively.

There is another method of measuring glare, which is rather easier especially when photographic recording apparatus only is available. This consists of measuring the absorbance of an even area of the specimen first with the normal large field and then with a field smaller than the evenly absorbing area selected. It is necessary to make certain that the exposure was the same for both measurements and to repeat the two measurements for a part of the clear field, since glare will also affect measurements of I_0.

Microscope glare can be reduced by using simple objectives, e.g., reflecting rather than refracting achromats, by eliminating secondary magnifiers, by stops between objectives and detector, and, perhaps most important, by illuminating only a small area of the specimen. This latter condition is easy to fulfill in photoelectric recording, but much more difficult in photographic, when the whole structure and background are normally photographed together.

3. *General Test*

A further procedure for estimating the effect of both monochromator stray light and microscope glare is to test the absorption law with objects small compared with the microscope field (King and Roe, 1953; Walker and Deeley, 1955). Such a model system is naphthalene dissolved in nonane or ethylene dichloride and suspended in a glycerol–water mixture of the same refractive index and density. Provided the densities have been correctly matched, a plot of absorbance against diameter should be linear for any given concentration of naphthalene. The result of such a test of Lambert's law for different condenser apertures is given in Fig. 5 for a recording microspectrophotometer. The main difficulty is in measuring the diameter of the smaller spheres, when refractive indices are closely matched.

Some of the tests outlined require the measurement of small transmissions, which may be difficult with photographic recording owing to the limited range of photographic emulsions in the ultraviolet. The easiest method is to use an ultraviolet-sensitive photomultiplier and correct for differences in sensitivity between the detectors. As already mentioned, this equipment may be expensive and unavailable and it may, therefore, be difficult to carry out these essential tests on photographic recording microspectrophotometers.

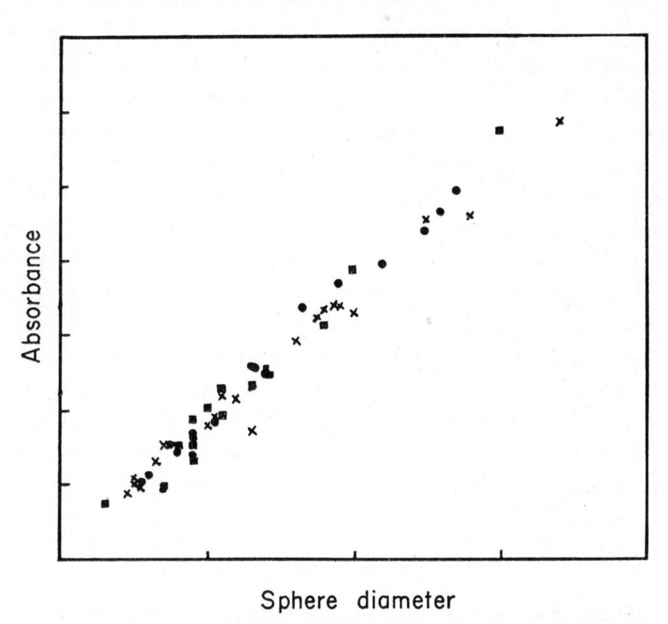

FIG. 5. Relation between diameter of spheres and their absorbance measured with different condenser apertures on a recording microspectrophotometer.

Additional tests may be desirable in certain instances, for example linearity with recording instruments, but as far as errors from stray light and glare are concerned, corrections are possible and they therefore need not be a limiting factor in the technique.

IV. CRITIQUE OF THE METHOD

A. ERRORS DUE TO THE BIOLOGICAL SPECIMEN

1. *Nonparallel Light*

We have to pass nonparallel light through the specimen in order both to define the detail and to ensure sufficient illumination. The mean path length is therefore increased by a factor determined by the numerical aperture mainly of the condenser and by the refractive index of the medium. This correction has been calculated by Blout *et al.* (1950) and replotted in a different form more suitable for microspectrophotometry by Walker (1956). This shows that with a numerical aperture of 1, a measured absorbance of unity will be about 12% too high for a living cell of refractive index $\sim$ 1.37, and 4% too high for a fixed cell of refractive index $\sim$ 1.53. For a lower aperture of 0.6 these errors are reduced to $\sim$ 5% and 1% respectively.

These maximum errors only apply to flat objects; spherical objects, on the other hand, show smaller errors from this cause (Caspersson, 1936).

Nonparallel light also causes the volume of the specimen actually contributing to the absorption to be inadequately delimited by that most useful fiction, the "projected area." This is defined as a cylinder through the thickness of the specimen, which is bounded by the measured area at the object plane, but the volume actually measured will be of a conical or biconical form. The error in determining the volume from the projected area is clearly proportional to the ratio of thickness to diameter, but the error in determining total absorbance, which is additional to that discussed in the last paragraphs, will depend on the relative absorbances of specimen and environment and may often be neglected. Large thin objects are better than small thick ones from this point of view, but in such thin specimens it is often difficult to measure thickness and to determine for example the exact amount of cytoplasm lying above and below the nucleus in a living tissue culture cell.

2. *Distributional Errors*

The uneven distribution of cellular material causes the most serious and least tractable errors in microspectrophotometry. In those living cells which may be observed in the phase microscope or photographed in the ultraviolet, there is often a markedly more homogeneous appearance than in corresponding fixed cells. This is primarily due to the contraction and aggregation of the proteins on denaturation and also possibly to the removal of matrix material. Thus the absorbing substances become more particulate and at the same time sharp gradients of refractive index appear between the dry protein and the medium in which the structures are immersed. This uneven distribution of the cellular material will cause errors of two types which are, however, interrelated.

The first type of inaccuracy is due to the inhomogeneity of the absorbing material within the measuring aperture, and results in an underestimation of the amount of absorbing substance. The second type is due to scatter or nonspecific light loss, caused by deflection of energy at the abrupt changes of refractive index.

a. Inhomogeneity of Absorbing Material.

If the same number of chromophores are concentrated into particles, the deviation from the true value of total absorbance obtained from an even distribution of material will depend on the absorbances of the particles and the relative projected areas of the measuring aperture and of all the particles. A two-dimensional treatment of this problem has been made by Glick *et al.* (1951), although the values they used for kcb ($= 0.434 \times kcl$ in our Eq. 1) for an evenly spread field would after concentration into particles be far higher than that normally found in cells. A plot of the error introduced by particles of a more usual absorbance is given in Fig. 6 for different ratios of measuring aperture to aggregate particle area.

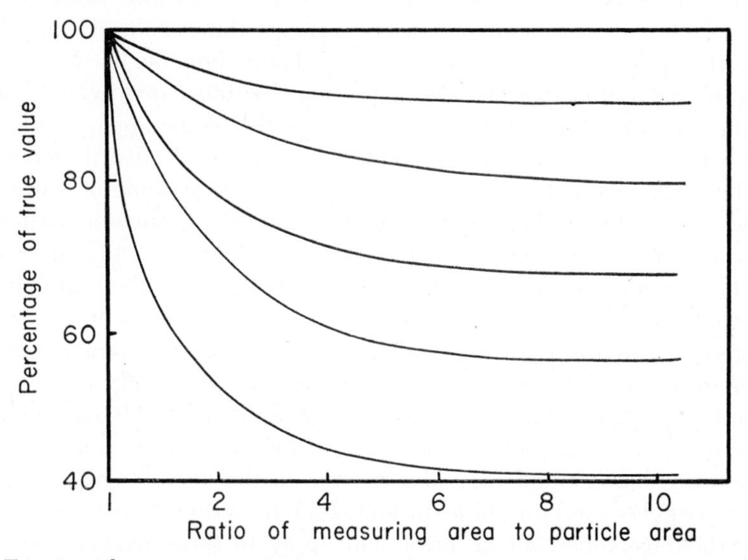

FIG. 6. The error, expressed as a percentage of the true value, that occurs when particles of various absorbances are measured with apertures larger than the particle size. A = 0.1, 0.2, 0.4, 0.6, and 1.0 for the top to bottom curves respectively.

A clear distinction should be made at this stage between the inaccuracies occurring with two different methods of measurement. First, there are those which are present when the absorbance is determined of a particle which is small compared with the measuring aperture, as in the aperture methods in Feulgen photometry. Second, there is the over-all error in measuring the total absorbance of a larger area, which may include local aggregations, when measurement has been made by scanning with a small aperture.

As an aid in assessing the magnitude of the error in the latter type of measurement, we can arbitrarily divide the absorbing material into three categories: (a) material evenly spread over areas larger than the measuring aperture; (b) material occurring in resolvable particles smaller than the measuring aperture but larger than about 0.1 λ; and (c) gradients of absorbance such as might be found at the boundaries of large or small absorbing areas. Any method of reducing the proportion of material in category (b) or the maximum absorbance differences in category (c) will clearly improve the accuracy. This may be done by decreasing the size of the measuring aperture or by fixing or other treatment that tends to prevent the local aggregation of the nucleoproteins. The smallest size of measuring aperture may be limited by the energy available, but it is also apparent that there is a lower limit set by the size of the diffraction patterns round images of small particles.

Unfortunately, it is difficult in practice to assign any values to the fraction of the chromophores that occur within each of the three categories suggested. It is, on the other hand, possible to give upper limits to the errors expected from particles within category b (Stokes, see Davies and Walker, 1953). It was shown that if the apparent absorbance was measured from an area containing randomly distributed small platelets, the percentage error in measuring the amount of substance would be 116 times the absorbance of the platelets. Since a particle of pure polymerized nucleic acid 0.1 μ thick will have an absorbance of 0.2, the maximum error would be about 22%. In the cell the particles will also contain proteins, sometimes lipids and, in living cells, water, which will reduce the absorbance and therefore errors of even larger particles to tolerable proportions. Another mitigating factor is that as the background absorbance due to very small chromophores $<0 \cdot 1\,\lambda$ increases, the error due to particles embedded within the background will decrease. The background may also be due to the out-of-focus images of larger particles, and in these instances more serious errors will occur since the 116 A percentage error will apply to these particles as well, and they may well have high absorbances. It will then be necessary to flatten the object, either by using a "crushing" condenser (Davies *et al.*, 1954) or by flattening cells prior to fixation (Rudkin, *et al.*, 1955).

A method exists for determining the extent of errors due to inhomogeneity, which entails the accurate comparison of the absorption curves of areas in the cell which are uniformly absorbing with those in which distributional error is expected (Ornstein, 1952; Patau, 1952). This method is particularly applicable to staining reactions, where the absorption curve would be expected to remain constant in different parts of a cell. In ultraviolet absorption studies, local changes in shape are probable, but it may be possible to use the method to detect large errors due to inhomogeneity.

The determination of the true absorbances of small structures ($< 1\,\mu$) is more seriously affected by objective aberrations than by distributional errors if the measuring aperture is kept small. The minimum size of object for which accurate measurements may be expected has been calculated by Caspersson (1936) in various ways, summarized in Davies and Walker (1953). It is found to be 3–4 λ in diameter, provided that an objective of at least 0.85 NA is employed together with a condenser of ½ to ⅔ the NA of the objective. This is in order that a sufficiently large part of the diffracted light contributes to the formation of the image.

In summary, the extent of the possible errors due to inhomogeneity can be reduced by using very small measuring apertures, and the extent of the remaining errors depends fundamentally on the absorbance of a given volume of nucleoprotein. In fixed cells any method which tends to swell the nucleoprotein or prevents its initial aggregation will reduce this error. Methods for obtaining structureless nuclei (except for nucleoli) have been described, for example, by Schneider (1955) and by Philpot and Stanier (1955), but they cannot be applied easily to tissues that require embedding and sectioning. Various media for mounting are discussed in the next section.

b. Scatter or Nonspecific Light Loss.

As we have mentioned, changes of refractive index cause the refraction and reflection of the energy traversing the specimen. Some of this energy in consequence will not be collected by the objective and a spuriously high absorbance will be recorded. The effect is wavelength dependent as in the relation

$$\epsilon = K\lambda^{-S} \tag{4}$$

where K is a constant and S some value between zero and four, depending on the size and nature of the particle and the numerical aperture of condenser and objective. For particles less than 0.1 λ and for parallel illumination, Rayleigh's law holds, where the exponent S will be four. This value was originally used by Caspersson (1936), when he extrapolated a scatter curve through the shorter wavelengths from the value for the apparent absorbance at 320 mμ where most cellular substances do not absorb. This is probably too high a value for S and Caspersson later (1950) proposed the experimental determination of the energy distribution round the portion of the specimen to be examined. This is difficult in the ultraviolet and would be impractical as a general routine procedure.

Now that microspectrophotometers are available which can record absorbance at all wavelengths from 320–500 mμ, a more reliable value for S may be obtained from the shape of the apparent absorbance curve in this region. If the scatter obeys Eq. 4, then a plot of log A against log λ should give a straight line, which may be extrapolated with more confidence than the single determination at 320 mμ. For several reasons, however, it would be helpful to have independent confirmation of the amount of scatter in the 240–260 mμ region. For example, small errors in determining the scatter relation at the longer wavelengths may cause quite serious errors at the shorter, particularly when cells with low specific absorbances must be investigated. A more fundamental criticism is that the sharp increase in protein absorptivity below 240 mμ will cause rapid changes in refractive index at slightly longer wavelengths (anomalous dispersion, Jenkins and White, 1937) and that this will cause changes in the value of S as the wavelength decreases, but it is probable that S will remain between 0–4.

There are several ways of obtaining this independent check. One method is to use a second parameter which has a constant relation to the true absorbance. Thus certain viruses can be obtained with a constant protein–nucleic acid ratio and we could therefore use the interference microscope to determine the mass of a particle and therefore the true value for the absorbance at any wavelength. Another way is to employ the various techniques of base line and curve analysis (Hiskey, 1955) which may be specially applicable to these instances where the shape of the nucleoprotein absorption curve is known. A third method of studying scatter is to measure cells in media of different refractive indices; this method will not give evidence about changes of S below 300 mμ, but it should overcome difficulties due to inaccuracy in measuring the apparent absorbance at longer wavelengths. Our preliminary experiments with these methods indicate that the scatter correction obtained from longer wavelengths will not introduce very serious errors.

An interesting method of estimating scatter has been recently devised by Rudkin and Corlette (1957a). This consists in measuring the light scattered by the object with a dark-field method of illumination and comparing this with the energy transmitted. These measurements are made at 275 and 257 mμ with immersion fluids consisting of glycerol 95%, saturated aqueous lanthanum acetate 5%, and enough zinc chloride to raise the refractive index at 540 mμ to various values between 1.46 and 1.52. It would be useful to compare the very low figures ($\sim$ 1%) for scatter which they obtained for salivary chromosomes with their data for non-specific light loss at 320 mμ.

So far we have emphasized methods of correction for scatter, but not its reduction or elimination. In order to eliminate scatter we should have to fill the interstices between the highly refracting protein with a substance of the same refractive index and dispersion. Such media, which must also have the additional property of ultraviolet transparency, are difficult to find. Köhler (1904) suggests saturated solutions of choral hydrate and Koenig *et al.* (1953) used a solution of zinc chloride in glycerol to give a medium of the right refractive index. The latter noted a shift of the absorption peak to 270 mμ, which is almost certainly due to the hydrolysis of DNA (Walker, 1957). Swift (1955) has embedded his sections in butyl methacrylate, which was polymerized with ultraviolet radiation, and Deeley and Walker (unpublished) have tried thiodiglycol ($n = 1.52$). A different approach is the useful method recommended by Caspersson (1950), in which the tissue is first frozen dried and then placed immediately in triple-distilled glycerol. This swells the protein so that it more nearly matches the refractive index of glycerol ($n = 1.47$). For a variety of reasons these methods are not entirely satisfactory.

We have ourselves attempted to use proteins and polypeptides as embedding media, although it is recognized that penetration of the larger molecules will be difficult. The method employed was to coat the frozen-dried section with a dilute solution of protein or polypeptide and then to dry it, either in the air or by freezing and dehydration with cold alcohol.

The first experiments with gelatin gave a very marked reduction in the scatter at long wavelengths (500 mμ) in frozen-dried tissue sections, compared with similar sections embedded in paraffin oil or glycerol, but some scatter was still present at 320 mμ. Gelatin absorbs at 280 and 240 mμ, and is therefore unsatisfactory as an embedding medium, but protamines or polypeptides which do not contain aromatic amino acids will eliminate this difficulty. Polyglutamic acid (Bricas and Fromageot, 1953) which can be obtained in a high degree of purity and has no specific absorption above 240 mμ, has been found most satisfactory, but some scatter still remains. It seems that the medium fills the surface irregularities and the larger holes in the tissue, but that it does not eliminate the molecular scatter due to the proteins, and in fact introduces a small amount of scatter itself.

The scatter of dried films of polyglutamic acid has been measured and is illustrated in Fig. 7. The lower curve and the two upper ones have been measured for thin and thick films, respectively, as determined by measurement with an interference microscope. The thin film curve does show some evidence of discontinuity at about 320 mμ, but it should be emphasized that the measurements are of apparent absorbances of < 0.01 for wavelengths longer than 320 mμ, and may therefore be subject to error. The thicker films however show a linear relation on this double-log plot and do not give evidence of any effects of anomalous dispersion.

In living cells it is comparatively unusual to find much scatter and this is probably due to the presence of gradients of refractive index around any local concentrations of substances. Certain fixed cells like mammalian sperm heads also do not scatter, and this is possibly due to the paracrystalline arrangement of the nucleoprotein. The methods of

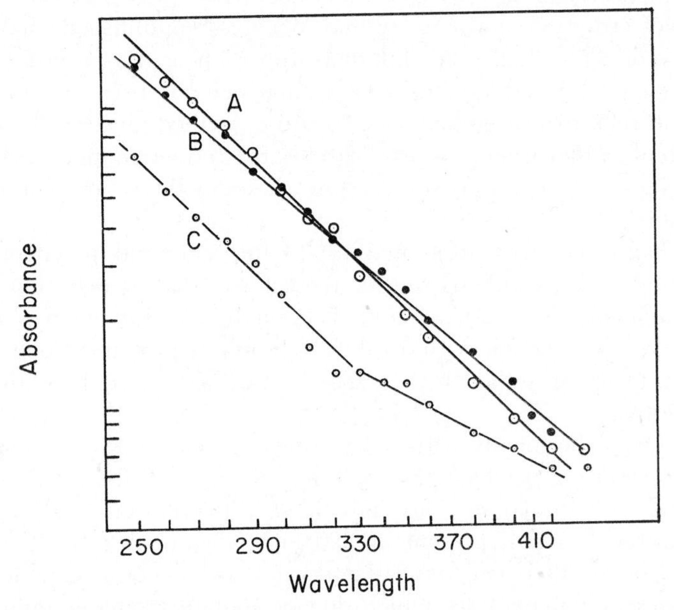

FIG. 7. Measurements of the apparent absorption of polyglutamic acid. Curves A and B, thick film; curve C, thin film.

eliminating or estimating scatter cannot be said to be very satisfactory, and it is therefore necessary to consider each cell type separately. In general, as in all cytological methods, the nearer we can approximate to conditions in living cells the more reliable the results.

3. *Dichroism of the Specimen*

There remains a further source of possible error in microspectrophotometry, which may be considered as having an intermediate status between errors due to the morphology of the specimen and those due to the molecular arrangement of the chromophores. This is due to the orientation of the chromophores in the tissue, which could result in a serious underestimation of the true absorbance and tend to give a maximum value of ~ 0.3 (Commoner, 1949). Commoner gave examples of results, which tend towards this value, but these were mainly from cytoplasmic RNA, which does not show evidence of spatial orientation. This orientation can be detected either by ultraviolet dichroism using polarized ultraviolet radiation, or by the accurate determination of birefringence in the visible region (Wilkins, 1950). Thorell and Ruch (1951) in particular have looked for evidence of dichroism in many cell types, but have failed to find it in animal cells except in certain elongated sperm heads, and in certain chromosomes as pointed out earlier by Caspersson (1940). Although in practice error due to molecular orientation does not seem to be serious with most cell types, it will always be preferable to test for it in new material.

4. *Changes Due to Radiation Damage*

It is well known that living cells are very sensitive to ultraviolet radiation damage, and that therefore care must be taken to ensure that this damage does not vitiate the measurements on such cells.

The conditions for obtaining the minimum radiation at the specimen have been discussed elsewhere (Walker, 1956) and may be summarized as follows:

(1) The specimen should be focused with visible radiation and not with a fluorescent eyepiece. (2) The specimen should only be irradiated during measurement; a shutter should therefore precede the specimen. (3) All optical components following the specimen must have the highest transmission possible. (4) In photographic recording the optimum combination of magnification onto the film and the grain size and sensitivity of the emulsion should be obtained for the particular amount of information that is required from the specimen. (5) In photoelectric recording only that part of the specimen which is being measured should be irradiated.

The general conclusion from work on tissue culture cells (Davies, 1950a, b; Walker and Davies, 1950) was that the amount of energy required for measurement at short wavelengths would eventually damage most cells, but that this damage does not become apparent immediately and that therefore there will be a short period during which valid measurements may be made.

When measurements are attempted on other cell types, this conclusion will obviously have to be tested afresh, but it seems that it would be unwise to base any results upon the behavior of "living" cells which had already been exposed to ultraviolet radiation for sufficiently long to obtain one high resolution photograph.

B. The Absorptivity of Cellular Constituents

If we assumed that all error due to instruments or to the specimen have been eliminated from microspectrophotometry, the results would still have to be interpreted with care. This is due to our present inadequate information about the intracellular environment and to our knowledge of the large number of factors which may influence the contour and height of absorption curves. The latter have been discussed by Scott (1955, pp. 150–160) and may be divided into two categories. There are first those conditions, chiefly degree of polymerization and configuration, but also redox potential and ionic environment, which may be expected to occur in living and fixed cells under normal circumstances, and second those like pH and temperature which may be controlled by the experimenter.

Comprehensive reviews of the spectroscopic properties of nucleic acids and their constituents have been made by Beaven *et al.* (1955) and of proteins by Beaven and Holiday (1952). Useful tables of the absorptivities of proteins, amino acids, and nucleic acids have been given by Nurnberger (1955), who also includes information on the amino acid composition of proteins. Perhaps the most striking spectroscopic property of nucleic acids is the hyperchromic effect, in which the molar absorptivity at 260 mμ of pure DNA *in vitro* expressed per gram atom of phosphorus (Chargaff and Zamenhof, 1948), as ϵ (P) rises from 6,800 to 10,800 on depolymerization. The same increase occurs, but apparently not to such a large extent, on denaturation (Thomas, 1954) and the effect is probably due to the interaction between the neighboring bases in the helical molecule of DNA. This increase was used as an indicator of the extent of denaturation by Thomas, when he treated the DNA with low concentrations of salts, changes in pH, and high temperatures (see also Shack *et al.*, 1953).

Di- and trinucleotides from DNA do not show the lower absorptivity (references in Beaven *et al.*, 1955) and Smith and Allen (1953) have a similar result from up to 20 bases. On the other hand, Sinsheimer, quoted in Scott (1955), has reported that the dinucleotide of deoxyguanylic acid has about 30% less absorptivity than the corresponding mononucleotide. Fortunately the question of the absorptivity of DNA in intermediate degrees of polymerization may not be of great importance to the cytochemist since there is no biochemical evidence for the occurrence of these stages in tissues, and the incorporation of mononucleotides directly into the DNA molecule could be accounted for by current theories of DNA synthesis (e.g., Watson and Crick, 1953). It may therefore be reasonable to recognize two absorptivities, one for the polymerized DNA (and RNA, Magasanik and Chargaff, 1951) and another average one for the labile mononucleotide fraction.

A related problem is the effect of concentration on the absorptivity. It is well known that some dyes do not obey Beer's law (e.g., Michaelis, 1947), and it has also been reported by Blout and Asadourian (1954) that the linear relation between absorbance and concentration breaks down at the very low concentration of 0.004%. It is possible to calculate the concentration of nucleic acids in fixed cells, from the concentration of dry protein (127 gm. per 100 cc. for dry gelatin) and the protein–nucleic acid ratio. The "concentration" of nucleic acid would be about 55 gm. per 100 cc. for fixed mammalian sperm heads and 10 gm. per 100 cc. for a fixed tissue cell. It would therefore be quite impossible to measure nucleic acids in fixed cells, unless the other cellular constituents, primarily proteins, prevented the interaction between adjacent DNA molecules which might result in lower absorptivity. Nurnberger (1955, pp. 4/16 and 17) has shown that DNA in concentrations up to 0.008% obeys Beer's law, but there is nothing in his experimental data to support his contention that Beer's law holds up to a concentration of 16 gm. per 100 cc.

The main problem of determining the absorbance of high concentrations of DNA and nucleoprotein is the difficulty of measuring the thickness of the optical cell. A very elegant method of doing this has been developed in Caspersson's laboratory (Svensson, 1955) which consists of forming the boundaries of the cell by a convex and a plane optical surface of quartz in contact at one point. The separation of these surfaces as a function of radius can be used to obtain the absorptivity or refractive index of a solution to a high degree of accuracy. A 4% solution of RNA has been measured and gave a value of A_{260} of 0.0808 for a 10 μ path length.

In living cells oxidation and reduction may be expected to alter the absorption of certain compounds such as diphosphopyridine nucleotides (Warburg and Christian, 1936) and ascorbic acid (Chayen, 1953). In addition it is possible that changes in ionic composition may occur locally in living cells causing spectroscopic variations.

Most of these possible causes of variation in absorption properties can be eliminated as sources of large error, if it can be shown that microspectroscopic measurements on both living and fixed cells are in agreement with those obtained independently by, for example, biochemical techniques. Such comparative measurements have been rather sparse, considering their importance. Leuchtenberger *et al.* (1952b) compared DNA measurements of certain fixed and isolated nuclei, which they had obtained biochemically and with a microspectrophotometer and found agreement between the mean values provided by both methods. Walker and Yates (1952) made similar comparisons for living mammalian sperm heads and for nucleated erythrocytes. Mellors *et al.* (1954), have published extensive measurements of total absorption in the normal and neoplastic nuclei of the mouse, which agree with relevant biochemical data and similar agreement has been found by Nurnberger (1955) for normal rat liver cells.

Davies (1954) in a careful study of the effects of fixation on the ultraviolet absorption of tissue culture nuclei, found about 20% decrease in the total absorbance of 265 mμ after fixing in neutral formalin or acetic alcohol. This may have been due to the changed physical and chemical environment of the cell, but the loss was ascribed to the removal of the extra DNA-absorbing material, which had been shown to exist in the nuclei of tissue culture cells (Walker and Yates, 1952). The ascription to loss of material was supported by Davies's finding that fixing by freeze-substitution caused no measurable loss. In the course of more recent work (Walker, 1957), it was found that the total absorbance of living and fixed sperm heads agreed closely, although of course the concentration changes on fixing are not so great with this material.

Within the experimental errors of the techniques involved, it appears that no serious error has been observed so far that may be attributed to the environmental effects mentioned. This may be due to the protein present in the cell. This view is supported by the experiments in varying the pH which are to be described next.

Only small changes in pH have been reported in living animal cells (e.g., Wiercinski, 1955) and it is unlikely that variations in absorptivity will occur *in vivo* from this cause. It is well known, however, that the ionization of the constituent bases of both DNA and RNA will cause marked changes in absorptivity. Not only must this be considered when procedures such as acid extraction are undertaken, but it may also be possible to use these changes in absorptivity as an experimental method for identifying cellular components.

As already noted treatment with strong acids or alkalis will denature DNA, causing an increase in absorptivity and thereafter reversible changes of smaller magnitude occur as the pH is altered. Different effects of pH might be expected when nucleoprotamine or nucleohistone are studied in the cell, and we have therefore (Walker, 1957), studied the behavior of the DNA-protein in mammalian sperm heads, which are very suitable objects optically and do not contain appreciable quantities of RNA.

The effect of $M/15$ buffers at various pH values on unhydrolyzed sperm heads is illustrated in Fig. 8. It was found that the changes in

A_{263} are small compared with those expected from pure DNA. This is supported by the small alteration in the contour of the absorption curve, which is shown by the variation in A_{242}/A_{260} ratio compared with that from pure DNA. It would appear that under these conditions the chemical and mechanical buffering action of the fixed protein can still prevent or delay the denaturation of DNA. This is also indicated by the effect of low salt concentration, which Thomas (1954) has shown to cause denaturation. Fixed sperm heads, when they are measured after passing through distilled water, show no difference in their DNA content from their living counterparts.

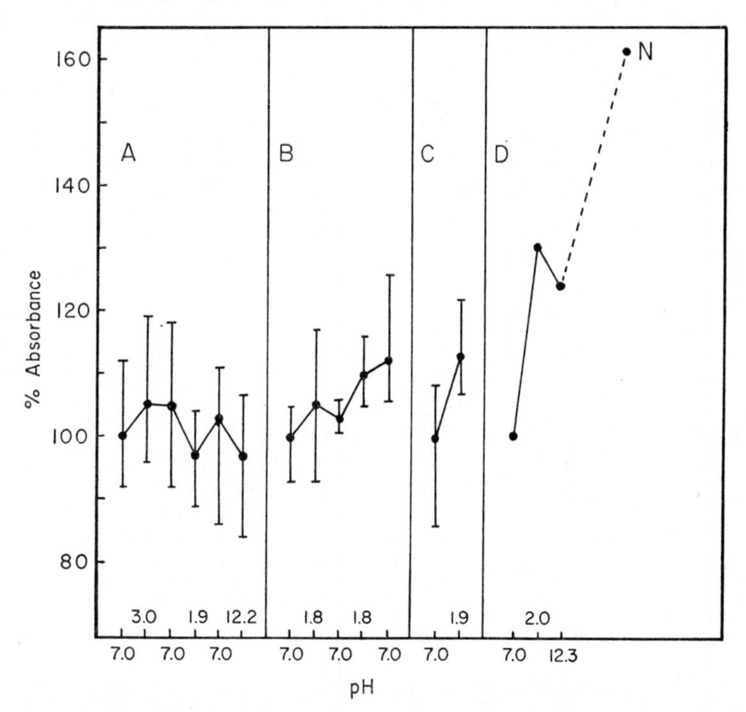

FIG. 8a. The effect of pH on (A, B, and C) the integrated absorbance of intact mammalian sperm heads and on the (D) absorptivity of DNA and its constituent bases (N). The values in A, B, and C give the means and the spread of values obtained. These are from Walker (1957). Those in D are from Beaven *et al.* (1955). Nucleohistone has the same absorptivity, as ε (P), as the intact DNA at pH 7 (Crampton *et al.*, 1954).

If the sperm heads are hydrolyzed so that some purines are removed and the DNA-protein bonding disrupted, two things occur. There is first an indication that the total absorbance actually increases during initial hydrolysis, probably due to the increased absorptivity on denaturation more than compensating for the loss due to the removal of the purines. In the second place the hydrolyzed cells are now much more sensitive to variations in pH (Fig. 8b). As the pH is increased from 7.0 to 12.2 the absorption curve changes shape (curves 1 and 2, Fig. 8b)

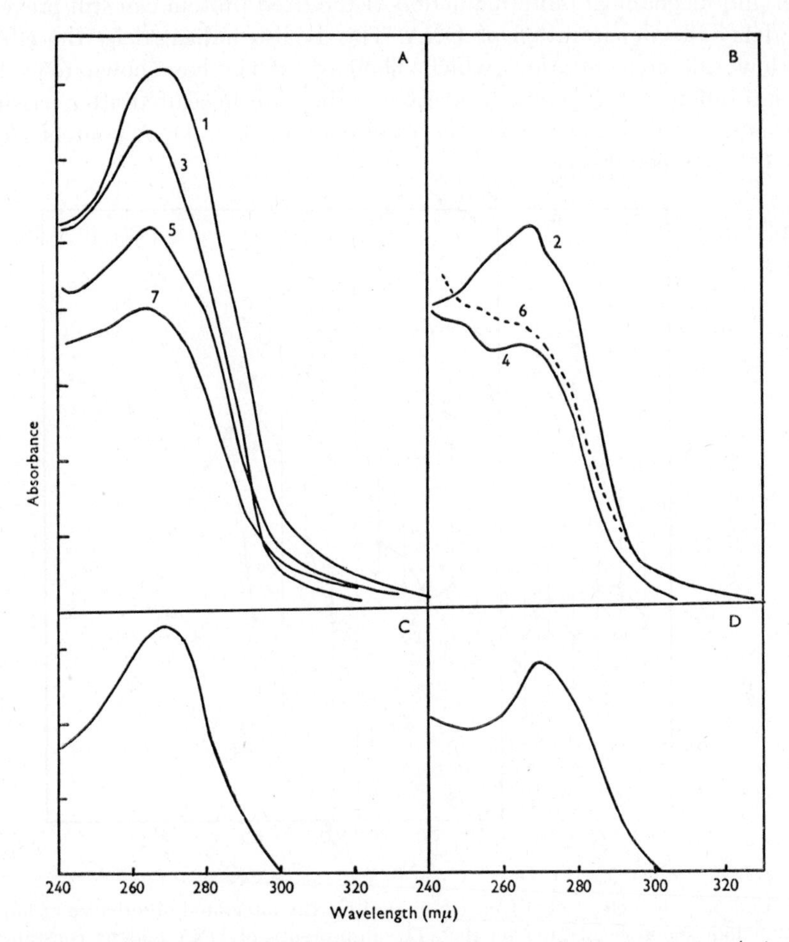

FIG. 8b. Absorption curves from hydrolyzed sperm heads at pH 7.0 (A) and pH 12.0 (B), compared with three parts cytidine and four parts thymidine at pH 7.0 (C) and pH 13.0 (D). The absorption curves in A and B were recorded in the order 1–7.

in a somewhat similar fashion to that of a mixture of cytidine and thymidine. This change is reversible (curve 3) although a progressive loss of absorption occurs, which is probably due to the alkali lability of the hydrolyzed DNA (Tamm *et al.*, 1953). This evidence for the increased absorptivity of the DNA in cells during the initial stages of hydrolysis was supported by an experiment in which unhydrolyzed sperm were placed in hot 10^{-2} M NaCl for 1 hour, which denatures DNA *in vitro*. It was found that the total absorbance increased by about 20%, which is similar to that found by Thomas (1954).

So far we have mainly considered the effects of changes in chemical environment on DNA, which has been much more amenable for isolation and study in an undenatured condition than other cellular constituents. RNA also has a lower absorptivity than its nucleotides, but it has not yet been studied from this viewpoint in the cell. Gordon and Nurnberger (1955) have shown, however, that the RNA isolated from rat liver by the method of Grinnan and Mosher (1951) gives an absorptivity (k in Eq. 1) of 17.6 compared with 24.6 for a commercial sample. They found that mild alkaline hydrolysis of their "native" RNA will cause its absorptivity to increase to the latter value. Since Tsuboi's (1950) RNA, prepared by a less mild method had an absorptivity of 25.5 and this increased to 33.5 after strenuous hydrolysis, Gordon and Nurnberger considered that Tsubois' preparation was in the same state as their commercial sample, and that therefore the absorptivity of RNA increased from 17.6 to 33.5 on complete hydrolysis, which is greater than that found for DNA. It may be safe to consider, as with DNA, that in cells the lower absorptivity (~ 18) can be used for RNA, but that a much higher value should be used for nucleotides. It is not however clear whether partially hydrolyzed RNA, with its intermediate absorptivity, represents a molecular state that ever occurs *in vivo*.

The absorption of proteins is even more difficult to investigate, since although the absorptivity of tyrosine and tryptophan is high, the proportion of these amino acids in cells is usually low (see table in Nurnberger, 1955, pp. 4/10 and 11). This small absorption is often difficult to measure accurately in cells and is much more subject to gross error due to nonspecific light loss.

There are indications however that anomalies are present in the absorption properties of these amino acids in proteins, particularly when they are associated with nucleic acids. Absorption curve analysis may be used to estimate the quantity of both tyrosine and tryptophan in proteins and gives results very similar to those obtained biochemically (Beaven and Holiday, 1952). Nurnberger (1955) has compared the absorption spectra of mixtures of nucleic acids and serum proteins with those obtained by adding the curves from the components and has found good agreement. On the other hand, he shows that his biochemical and microspectrophotometric results for protein from rat liver cells do not agree. Walker (1956) discussed Moberger's (1954) results and concluded that the absorption due to tyrosine and tryptophan cannot be used to estimate proteins in cells containing more than about 5% nucleic acids, as was attempted by Caspersson and Santesson (1942) in their two-wavelength method. The protein-nucleic acid ratios show a wide variation compared with the values calculated from the dry mass and A_{260} data alone. The wide variation in these ratios made it impossible to give any value for the deviation of the A_{280} values from those expected from a mixture of nucleic acid and protein, except to say that the $A_{260}/_{280}$ ratio is usually too low.

A related effect was discovered by Davies (1954), when he compared the absorption spectra from living and fixed nuclei of tissue culture cells. He found that the shape of the absorption spectra from living nuclei is very similar to that obtained from pure DNA and the ratio A_{265}/A_{280} was 1.8. On fixing the cells this ratio fell to 1.35, which corresponded to a protein–nucleic acid ratio of about 15. In subsequent measurements on sperm heads with a recording microspectrophotometer it was found (Walker, 1957) that no change in the A_{260}/A_{280} ratio occurred on fixing these cells, while the change in the nuclei from tissue culture cells was confirmed. The difference in the behavior of the sperm heads is probably due to their differing chemical constitution.

These anomalies illustrate the difficulties of making accurate measurements of proteins in cells from ultraviolet measurements. Interference microscopy is so much a better method of obtaining the total amount of protein, that it is unlikely that ultraviolet measurements will in the future be used for this purpose. The anomalies that exist in amino acid absorption may eventually provide useful information however about nucleic acid–protein relationships in living cells, which may be difficult to obtain in other ways.

V. COMMENT ON RECENT BIOLOGICAL RESULTS

In this short concluding section, we will only consider a small number of recent papers, which illustrate different methods both of instrumental and of biological approach. Earlier papers have been reviewed by Caspersson (1950), Blout (1953), and Nurnberger (1955), among others. Compared to the vast and increasing literature on microphotometry with visible light, particularly with the Feulgen reaction, the number of studies with the ultraviolet microscope have been relatively few. This may not be a disadvantage since those who have been prepared to surmount the greater technical difficulties might be expected to be well aware of the difficulties of interpretation and of the many other hazards of inaccuracy.

Some of Gordon and Nurnberger's (1955) results from *in vitro* experiments have already been described. For their measurements on frozen-dried rat liver cells, they used a photographic method similar to that described by Caspersson (1936) with a quartz monochromatic objective. They state that they cannot detect any scatter by measurements at 325 mμ, although they recognize that this may be in some degree due to the spherical aberration in the objective, when it is used so far from its designed wavelength (275 mμ). Their most interesting result is that cold shock causes an increase in the concentration of RNA in rat liver cells as measured by a modified Schmidt-Thannhauser technique, compared with an apparent decrease when it is determined with a microspectrophotometer. They interpret this difference as showing possibly that the cold shock causes an increase of polymerized RNA at the expense of a less polymerized form which has a higher absorptivity. This less polymerized RNA could be ribonucleotides, if many of these are left after freeze-drying and subsequent procedures, but we have insufficient knowledge of the molecular structure of RNA to determine whether it can exist in a form in which the bases do not interact to give the lower absorptivity value. These results do show however the way in which new information about chemistry of the cell may be obtained by parallel measurements with different techniques.

Rudkin *et al.* (1955) have described in detail the photographic method of recording, which was used in their study of haploid and diploid strands of salivary gland chromosomes. They also employed monochromatic objectives in a Caspersson type of equipment, and show that in a triploid tissue equivalent bands of the diploid and haploid strands, which had locally failed to pair, had the expected ratio of total absorbance of two to one. If, therefore, we can assume that the DNA content of a diploid chromosome is indeed twice that of the haploid, then within the concentrations and thickness of material measured, the absorption law holds. These workers also give absorption spectra measured at 7 wavelengths between 230 and 320 mμ, which show that the apparent absorbance at 320 mμ is usually less than 10% of that at 257 mμ. No attempt was made to correct for scattering but it tended to be proportionally the same for both diploid and haploid strands, and this work therefore illustrates how a limited cytochemical result can be quite convincingly demonstrated despite the presence of systematic error.

Rudkin and Corlette (1957b) have also measured the increase in absorption which occurs during the process of "puffing" in the salivary chromosomes of *Rhynchosciara angelae*. They find that while the ordinary segments of the chromosomes double their content of ultra-violet absorbing substances between the pre- and post-puff stages, there is a fourfold increase in the segments with puffs. They consider that this extra material is DNA on the basis of treatment with appropriate enzymes and hot trichloroacetic acid. The nature of the measurements indicate the very great difficulty of this kind of characterization. In this preliminary communication absorption curves have not been published and no correction for scatter has been made, since their method of measuring scatter (Rudkin and Corlette, 1957a) indicates that it will not introduce appreciable error.

A similar comment may be made on Lin's (1955) results, in which the amount of RNA in the nucleoli of maize is shown to be proportional to the number of nucleolar organizers within the nucleus. Lin used a simple photoelectric method, with a reflecting objective and a quartz refracting condenser, which caused problems in focusing. The apparent absorbance at 320 mμ was about 30% of that at 265 mμ, but a correction for light scatter was obtained by extracting the RNA from similar cells with cold perchloric acid. This procedure has been used by other workers and may be justifiable, although some doubt will remain that such treatment might alter the scattering properties of the protein. An examination of Lin's absorption curves appears to show the presence of monochromator stray light, since his A_{245} values are a rather small fraction of the A_{265}, after scatter has been subtracted from both values. In the perchloric acid extracted nucleoli themselves, the subtraction of a normal protein absorption curve would indicate a positive wavelength exponent of scatter.

Grun (1956) has recently published measurements obtained with Caspersson's latest photoelectric equipment from nuclei of *Tradescantia.* Plant cells are notoriously more difficult for ultraviolet absorption studies than animal cells, partly because of the increased scatter due to cell walls and also because of the presence of absorbing substances other than nucleic acids and proteins (Chayen, 1952). A modification of Caspersson's classic two-wavelength method of determining protein–nucleic acid ratios was used in which the amino acid absorption is measured as a mixture of 1.7 parts tryptophan and 2.6 parts of 25% dissociated tyrosine. This percentage of dissociation of the tyrosine was selected as that giving the best fit to the observed data at wavelengths other than 265 and 280 mμ.

We have already commented on the anomalies in protein absorption, which makes confirmation of these results by an independent technique such as interference microscopy particularly desirable. Parallel measurements of a Feulgen stained preparation would also be helpful, as this should definitely confirm whether the cells measured from the "mitotic" series can be considered as all belonging to one dividing population, or whether different degrees of polyploidy occur. In addition Feulgen measurements would indicate the amount of RNA or other absorbing material that was present.

Attardi (1953) has also used Caspersson's apparatus for studying the cytoplasmic absorption of Purkinje cells. In these experiments the non-specific light loss was high ($\sim$ 40% of A_{260}) and correction was made by subtracting the A_{315} from the A_{260} values (λ^0). It was found that the absorbance per unit projected area fell by 20–30% after distilled water extraction and by 70–90% after ribonuclease. Estimation of nerve cell volumes is difficult and was not attempted in this instance, and therefore the differences reported may not exactly reflect the actual loss of material. Nevertheless, the errors here were probably small, unlike measurements in which experimental treatments preceded the fixation of material, and in which differences in cell volume cannot be eliminated.

Mellors *et al.* (1954) in the course of a large-scale investigation of normal and neoplastic tissues of the mouse have published measurements of the total amount of absorbing material from a number of cell types. They used a reflecting objective and condenser and a photographic method of recording, and found that their isolation procedure resulted in nuclei with very little light scatter and gave absorption curves which closely corresponded with those from a nucleoprotein. A frequency distribution was given for the average absorbance, nuclear area, and the total nucleic acid content of each of the cell types studied. Compared to the normal tissues, all the nuclei from those of neoplastic origin had a very wide spread in nucleic acid content.

Their measurements include nuclear RNA, which may occur in quite high proportion in tumor cells (Leuchtenberger *et al.,* 1952a), but unfortunately it is impossible with their technique to make accurate measurements on mitotic stages as may be achieved in Feulgen photometry with certain instrumental methods (Richards *et al.,* 1956). We could then decide how much weight to give to variation in chromatid number, to nucleic acid synthesis in dividing tissues, and to experimental error as possible causes of the wide variation reported here for neoplastic tissues.

Although it is quite undesirable to lay down any hard and fast rules for any type of research, it may be justifiable to comment in general on the results just discussed. In most of the publications discussed, details of the apparatus used are meager, and the results of any tests for instrumental errors, particularly monochromator stray light and microscope glare entirely lacking. Complete absorption curves should also be given if possible, and results based on measurements at two or three wavelengths avoided. It is true that in the past the biological problems have often been of such a nature that high accuracy was not essential, but this stage in the development of the subject should be passing.

In considering specimen errors it may not always be possible to reduce scatter to an acceptably low level ($< 5\%$), and it is useful if corrected absorbances at 260 mμ can be given for different values of the wavelength exponent of scatter, unless it is possible to determine this value by other means. Specimen heterogeneity should also be estimated and high accuracy not expected from very particulate material, and in addition the possibility of changes in absorptivity of cellular constituents during treatment must also be considered.

In general, measurements of absorbance per unit area have less significance than those of the total amount of absorbing material in a defined biological structure, such as nucleus, nucleolus, or cell, which can be validly compared with similar structures elsewhere. In this way chance variations in the size of the structures, due to either chemical environment or position in the tissue, can be eliminated. It is also abundantly clear that ultraviolet absorption measurements should not be made in isolation. Collateral measurements by other methods will nearly always illuminate and qualify the results.

Despite many drawbacks ultraviolet microspectrophotometry remains one of the very few methods that can be used to identify and measure a class of important biological constituents in either their natural state or with the least possible interference by chemical procedures. It is hoped that by describing various types of equipment and by discussing the errors in the method, it has been possible to give a general outline of the usefulness and limitation of ultraviolet microspectrophotometry.

VI. APPENDICES

Appendix 1: Basic Photographic and Photoelectric Instruments

The basic photographic and photoelectric instruments are illustrated in Figs. 9 and 10. It should be emphasized that these designs are not to be considered as in any way mandatory, and in particular the opportunity has been taken to vary certain details in each drawing so that a wider range of methods are illustrated. For example, different methods of illuminating the input slit are shown in Figs. 9, 10, and 19, but they are not confined to use with instruments there illustrated and often may be interchanged.

The photographic equipment (Fig. 9) is similar to that used by Chayen (personal communication), in which the microscope, auxiliary eyepiece, and camera are all combined in the reflecting microscope stand made by R. and J. Beck. The dispersion of the monochromator in this instance will be governed by the size of field which can be filled by one line of the mercury arc spectrum, without overlapping by the neighboring lines. The monochromator illustrated in Fig. 9, has a very high dispersion and gives a large field area suitable for photographic recording. It was designed by Seeds and Wilkins (see Seeds, 1951) and is

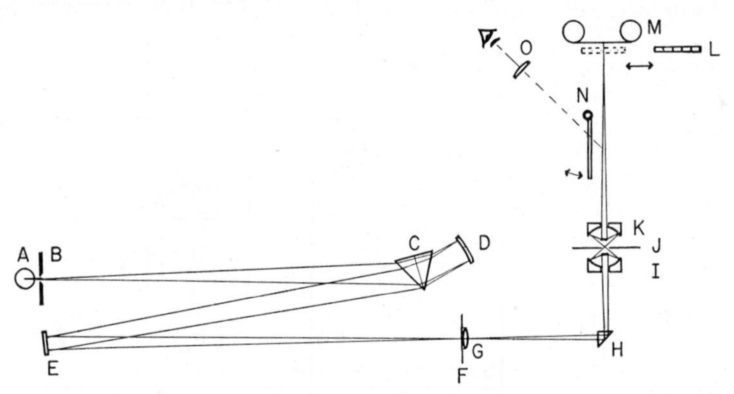

FIG. 9. Basic equipment for photographic recording. A, source; B, monochromator input slit; C, prism; D, concave mirror; E, plane mirror; F, output slit; G, lens; H, prism; I, condenser; J, object plane; K, objective; L, movable step wedge; M, camera; N, movable mirror; O, auxiliary viewing eyepiece.

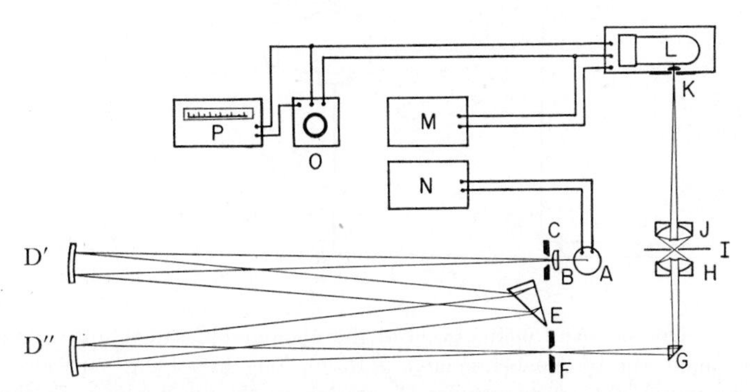

FIG. 10. Basic equipment for photoelectric recording. A, source; B, lens; C, monochromator input slit; D, concave mirrors; E, dispersing prism; F, output slit; G, prism; H, condenser; I, object plane; J, objective; K, aperture plate with lens behind; L, photomultiplier; M, stabilized voltage supply for photomultiplier; N, stabilized voltage supply for lamp; O, Ayrton shunt; P, galvanometer.

more fully described later (Fig. 18). In this relatively simple instrument the film is calibrated by photographing a blank field with the step wedge inserted in front of the film.

The photoelectric instrument illustrated in Fig. 10 has a Czerny-

Turner type of monochromator, which for a similar size has about half the dispersion of that illustrated in Fig. 9. Since only a small field at I is normally required, this does not matter. This monochromator also gives much less aberration because the radiation passing through the prism is properly collimated. Commercial monochromators of the Littrow form are also available (see Fig. 19) but they normally have a

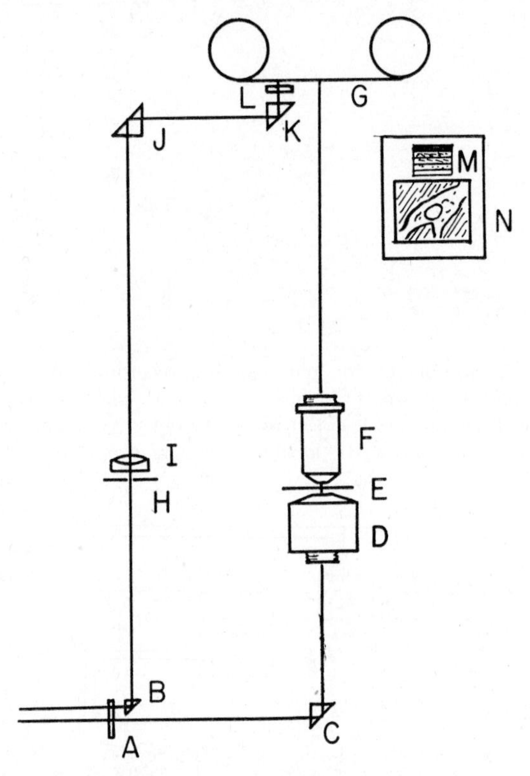

FIG. 11. Double-beam modification of the photographic apparatus described in FIG. 9. A, input slit of monochromator with its long axis lying vertically in the plane of the paper; B,C, fixed prisms; D, condenser; E, object plane; F, objective; G, camera; H, adjustable diaphragm; I, lens, J,K, prisms; L, fixed step wedge; M, wedge shadow on negative; N, image of object on film.

low f/number (the distance DF over the diameter illuminated at D″) and will therefore require relay lenses or mirrors to match this f/number to that of the condenser (the distance FGH over the diameter of the small mirror at H) which is usually about f/80 for reflecting optics. Commercial monochromators often use only one mirror for D′ and D″ and arrange the slits C and F above each other. This makes it difficult to provide effective baffles to reduce stray light. Since in the arrangement given in Fig. 10, the distance DF can be about 150 cm. the angles of reflection at D′, E, and D″ can consequently be made sufficiently small to reduce aberration and still provide positions for suitable baffles.

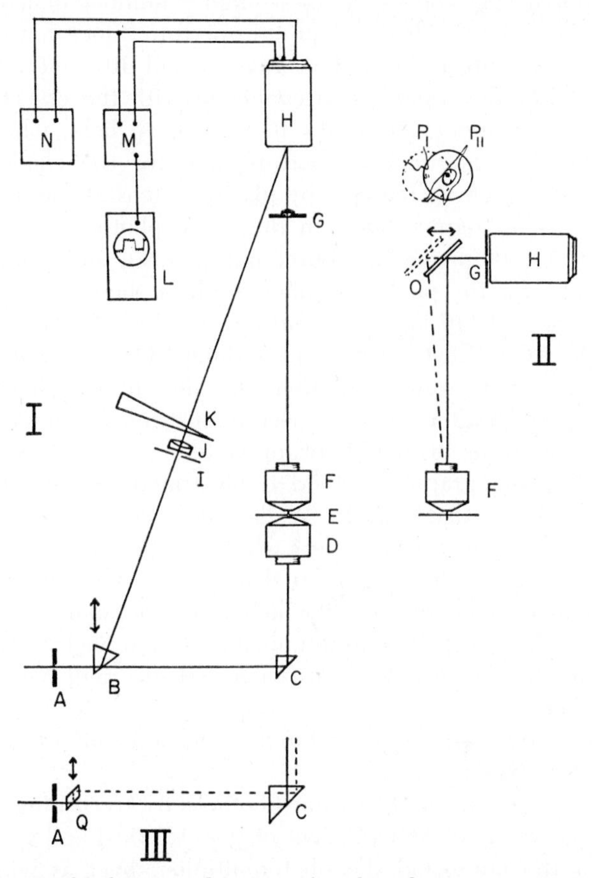

FIG. 12. Double-beam modifications for photoelectric measurements. Modification I: A, output slit of monochromator; B, vibrating prism; C, prism; D, condenser; E, object plane; F, objective; G, aperture plate; H, photomultiplier; I, adjustable diaphragm; J, lens; K, optical attenuator; L, oscilloscope photomultiplier signal; M, amplifier; N, smoothed power supply. Modification II: O, Vibrating mirror; P_1 and P_{11}, two positions of the image on the aperture plate G. Modification III: Q, vibrating prism. The vibrating prisms B and Q, should move at right angles to the plane of the drawing.

It will be noted that a lens (B) is used at the input slit to illuminate the mirror (D′). This lens should have a sufficiently short focal length so that the required diameter at D′ will be evenly filled by the particular source used. The lens B can be achromatic, but provided the size of the slit C is small, it will be satisfactory to use a quartz lens correctly focussed at the shorter wavelengths, when all available energy is required. It is also possible with some mercury and hydrogen arcs to place the source near the slit in these high *f*/number monochromators (as in Fig. 9) and still fill the required aperture. Since however in the particular instruments illustrated in Figs. 9 and 10 (but not those in Figs. 11 and 12) the input slit is conjugate with the object field, any unevenness at the input slit results in an uneven field, which is often less desirable than an uneven aperture stop for the condenser. This unevenness may often be reduced by placing a lens at the input slit.

The other main requirements in this basic photoelectric instrument are a method of stabilizing the source output (N), and a stable supply of high voltage for the photomultiplier. Both of these requirements are discussed in more detail later. The output of the photomultiplier is read on a suitable galvanometer (P), and a shunt (O) is required to compensate for the wide signal variations at different wavelengths.

Both the photographic and photoelectric instruments described would benefit if some form of double-beam system was employed. Such a variant of the photographic method is illustrated in Fig. 11, which is essentially similar to that used by Walker and Davies (1950). The output slit (A) of the monochromator, which is also the microscope field stop, is divided along its length, so that only part of the energy falls on the specimen. The remainder which is of the same wavelength is directed by the prisms B, J, K to illuminate the step wedge (L), situated contiguous to the image area. Thus each exposure will have a shadow of the step wedge (M) next to the specimen image (N) and errors due to variations in development or differences in the conditions of exposure will be much reduced.

The main object in double-beam photoelectric recording is to make measurements which are independent of the electrical and spectral characteristics of the lamp and the photomultiplier, over as wide a range of wavelengths as possible. In the main drawing (I) of Fig. 12, the energy leaving the monochromator alternately falls on the microscope condenser (D) or is directed by the vibrating prism (B) through the external light path (I, J, K) to the photomultiplier (H). The signal is amplified (M) and displayed on the oscilloscope (L). If two photomultipliers had been used as in Caspersson's latest apparatus (Caspersson, 1955), or if some method is available of only attenuating the photomultiplier signal during the half-cycle during which it is receiving energy from the external beam, it is possible to measure I and I_0 as the ratio of the signals produced at the two positions of the specimen. It is simplest however to use an optical attenuator (K) in the external beam and measure I and I_0 from the position of the attenuator when the signals from both beams balance in the oscilloscope. An attenuator made by depositing rhodium or platinum onto quartz is simplest to use, but they are not wavelength independent (Walker, 1956), and a correction for each position of the attenuator will be required for all wavelengths. An aperture form of attenuator which is wavelength independent was used by Walker and Deeley (1956) but it is difficult to provide the even illumination of the aperture which is so essential.

An external light path of this form has the great advantage over the internal forms illustrated in Figs. 12, II and III, in that it is unnecessary to reserve a fixed part of the specimen area for the measurement of I_0. They have the disadvantage, however, that the transmissions of both beams may be difficult to match at all wavelengths, particularly when low transmission refracting optics are employed in the microscope. The optically matched double-beam system of Pollister and Ornstein (personal communication) overcomes this difficulty.

The two internal double-beam systems illustrated can only be used with electrical attenuation, and will also require a shutter to cut off the energy during every half-cycle, in order that a reference voltage may be obtained from the photomultiplier. In the method shown in Fig. 12, II and designed in Thorell's laboratory, a vibrating mirror (O) causes the image of the specimen to alternate between two positions (P_1, P_{11}) relative to the aperture plate (G). In the other variant (Fig. 12, III) a small vibrating prism (Q) or mirror (see Haggis, 1956) causes the illuminating aperture at the monochromator exit slit to alternate between two positions. If the aperture plate (G) has two apertures suitably placed in the image plane, the photomultiplier will then receive alternating signals of I and I_0 with the necessary dark portion in between. The chief disadvantage of these internal systems is that the measurements are not made at the same central portion of the optical system, but provided the relative transmissions of these two parts remain constant, this should not be a serious drawback.

Appendix 2: Microscope Followed by a Spectrograph

An alternative arrangement of microscope and spectrometer is to arrange that all the energy from the source passes through the microscope before being analyzed by a monochromator or spectrograph. This is the method commonly used in infrared microspectrometry, but it has not often been used in the ultraviolet region.

It may be employed to obtain adjacent images of an area of the specimen with a line source (e.g., Hg arc) but difficulties may be encountered due to the varying intensities and multiplet natures of the spectral lines. Alternatively an absorption spectrum can be recorded from points along a line across the specimen, if a continuous source such as a hydrogen arc is used.

Seeds (1951) used this kind of instrument with a mercury arc for studying the dichroism of tobacco mosaic virus crystals. More recently Ruch (personal communication) has used the instrument illustrated in Fig. 13 for the study of plant cells.

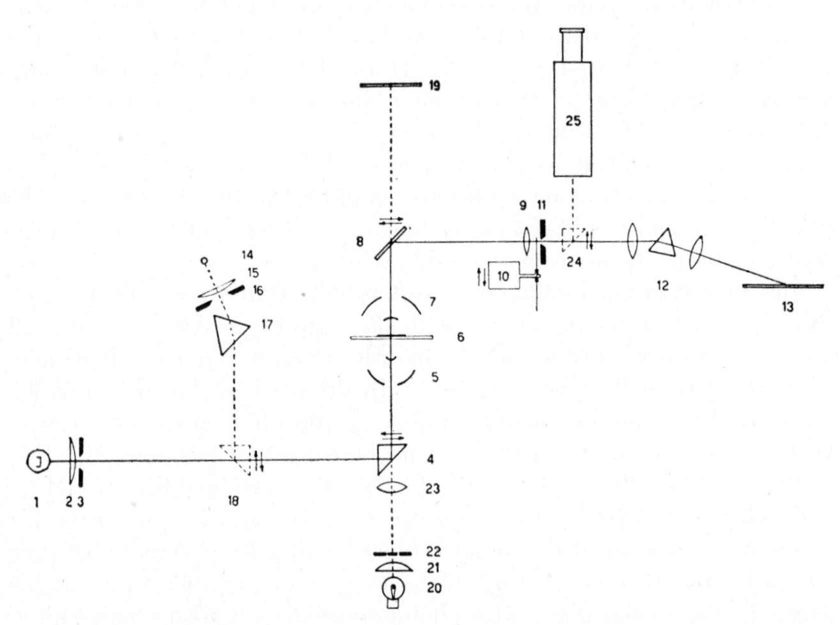

FIG. 13. A photographic recording instrument which has been designed by Ruch, and incorporates a spectrograph following the microscope. 1, hydrogen arc; 2, field lens; 3, field stop; 4, movable prism; 5, reflecting condenser; 6, specimen; 7, reflecting objective; 8, movable mirror; 9, field lens of spectrograph; 10, calibrating sector; 11, spectrograph input slit; 12, spectrograph; 13, spectrograph plate; 14–19, auxiliary light train with a monochromator for taking photographs at one wavelength; 20–25, phase contrast illuminating and viewing system (see Fig. 22).

APPENDIX 3: A "MINIMUM" ULTRAVIOLET MICROSCOPE

It is sometimes necessary to make a simple apparatus for illuminating a specimen with reasonably pure monochromatic ultraviolet radiation, and the design illustrated (Fig. 14) may then be useful. The dispersion element is a Féry (1910) prism, which consists of two spherical surfaces arranged at an angle to each other. This prism both disperses the energy and forms an image of the slit (B) at the condenser mirror (D). A prism with a 15° angle and radii of 100 cm. can disperse the 240–600 mμ spectral region over about 5 cm. if CD is 200 cm. For small aperture ratios the prism only needs rotating. The specimen is placed at E, and the objective (F) can be the 0.2 NA ultraviolet achromat designed by Johnson (1939).

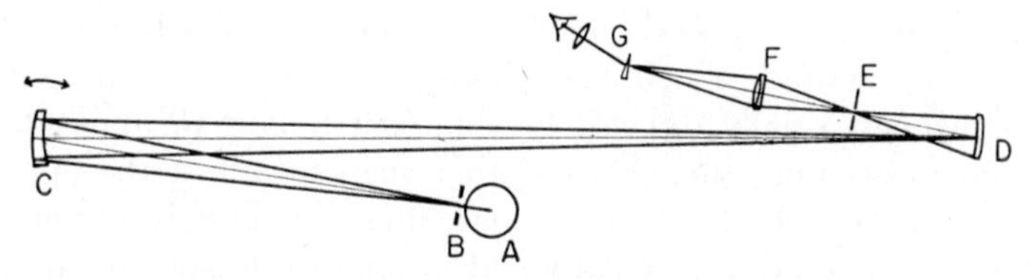

FIG. 14. A "minimum" ultraviolet microscope. A, source; B, monochromator input slit; C, Féry prism; D, concave mirror acting as monochromator output slit and condenser; E, object plane; F, achromatic objective lens; G, Köhler eyepiece.

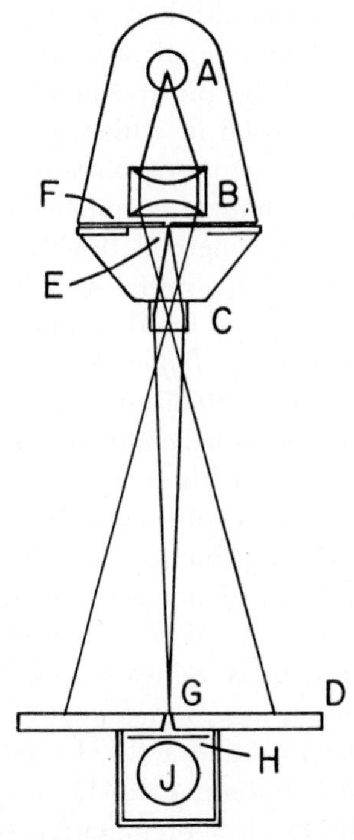

FIG. 15. An enlarger adapted for use as a microdensitometer. A, source; B, condenser; C, lens; D, screen; E, aperture plate; F, film; G, aperture; H, filter; J, photomultiplier.

APPENDIX 4: MICRODENSITOMETERS

Means of measuring photographic plates in spectrography have been fully discussed by Harrison (1937) and the modifications desirable in microspectrographic procedures have been mentioned by Walker

(1956). It is intended to illustrate here only rather simple designs, which may be constructed in the laboratory, since automatic recording instruments are now commercially available.

(*a*) *Enlarger Method* (*Fig. 15*). Instruments based on the ordinary photographic enlarger have been described by Hesthal and Harrison (1929). The image is viewed on a white screen (D), with a small central aperture (G) behind which the photomultiplier (J) is placed. Since field evenness is usually difficult to obtain, it is best to move the negative, when positioning the object, rather than to adjust the aperture or objective positions. Errors due to glare may be largely eliminated by arranging just behind the negative (F) a movable screen, in which a small aperture (E) is optically aligned with the photomultiplier aperture. Alternatively a fixed, preferably red, filter with a similarly placed aperture can be employed and a complementary filter placed between the viewing screen and photomultiplier. Some difficulty may be experienced in this instrument if very small apertures are required since the object hole size may have to be inconveniently small. This difficulty is overcome if a microscope is used in which the compound stage is used as the film holder; the field stop diaphragm of conventional Köhler illumination limits the area illuminated on the film. The image may be viewed on a screen, as in the method described by Lison (1950) for measuring stained specimens, or special viewing attachments, such as those illustrated in Fig. 17, can be fitted compactly onto the microscope.

(*b*) *Double-Beam Recording Type.* A simple double-beam instrument described by Swann and Mitchison (1950) is illustrated (Fig. 16), as it shows how standard laboratory components can be used in a microdensitometer. Clearly such an instrument is similar in principle to those required for the photoelectric recording of the absorption in cells. The following simplifications are possible for measuring the density of photographic film: (1) the light need not be monochromatic; (2) ordinary microscope objectives may be used; (3) low-voltage tungsten filament lamps, run off accumulators, may be used as a stable light source.

(*c*) *Viewing and Measuring Attachments.* These are attachments which allow the operator to see exactly the position on the specimen from which measurement is to be made. The first (Fig. 17, I) is suitable for instances in which the level of illumination is high. Another method (Fig. 17, II) is that employed by Pollister (1952) for stain measurements in which a comparatively large area is measured. The lens placed behind the aperture is useful with such large apertures to ensure that all the energy falls on the same part of the photocathode so that variations in its sensitivity do not introduce errors. This lens is arranged to focus the exit pupil of the microscope objective on to the

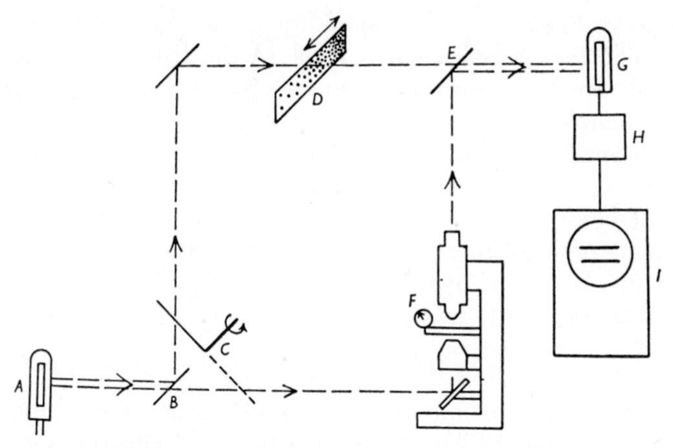

FIG. 16. Double-beam microdensitometer using standard components. A, source; B, fixed beam splitter; C, beam chopper; D, density wedge; E, partially reflecting plate; F, movement indicator; G, photo-cell; H, amplifier; I, oscilloscope.

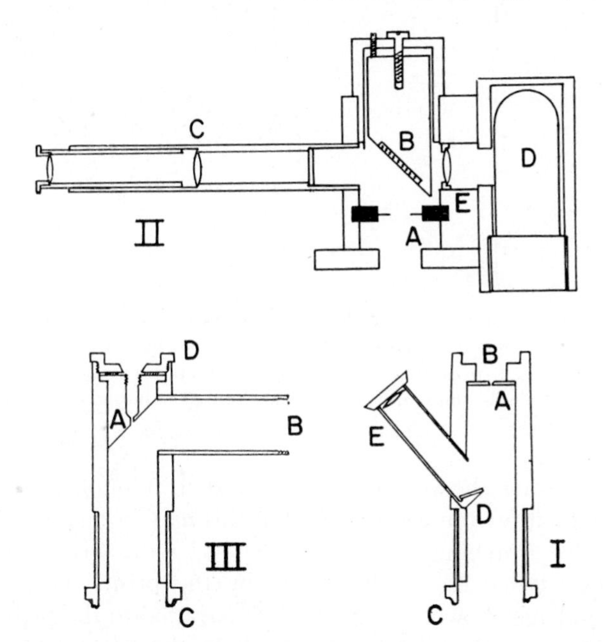

FIG. 17. Viewing attachments. I: Simple attachment for use in microdensitometers, where the energy available is large. A, matt-white plate with aperture; B, adapter for photomultiplier; C, microscope eyepiece holder; D, small plane mirror, E, eyepiece. II: Attachment suitable for employment with larger measuring apertures. A, adjustable aperture; B, rotatable mirror; C, eyepiece; D, photomultiplier; E, lens. III: Attachment for use with a small fixed aperture. A, aperture; B, fitting for viewing eyepiece; C, microscope eyepiece holder; D, photomultiplier adapter.

photocathode. Provision is also made for photographing the exact position at which the measurement is made. A third method (Fig. 17, III) suitable for small measuring apertures, which we have employed is somewhat simpler and consists of a stainless steel mirror arranged at 45° to the optic axis with the measuring aperture situated on the axis. The image is viewed by a simple microscope such as that in the R, and J. Beck demonstration eyepiece. The mirror may be turned about its axis if photographs of the measured area are required. The field of view on these attachments is usually limited; it is therefore often convenient to fit a commercial side tube viewing device for surveying larger fields.

Appendix 5: A High-Dispersion Monochromator

We have already mentioned the high-dispersion monochromator designed by Seeds and Wilkins for use particularly with the photographic method of recording. Figure 18 shows the design in greater detail. Since

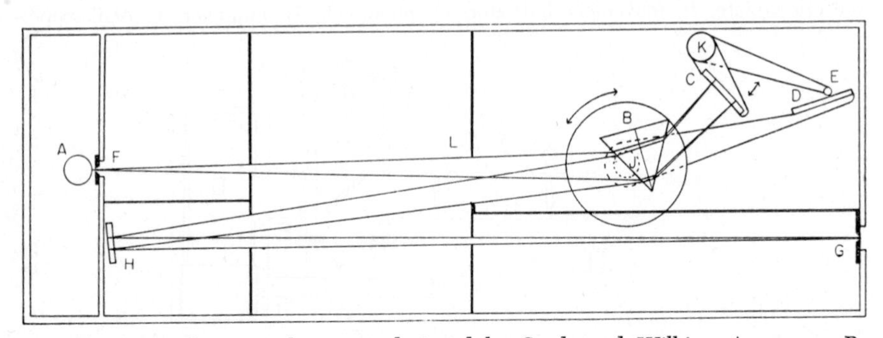

Fig. 18. The monochromator designed by Seeds and Wilkins. A, source; B, prism; C, spherical mirror; D, radial plate; E, hardened ball; F, input slit; G, output slit; H, plane mirror; J, prism axis; K, mirror axis; L, position of aperture mentioned in text. The thick single lines indicate the position of suitable baffles and stops for reducing stray light.

the rays traversing the prism are not parallel, minimum deviation can only be obtained for the central ray, but this may be maintained for any wavelength by arranging that the mirror (C) with axis of rotation (K) turns through twice the angle moved by the prism. This is done by the simple linkage shown. The plate (D) is radial to the prism axis (J), and the distances JK and KE are equal. In the monochromators used in this laboratory, the mirrors are made of quartz with both surfaces curved and the back one aluminized. In a city atmosphere, this prevents the rapid deterioration of the reflectivity and it is possible to arrange that the inevitable ghost images are aligned with the main image of the slit.

A disadvantage of this design is that the image of the source moves across the mirror face (C) as the wavelength is changed, and it is therefore impossible to place a diaphragm controlling the condenser aperture at this mirror. An aperture at the position (L), the baffles indicated in the figure, and lining the box with black velvet paper should ensure that monochromator stray light is acceptably low.

Appendix 6: Optical Matching of a Commercial Monochromator to a Reflecting Condenser

The monochromators which we have already described in the appendices have been specially designed for use with reflecting optics, but the facilities for constructing them may not always be available. Com-

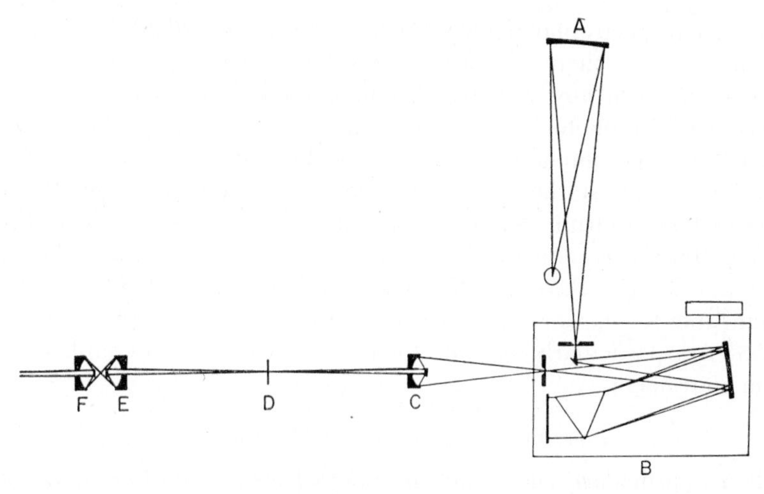

Fig. 19. Optical matching of a commercial monochromator of low f/number to a reflecting condenser. A, Mirror focusing source onto the input slit of the monochromator; B, commercial monochromator; C, relaying mirror; D, microscope field stop; E, reflecting condenser; F, reflecting objective.

mercial monochromators, on the other hand, usually have too high an aperture ($f/10$) to illuminate a reflecting condenser without serious loss of energy. Additional optical components are therefore required to match the aperture ratios (Loofbourouw, 1950). A neat solution to this problem is due to Pollister and Ornstein (personal communication) who used the method illustrated in Fig. 19. It will be seen that the reflecting "objective" (C) effectively matches the monochromator NA to that of the condenser. In this instrument a beam splitter was placed at D, which is the microscope field stop. It would also be possible to place the condenser at the focus of C, the small mirror of which would then act as the field stop. Figure 19 also indicates how a mirror (A) can be used to illuminate the entrance slit of the monochromator (B).

A similar optical system to that shown is used in commercial monochromators which have a diffraction grating instead of a quartz prism as their dispersing element. The advantage of a grating is that its dispersion is substantially linear over the visible and ultraviolet spectrum. The grating should be blazed so that the maximum energy is available in the short-wave ultraviolet region, but it will also require an auxiliary prism to remove unwanted spectral orders and therefore single-prism monochromators will usually have a higher transmission at these wavelengths.

Appendix 7: Slit Designs

In microspectrophotometers such as those described here (except that in Fig. 13) there is a fixed exit slit to the monochromator, which may be made in any way which eliminates reflections from the edges of the slit. In photoelectric recording, when a source which gives a continuous spectrum and a prism monochromator are used, it will be necessary to adjust the input slit to give a constant band width at all parts of the spectrum. Suitable adjustable slits are described by Fassin (1937), but it is unnecessary in these instruments to use the more elaborate designs. Razor blades make good jaws and the Fassin (1933) design and the newer and ingenious design of Crosswhite and Fastie (1956) are particularly suitable for operation by a cam coupled to the prism movement.

Appendix 8: Micrometer Moving Stage

It is particularly necessary in photoelectric recording to position the cell accurately on the measuring aperture. This is difficult with the ordinary commercial mechanical stage of the microscope. The gliding stage made by Zeiss is a great improvement in this respect, but exact return to the same position after each measurement would not be easy. Caspersson designed an elegant hydraulic stage for this purpose, which can be positioned with very great accuracy. Mechanical stages actuated by micrometers are commercially available, but these were specially designed for the study of photographic plates in nuclear physics. Their range is unnecessarily large and they are expensive. The stage shown in Fig. 20 has a limited range (8×8 mm.) but it is cheap and easy to make and can usually replace an ordinary stage without displacing the alignment of other components. The top plate carrying the specimen is free to roll on the three balls parallel to the right-hand micrometer, and it can also rotate round the ball (F), when the other micrometer is adjusted.

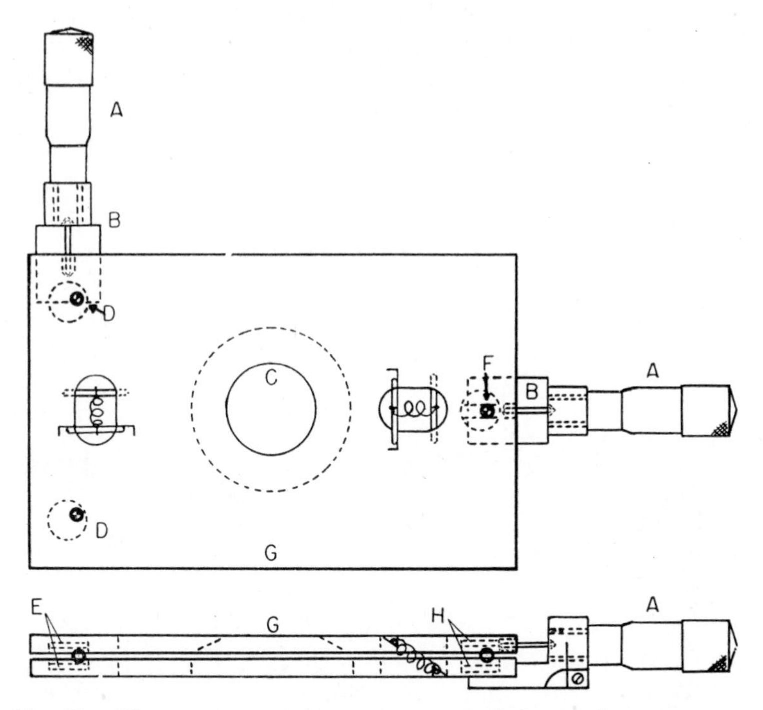

FIG. 20. Micrometer stage. A, micrometers; B, links; C, objective aperture; D, hardened steel balls running between circular hardened plates (EE); F, hardened ball running between hardened plates (HH), which have a parallel V-groove in each; G, position of bolting face for microscope stand.

APPENDIX 9: METHODS OF USING PHASE CONTRAST ILLUMINATION WITH REFLECTING MICROSCOPES

It is often a great advantage to have a method of viewing the specimen in phase contast when reflecting objectives are employed. In this way the fullest advantage may be taken of the achromatic properties of these objectives in searching and focusing the specimen. Unfortunately the usual position for the phase annulus and retardation plate is near the small mirror, and therefore the removal of these components before measurements are made will be inconvenient. One method of overcoming this (Walker and Davies, 1950; Norris *et al.*, 1951) is to place the retardation plate outside the objective (E, Fig. 21) and to image the illuminating annulus (A) into the conjugate position within the condenser (C) by means of the lens (B). In this method the focusing and alignment of the two phase components is critically dependent on the positions of C and D, but this is not a disadvantage if the stage only is moved to focus the specimen.

Another method which in effect places the retardation plate at the

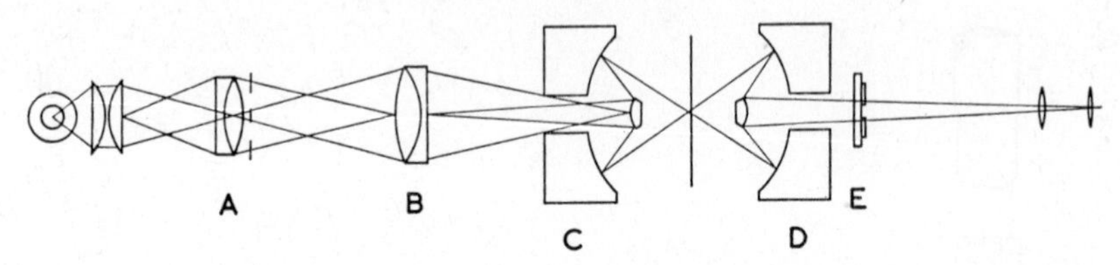

FIG. 21. Arrangement for visible phase contrast illumination. A, phase annulus; B, field stop; C, condenser; D, objective; E, phase plate ($\lambda/4$).

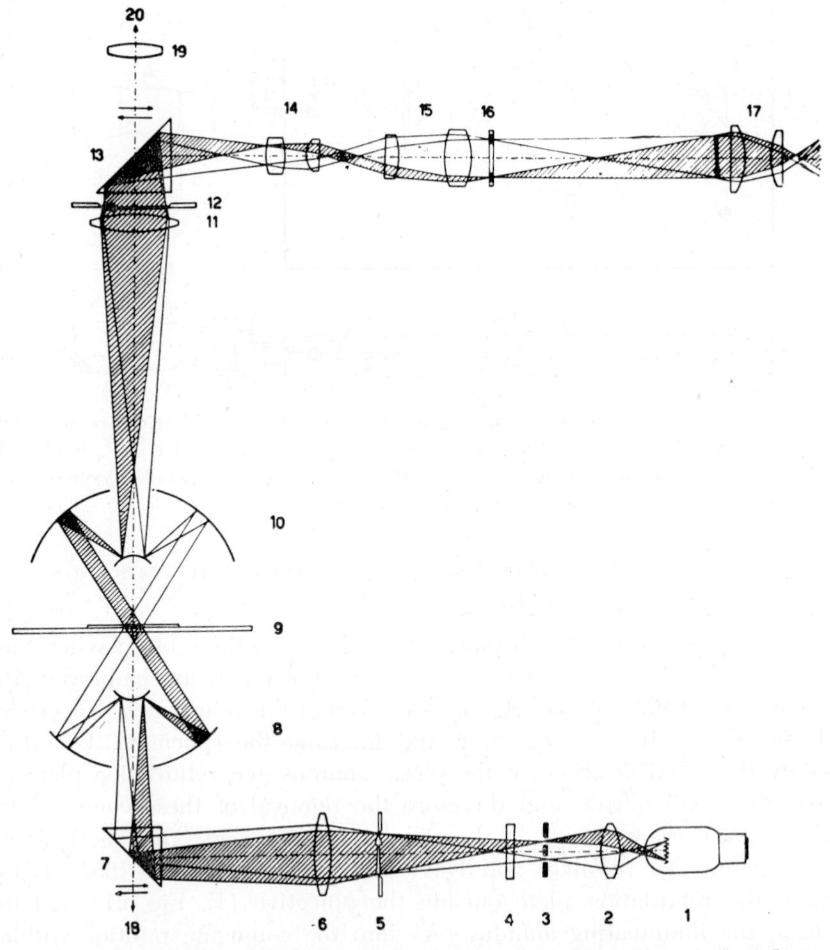

FIG. 22. Ruch's method of using phase contrast illumination with reflecting objectives. 1, tungsten lamp; 2, field lens; 3, illuminating annulus; 4, filter; 5, microscope; 6, field lens; 7, movable prism; 8, reflecting condenser; 9, specimen; 10, reflecting objective; 11, field lens; 12, image stop; 13, movable prism; 14, 15, 17, condenser, objective, and eyepiece of the auxiliary microscope; 16, phase retardation plate, which is conjugate with 3; 18–20, light path for ultraviolet recording.

correct position is to use one of the special phase systems (e.g., Payne, 1954) in which the exit pupil of the objective is reimaged by a mirror to a position where the phase plate may be conveniently placed. This method was at first used by Walker and Deeley (1956) in their microspectrophotometer, but it had the disadvantage that a number of optical components had to be inserted and removed from the optical path, and consequently it was difficult to maintain correct alignment.

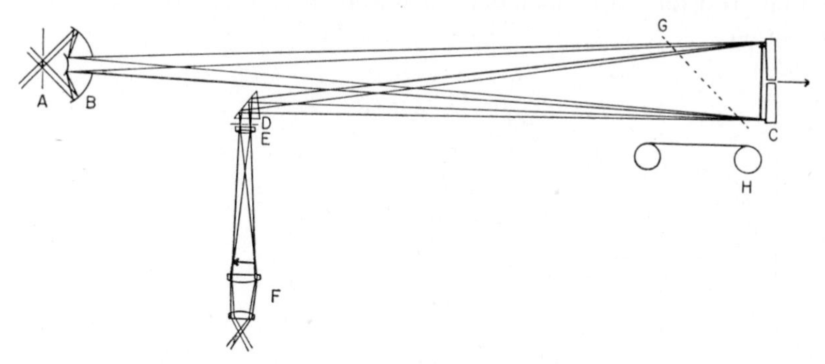

FIG. 23. Method of using phase contrast observations with a reflecting objective. A, specimen; B, objective; C, mirror with measuring aperture; D, phase retardation plate; E, lens; F, eyepiece; G, moveable mirror; H, camera.

Ruch (personal communication) has used an auxiliary refracting microscope (Fig. 22, nos. 14, 15, 17) instead of a mirror for obtaining an external image of the primary focus of the objective. Since the ultraviolet radiation by-passes this auxiliary microscope, normal visible objectives of good quality can ensure that there is the minimum loss of definition due to the introduction of these additional components.

We have recently made an improvement (Fig. 23) on Payne's system, in which the limiting aperture in photoelectric recording is placed at the mirror (C). The specimen is seen focused and in phase contrast illumination at this mirror by means of the fixed auxiliary microscope (F, E). This system has no moving parts and is very simple, but has the disadvantage that the mirror (C) must be very well figured to avoid its irregularities being so enhanced by the phase system as to affect the clarity of the image. This method could be used in photographic recording if the mirror could be easily replaced by the camera or if a second moveable mirror (G) were used.

APPENDIX 10: POWER STABILIZATION

We have already mentioned the need to stabilize the power supplies to photomultipliers in photoelectric recording, owing to their extreme sensitivity to changes in the voltage of the power supply. Perhaps the simplest method of achieving this is to use a stack of high-tension batteries as shown in Fig. 24. Provided that batteries with a long shelf-life are chosen, the drain on them is so small that they can be used for 2 or 3 years without replacement. Figure 24 also shows a convenient mounting for the RCA 1P28 type of photomultiplier, which may be easily used with the viewing attachments described earlier. The correct pin connections are shown in the circuit and it will be noted that current-limiting resistors are included between each battery as a safety precaution.

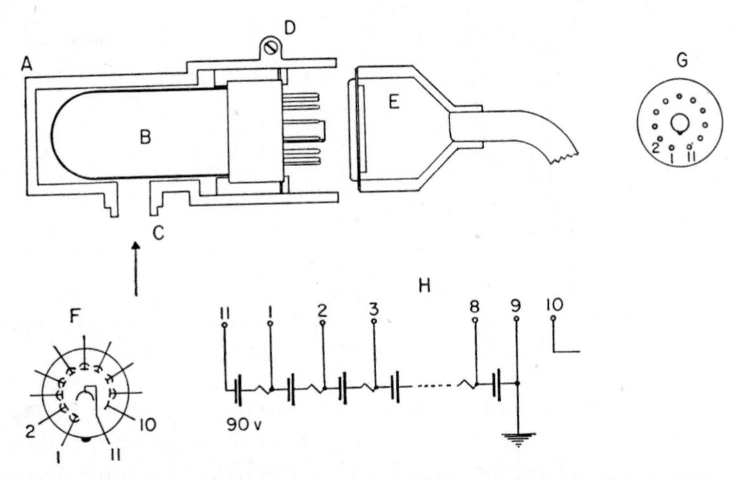

FIG. 24. Adapter for RCA 1P28 type photomultipliers; A, housing; B, photomultiplier; C, adapter to suit for example the viewing attachments illustrated in Fig. 17; D, clamping screw; E, base and cable connections; F, arrangement of electrodes; G, base connections; H, suggested circuit for battery operation with limiting resistances included for safety purposes.

Alternatively electronic stabilization may be used and suitable circuits for this can usually be obtained from the manufacturers of the photomultipliers.

It is also possible to compensate simultaneously for both voltage changes in the high-tension supply and variations in the energy emitted by the light source by a circuit such as that illustrated in Fig. 25. Both photomultipliers are supplied by the same power-pack (E), while one photomultiplier receives energy from the microscope (L_2), the other receives a small fraction of the energy leaving the monochromator (L_1). This small fraction can be obtained by putting a quartz coverslip in the beam between source and microscope condenser, and arranging an aperture and collecting lens in front of the photomultiplier which receives the same angular distribution of energy as this condenser. This circuit maintains a steady anode current in the left photomultiplier, by altering the voltage across the dynodes. Provided the two photomultipliers can be matched, the voltage appearing at F will be independent of source energy and fluctuations in the output of the power-pack.

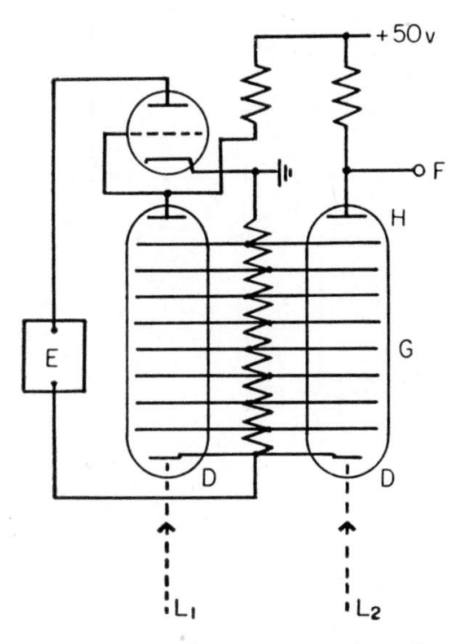

FIG. 25. Automatic gain control circuit using two photomultipliers. D, matched photomultipliers; E, high tension supply; F, output; G, photomultiplier dynodes; H, photomultiplier anodes; L_1 and L_2, compensating and measuring light paths respectively.

APPENDIX II: LAMPS AND LAMP STABILIZATION

In the last section a method of eliminating the need for lamp stabilization was described, but in some applications this circuit may not be desirable and the lamp may have to be stabilized independently. If hydrogen arcs can be used, current stabilization is satisfactory and has been often used in commercial single-beam spectrometers. With a mercury arc such as the British Thomson Houston MB/D or the American AH/4 designs, arc movement makes brilliance stabilization necessary. This can be achieved (Deeley, 1955) by using a second photomultiplier, which need not be the same as the main photomultiplier, to control the current through the lamp. The circuit shown in Fig. 26 can eliminate all comparatively short-term fluctuations in the energy available, so that I and I_0 can be measured consecutively, but of course the effect of long-term variations in photocathode response cannot be eliminated in these two-multiplier designs.

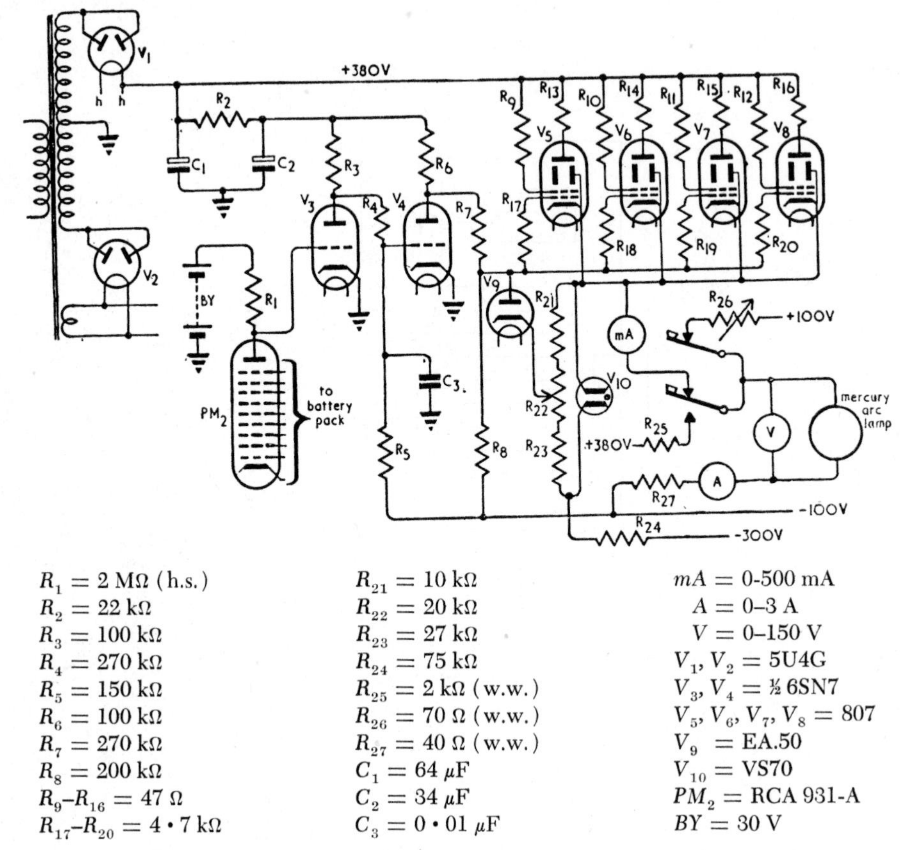

$R_1 = 2$ MΩ (h.s.)	$R_{21} = 10$ kΩ	$mA = 0$-500 mA
$R_2 = 22$ kΩ	$R_{22} = 20$ kΩ	$A = 0$–3 A
$R_3 = 100$ kΩ	$R_{23} = 27$ kΩ	$V = 0$–150 V
$R_4 = 270$ kΩ	$R_{24} = 75$ kΩ	$V_1, V_2 =$ 5U4G
$R_5 = 150$ kΩ	$R_{25} = 2$ kΩ (w.w.)	$V_3, V_4 =$ ½ 6SN7
$R_6 = 100$ kΩ	$R_{26} = 70$ Ω (w.w.)	$V_5, V_6, V_7, V_8 = 807$
$R_7 = 270$ kΩ	$R_{27} = 40$ Ω (w.w.)	V_9 = EA.50
$R_8 = 200$ kΩ	$C_1 = 64$ μF	V_{10} = VS70
R_9–$R_{16} = 47$ Ω	$C_2 = 34$ μF	PM_2 = RCA 931-A
R_{17}–$R_{20} = 4 \cdot 7$ kΩ	$C_3 = 0 \cdot 01$ μF	$BY = 30$ V

FIG. 26. Mercury arc lamp stabilizing circuit. This circuit is designed for use with a +100v and −100v DC supply to the lamp (Type MB/D or AH/4), and the press switch shown is for starting the arc.

REFERENCES

Attardi, G. (1953). *Experientia* **9,** 422.

Baird, D. K., Haworth, W. N., Herbert, R. W., Hirst, E. L., Smith, F., and Stacey, M. (1934). *J. Chem. Soc.,* p. 62.

Barnard, J. E., and Welch, F. V. (1936). *J. Roy. Microscop. Soc.* **56,** 365.

Beaven, G. H., and Holiday, E. R. (1952). *Advances in Protein Chem.* **7,** 319.

Beaven, G. H., Holiday, E. R., and Johnson, E. A. (1955). "The Nucleic Acids" (E. Chargaff and J. N. Davidson, eds.), Vol. I, p. 493. Academic Press, New York.

Blaisse, B. S., Bouwers, A., and Bulthuis, H. W. (1952). *Appl. Sci. Research* **B2,** 453.
Blout, E. R. (1953). *Advances in Biol. and Med. Phys.* **3,** 285.
Blout, E. R., and Asadourian, A. (1954). *Biochim. et Biophys. Acta* **13,** 161.
Blout, E. R., Bird, G. R., and Grey, D. S. (1950). *J. Opt. Soc. Am.* **40,** 304.
Bricas, E., and Fromageot, C. (1953). *Advances in Protein Chem.* **8,** 4.
Burch, C. R. (1947). *Proc. Phys. Soc.* **59,** 41.
Caspersson, T. (1936). *Skand. Arch. Physiol.* **73,** Suppl. 8.
Caspersson, T. (1940). *Chromosoma* **1,** 605.
Caspersson, T. (1950). "Cell Growth and Cell Function." Norton, New York.
Caspersson, T. (1955). *Experientia* **11,** 45.
Caspersson, T., and Santesson, L. (1942). *Acta Radiol. Suppl.* **46.**
Chargaff, E., and Zamenhof, S. (1948). *J. Biol. Chem.* **173,** 327.
Chayen, J. (1952). *Symposia Soc. Exptl. Biol.* **No. 6,** 290.
Chayen, J. (1953). *Intern. Rev. Cytol.* **2,** 77.
Commoner, B. (1949). *Science* **110,** 31.
Crampton, C. F., Lipshitz, R., and Chargaff, E. (1954). *J. Biol. Chem.* **206,** 499.
Crosswhite, H. M., and Fastie, W. G. (1956). *J. Opt. Soc. Am.* **46,** 110.
Davies, H. G. (1950a). *Discussions Faraday Soc.* **9,** 442.
Davies, H. G. (1950b). *Discussions Faraday Soc.* **9,** 397.
Davies, H. G. (1954). *Quart. J. Microscop. Science* **95,** 433.
Davies, H. G., and Walker, P. M. B. (1953). *Progr. in Biophys. and Biophys. Chem.* **3,** 195.
Davies, H. G., and Wilkins, M. H. F. (1950). *J. Roy. Microscop. Soc.* **70,** 280.
Davies, H. G., Wilkins, M. H. F., and Boddy, R. G. H. B. (1954). *Exptl. Cell Research* **6,** 550.
Deeley, E. M. (1955). *J. Sci. Instr.* **32,** 263.
Fassin, G. (1933). *J. Opt. Soc. Am.* **23,** 186.
Fassin, G. (1937). *In* "Measurement of Radiant Energy" (G. E. Forsythe, ed.), p. 147. McGraw-Hill, New York.
Féry, C. (1910). *J. phys.* **9,** 762.
Foster, L. V., and Thiel, E. M. (1948). *J. Opt. Soc. Am.* **38,** 689.
Glick, D., Engström, A., and Malmström, B. G. (1951). *Science* **114,** 253.
Gordon, M. W., and Nurnberger, J. I. (1955). *Exptl. Cell Research* **8,** 279.
Grey, D. S. (1950). *J. Opt. Soc. Am.* **40,** 283.
Grinnan, E. L., and Mosher, W. A. (1951). *J. Biol. Chem.* **191,** 719.
Grun, P. (1956). *Exptl. Cell Research* **10,** 29.
Haggis, G. H. (1956). *J. Sci. Instr.* **33,** 481.
Harrison, G. R. (1937). *In* "Measurement of Radiant Energy" (W. E. Forsythe, ed.), p. 283. McGraw-Hill, New York.
Hesthal, C. E., and Harrison, G. R. (1929). *Phys. Rev.* **34,** 543.
Hiskey, C. F. (1955). *In* "Physical Techniques in Biological Research" (G. Oster and A. W. Pollister, eds.), Vol. I, p. 73. Academic Press, New York.
Holiday, E. R., and Beaven, G. H. (1950). *Photoelec. Spectometry Group Bulletin* **3,** 53.
Jenkins, F. A., and White, H. E. (1937). "Fundamentals of Physical Optics," p. 297. McGraw-Hill, New York.
Johnson, B. K. (1939). *Proc. Phys. Soc. (London)* **51,** 1034.
King, R. J., and Roe, E. M. F. (1953). *J. Roy. Microscop. Soc.* **73,** 82.
Köhler, A. (1904). *Z. wiss. Mikroskop.* **21,** 129.

Koenig, H., Schildkraut, D., and Galler, E. (1953). *J. Histochem. and Cytochem.* **1,** 384.
Leuchtenberger, C., Klein, G., and Klein, E. (1952a). *Cancer Research* **12,** 480.
Leuchtenberger, C., Leuchtenberger, R., Vendrely, C., and Vendrely, R. (1952b). *Exptl. Cell Research* **3,** 240.
Lin, M. (1955). *Chromosoma* **7,** 340.
Lison, L. (1950). *Acta Anat.* **10,** 333.
Loofbourow, J. R. (1950). *J. Opt. Soc. Am.* **40,** 317.
Ludford, R. J., and Smiles, J. (1950). *J. Roy. Microscop. Soc.* **70,** 186.
Magasanik, B., and Chargaff, E. (1951). *Biochim. et Biophys. Acta* **7,** 396.
Mellors, R. C., Hlinka, J., Kupfer, A., and Sugiura, K. (1954). *Cancer* **7,** 771.
Michaelis, L. (1947). *Cold Spring Harbor Symposia Quart. Biol.* **12,** 131.
Moberger, G. (1954). *Acta Radiol. Suppl.* **112.**
Norris, K. P., Seeds, W. E., and Wilkins, M. H. F. (1951). *J. Opt. Soc. Am.* **41,** 111.
Nurnberger, J. I. (1955). *In* "Analytical Cytology" (R. C. Mellors, ed.), Chapter 7. McGraw-Hill, New York.
Opler, A. (1950). *J. Opt. Soc. Am.* **40,** 401.
Ornstein, L. (1952). *Lab. Invest.* **1,** 250.
Patau, K. (1952). *Chromosoma* **5,** 341.
Payne, B. O. (1954). *J. Roy. Microscop. Soc.* **74,** 108.
Philpot, J. St. L., and Stanier, J. E. (1955). *Biochem. J.* **63,** 214.
Photoelec. Spectrometry Group Bull. (1950). **3,** 39.
Pollister, A. W. (1952). *Lab. Invest.* **1,** 106.
Richards, B. M., Walker, P. M. B., and Deeley, E. H. (1956). *Ann. N. Y. Acad. Sci.* **63,** 831.
Rudkin, G. T., Aronson, J. F., Hungerford, D. A., and Schultz, J. (1955). *Exptl. Cell Research* **9,** 193.
Rudkin, G. T., and Corlette, S. L. (1957a). *J. Biophys. Biochem. Cytol.* **3,** 821.
Rudkin, G. T., and Corlette, S. L. (1957b). *Proc. Nat. Acad. Sci.* **43,** 964.
Schneider, R. M. (1955). *Exptl. Cell Research* **8,** 24.
Scott, J. (1955). *In* "Physical Techniques in Biological Research" (G. Oster and A. W. Pollister, eds.), Vol. I, pp. 150–166. Academic Press, New York.
Scott, J., and Sinsheimer, A. (1950). "Medical Physics," Vol. **2,** pp. 537. Year Book, Chicago, Illinois.
Seeds, W. E. (1951). Ph.D. Thesis, London University, p. 84.
Shack, J., Jenkins, R. J., and Thompsett, J. M. (1953). *J. Biol. Chem.* **203,** 373.
Smith, K. C., and Allen, F. W. (1953). *J. Am. Chem. Soc.* **75,** 2131.
Steel, W. H. (1953). *Rev. opt.* **32,** 4.
Svensson, G. (1955). *Exptl. Cell Research* **9,** 428.
Swann, M. M., and Mitchison, J. M. (1950). *J. Exptl. Biol.* **27,** 226.
Swift, H. (1955). *In* "The Nucleic Acids" (E. Chargaff and J. N. Davidson, eds.), Vol. II, p, 51. Academic Press, New York.
Tamm, C., Shapiro, H. S., Lipshitz, R., and Chargaff, E. (1953). *J. Biol. Chem.* **203,** 673.
Taylor, W. (1953). *J. Opt. Soc. Am.* **43,** 299.
Thomas, R. (1954). *Biochim. et Biophys. Acta* **14,** 231.
Thorell, B., and Ruch, F. (1951). *Nature* **167,** 815.
Tsuboi, K. K. (1950). *Biochim. et Biophys. Acta* **6,** 202.
Walker, P. M. B. (1956). *In* "Physical Techniques in Biological Research" (G. Oster and A. W. Pollister, eds.), Vol. III, p. 401. Academic Press, New York.

Walker, P. M. B. (1957). *Exptl. Cell Research Suppl.* **4,** 86.
Walker, P. M. B., and Davies, H. G. (1950). *Discussions Faraday Soc.* **9,** 461.
Walker, P. M. B., and Deeley, E. M. (1955). *Photoelec. Spectrometry Group Bull.* **No. 8,** 192.
Walker, P. M. B., and Deeley, E. M. (1956). *Exptl. Cell Research* **10,** 155.
Walker, P. M. B., and Yates, H. B. (1952). *Proc. Roy. Soc.* **B140,** 274.
Warburg, O., and Christian, W. (1936). *Biochem. Z.* **287,** 291.
Watson, J. D., and Crick, F. H. C. (1953). *Nature* **171,** 964.
Wiercinski, F. J. (1955). "Protoplasmatologia" (L. V. Heilbrunn and F. Weber, eds.), Vol. II, Part B, 2, C.
Wilkins, M. H. F. (1950). *Discussions Faraday Soc.* **9,** 363.
Wilkins, M. H. F. (1953). *J. Roy. Microscop. Soc.* **73,** 77.
Wyckoff, H. (1952). *Lab. Invest.* **1,** 115.

QUANTITATIVE DETERMINATION OF DNA IN CELLS BY FEULGEN MICROSPECTROPHOTOMETRY

CECILIE LEUCHTENBERGER

Institute of Pathology, Western Reserve University, Cleveland, Ohio

I. INTRODUCTION

Ever since the microscope was discovered and staining methods developed which allowed the studies of tissues and cells, questions as to the chemical makeup of cells and their changes under different physiological conditions have been raised. The questions have not only been asked by those who used the microscope to study morphology of tissue and cell structures, such as the histologist, cytologist, geneticist, and pathologist, but physiologists and biochemists have been equally concerned with these problems.

The interest was especially stimulated by the brilliant work of Miescher (1897). In 1871, Miescher not only isolated cell nuclei for the first time, but he also showed that the nucleoproteins are most important chemical components of all the cells of all living organisms. Thus Miescher started the field of chemical cytology and his discovery led to the early recognition and chemical characterization of the two types of nucleic acids, deoxyribosenucleic acid (DNA) and ribosenucleic acid (RNA) (Hammarsten, 1924; Kossel, 1891; Levene and Bass, 1931). Since the work of the early investigators, considerable progress has been made in our understanding of the chemical nature and biological importance of the nucleic acids for cell metabolism. The continuous active interest in the nucleic acids and the enormous strides made in this field in the last two decades are well exemplified by the wealth of material which Chargaff and Davidson have covered in their two volumes dealing with the chemistry and biology of the nucleic acids (1955).

Our present day concept that RNA in cells is metabolically highly active and probably plays a role in protein synthesis, while DNA is metabolically stable and possibly plays a role in genetic processes is based on the integrated knowledge derived from nearly all branches in the fields of chemistry and biology. The recognition of the great biological importance of DNA as a constant chromosomal and nuclear component and its probable participation in genetic mechanisms (for review, see Hotchkiss, 1955) has stimulated and led to the development of many methods for the study of DNA. One of the recent cytochemical approaches which has been especially useful in the study of the *quantitative* aspects of DNA in the nuclei of cells is Feulgen microspectrophotometry, a method which will be discussed in this chapter.

II. THEORY OF THE METHOD

The development of Feulgen microspectrophotometry for the quantitative determination of DNA in cells is one of the classic examples of the successful integration of knowledge derived from cytology, genetics, chemistry, and physics. This cytochemical technique attempts to determine the DNA content in nuclei of single cells *in situ* directly under the microscope and thus permits a direct correlation between the morphological appearance and chemical composition of the same cell structure. The technique which was first described by Pollister and Ris in 1947 is based on the combination of two major advances, one made by Feulgen and Rossenbeck over 30 years ago (1924) and one made by Caspersson, 20 years ago (1936).

In 1924, Feulgen and Rossenbeck reported the microscopic chemical identification of DNA in nuclei by means of the "nucleal reaction," that is a red-violet staining of DNA with fuchsin-sulfurous acid after the purines had been split off by mild acid hydrolysis. This specific staining reaction for DNA, now generally known as the Feulgen reaction and which can be carried out successfully on microscopic sections, has become one of the most important criteria in providing information on the precise localization of DNA in chromatin and chromosomes of cells.

The contribution of Caspersson which serves as a basis for determining the quantity of DNA in cells by Feulgen microspectrophotometry is of special significance because it has opened completely new pathways for the quantitative studies of intracellular components. In his brilliant thesis "Über den chemischen Aufbau der Strukturen des Zellkernes" (1936), Caspersson presented an ingeniously new type of microscopy, closely resembling a photometric chemical analysis of solutions but with the added advantage that the chemical analysis can be made in a single cell or even a cell part such as the nucleus, nucleolus, or cytoplasm. By combining the microscope with a photometric device and measuring the natural absorption of ultraviolet light by intracellular substances at specific wavelengths, such as the absorption of the nucleic acids at 2570 Å, Caspersson determined the quantity of nucleic acids in the cell structure. He thus established the basic principle that light absorption by intracellular substances can be measured in a single cell in microscopic sections directly under the microscope, and that the light absorption can be used as a basis for computing the quantity of the substance contained in the cell.

In Feulgen microspectrophotometry the DNA determinations in single cells are also a result of light absorption measurements in cells under the microscope. However, in this technique the DNA is first stained by the Feulgen reaction. Then the amount of light absorbed by the Feulgen-stained cell structure is determined and from this light absorption, the relative amount of DNA in the cell is computed. Therefore we are dealing with two important steps in this technique, first, the *qualitative* identification of DNA by the Feulgen reaction and second, the *quantitative* measurement of DNA by Feulgen microspectrophotometry. In the interest of clarification, the theories of the two procedures are discussed independently.

A. Theory of the Chemical and Qualitative Aspects of the Feulgen Reaction for DNA

According to Feulgen and Rossenbeck (1924), the basis for the Feulgen reaction is that mild acid hydrolysis breaks the purine sugar glycoside linkage in the DNA molecule, thus liberating aldehydes. These aldehydes react with colorless fuchsin-sulfurous acid (Schiff reagent) (1866) to give a magenta-colored compound. The same reaction can be produced in histological sections where, after mild acid hydrolysis, the aldehydes can be demonstrated *in situ* as components of the deoxyribonucleoproteins in the nucleus (Feulgen and Voit, 1924a).

The mechanism of aldehyde reactions with Schiff reagent was thoroughly investigated by Wieland and Scheuing (1921). According to their theory the Schiff reagent itself is formed in two successive steps:

(1) The principle component of basic fuchsin, triaminotriphenylmethane chloride is converted to the colorless leucosulfonic acid by adding sulfurous acid in the 1:6 position to the quinoid nucleus of the dye.

(2) Addition of sulfur dioxide to the leucosulfonic acid results in the production of the N-sulfinic acid of *p*-fuchsinleucosulfonic acid (Schiff reagent).

Each mole of Schiff reagent will condense with 2 moles of the aldehyde released after acid hydrolysis has split off the purines from the DNA molecule. This combination leads to the formation of a magenta-colored compound which therefore indicates the presence of DNA. This magenta-colored compound is a different dye from the

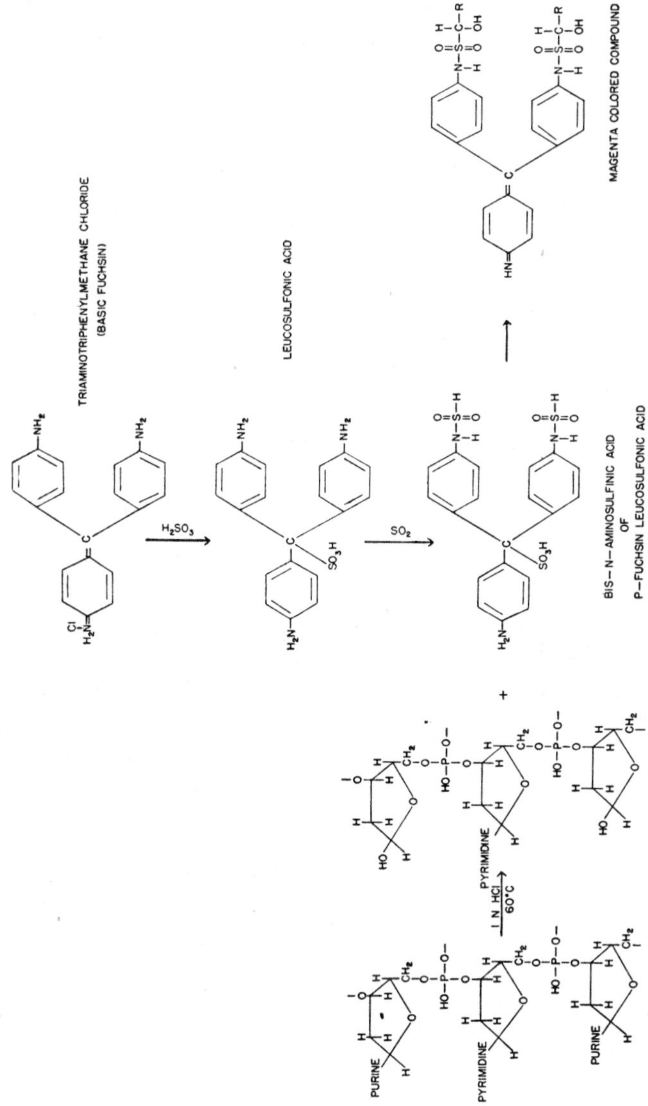

FIG. 1. The chemistry of the Feulgen reaction for DNA.

original basic fuchsin (Wieland and Scheuing, 1921; Feulgen and Rossenbeck, 1924). The formulae shown in Fig. 1 give the essential chemistry of the above reactions.

Although a considerable amount of work has been done on the chemistry of the Feulgen reaction (for detailed list of references, see Milovidov, 1954, and a recent review by Lessler, 1953) the exact chemical processes involved in the Feulgen reaction are by no means completely understood. Nevertheless, most of the recent work concerned with the chemistry of the Feulgen reaction appears to be in agreement with the theory originally postulated by Feulgen and his school (1924) and by Wieland and Scheuing (1921).

Recent investigations dealing with the mechanism of the Feulgen reaction in solutions, isolated cells, and tissues (Baker, 1942; Stowell and Albers, 1943; Stowell, 1945; DiStefano, 1948; Overend and Stacey, 1949; Delamater *et al.*, 1950; Lessler, 1951, 1953) seem to support at least three essential points:

(1) After mild acid hydrolysis, breaking of the linkages attached glycosidically to purine bases leads to liberation of aldehyde groups on the deoxy sugar.

(2) These aldehydes then react with the Schiff reagent in the manner suggested by Wieland and Scheuing (1921) giving an insoluble magenta-colored compound.

(3) The Feulgen reaction does locate the precise site of DNA.

That the purines are split off from the DNA in cells by mild acid hydrolysis was first demonstrated by DiStefano (1948, 1952). In nuclei of the frog, he found that the ultraviolet absorption of the purines and pyrimidines decreased in a straight line curve with varying times of hydrolysis (1 *N* HCl at 60° C., from 0–24 minutes) and that the Feulgen dye–DNA complex reached its maximum at 12 minutes hydrolysis. After this period, the Feulgen dye–DNA complex decreased rapidly (Fig. 2).

Support for the occurrence of aldehydes was brought forth by Lessler (1951) who showed that aldehyde coupling reagents applied after acid hydrolysis of DNA blocked the Feulgen reaction completely. That the magenta-colored compound (Feulgen dye) represents a different type from the original basic fuchsin was conclusively demonstrated by Stowell and Albers (1943). These workers found that absorption curves of basic fuchsin gave a maximal absorption of 560–570 mμ and a secondary absorption at 400 mμ, while Schiff reagent treated with formaldehyde or Feulgen-stained thymus had a maximal absorption only at 550 mμ. Moses (1951) who made absorption studies on Feulgen-stained hydrolyzed solutions of DNA and on Feulgen-stained hydrolyzed plant cells found for both also a peak of 550 mμ, but reported, in addition, a second maximum between 575 and 580 mμ. Recently, Kasten (1956b) in a thorough systematic study of Feulgen-DNA absorption curves showed that the absorption peak was 570 mμ for animal nuclei. The major peak and the type of absorption curve remained unchanged regardless of the fixatives used or whether the measurements were made in nuclei from paraffin embedded sections or in isolated nuclei (Figs. 3 and 4).

It is obvious that before the Feulgen reaction could be accepted as

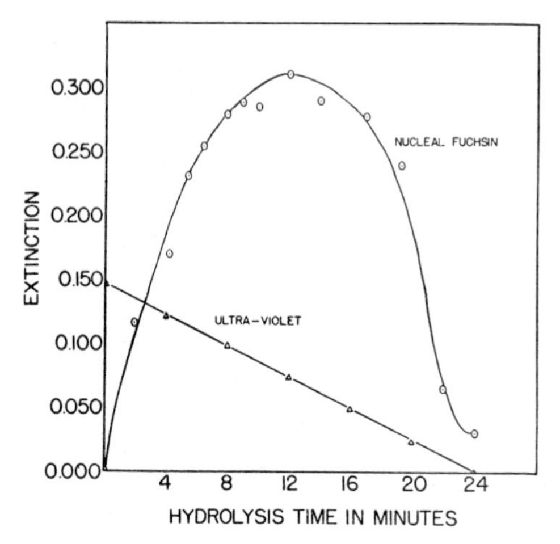

FIG. 2. Absorption measurements of cartilage cell nuclei from the head region of frog tadpoles at various times of hydrolysis (from DiStefano, 1948).

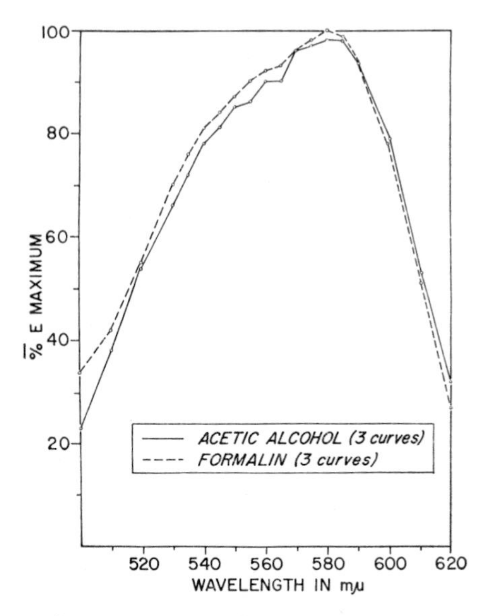

FIG. 3. Feulgen absorption curves from paraffin sections of mouse liver fixed in 50% formalin and in acetic alcohol (1:3) (courtesy of F. H. Kasten, 1956b).

an analytical tool for the determination of DNA in cells its specificity for DNA had to be investigated. This question was all the more pertinent since the Feulgen reaction does not visualize the whole DNA molecule, but the aldehydes liberated from the deoxyribose sugars after hydrolysis of DNA. Consequently the possible interference of other aldehydes

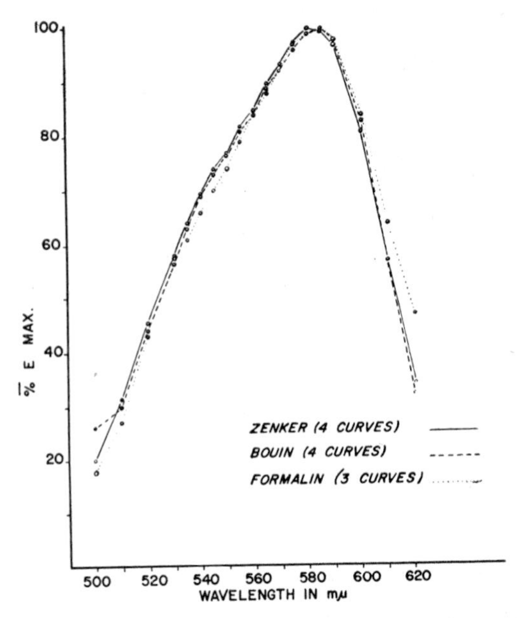

FIG. 4. Average Feulgen absorption curves are shown for nuclei that had been isolated in sucrose, fixed in different solutions, placed on the same slide and stained together (courtesy of F. H. Kasten, 1956b).

present in tissues had to be considered. In an extensive series of experiments, Bauer (1933a, b), Feulgen (1932), and Feulgen's school (Feulgen and Rossenbeck, 1924; Feulgen and Voit, 1924b) examined this question critically. They brought forth evidence to show that the very few aldehydes naturally occurring in tissues and which may also form the magenta-colored compound with the Schiff reagent, will not interfere with the detection of DNA. These substances combine with the Schiff reagent without hydrolysis, and can be readily differentiated from DNA which will react with the Schiff reagent only after acid hydrolysis. Furthermore, aldehydes formed from "plasmalogen" can be readily extracted with alcohol while DNA is not affected by the same alcoholic extraction. A critical study of the various aldehydes occurring in tissues and their cytochemical characterization is found in Danielli's recent book (1953).

Feulgen and Rossenbeck (1924) also noted that in contrast to DNA, RNA was Feulgen-negative. Whatever the explanation for this phenomenon may be, the fact that RNA is Feulgen-negative, serves as an excellent criterion for differentiating DNA from RNA by the Feulgen reaction (see also Carr, 1945; Ely and Ross, 1949).

Further support for the claim of specificity of the Feulgen reaction for DNA in tissues comes from experiments in which the DNA has been removed. Either digestion of DNA by a specific deoxyribonuclease (McCarty, 1946; Brachet, 1947) or extraction of DNA by hot trichloroacetic acid (Schneider, 1945) will abolish the Feulgen reaction. On the other hand, digestion with other enzymes such as RNase (Kaufmann *et al.*, 1951) and proteases (Catcheside and Holmes, 1947; Gomori, 1952) will not produce a change in the Feulgen reaction.

On the basis of these results and in view of the extensive Feulgen data obtained in recent years on DNA in a great variety of biological materials (viruses, yeasts, bacteria, plant, and animal cells), it is justified to state that the Feulgen reaction when carried out under standardized and controlled conditions can be used as a specific method for DNA. Although the precise chemical steps involved in the Feulgen reaction are not completely known, its specificity for the determination of DNA has been widely accepted (for recent reviews see Lessler, 1951; Moses, 1952; Swift, 1953, 1955; Davidson, 1953; Leuchtenberger, 1954a; Pollister and Ornstein, 1955), and thus a basis has been provided for considering the quantitative estimation of DNA by the Feulgen reaction.

B. Theory of the Feulgen Reaction for Quantitative Determination of DNA (Feulgen Microspectrophotometry)

The basic principle underlying Feulgen microspectrophotometry is actually a very simple one since it is an application of a typical photometric chemical method. It involves the quantitative analysis of a colored compound (Feulgen dye–DNA complex) by measuring the amount of monochromatic light absorbed by this compound.

The use of such optical methods for determining the quantity of colored substances in *solutions* is widespread and extensively used in biological work (Brode, 1949; Gibb, 1942; Miller, 1939). In practice the intensity, I, of monochromatic light which has passed through a colored solution is compared with the intensity I_0 of the same monochromatic light which has passed through a colorless solution (blank) of similar refractive index in the same cuvette. The blank is absolutely necessary since it allows correction for possible unspecific light loss caused by scatter and reflection as contrasted with the light loss due to the specifically absorbing colored substance. The ratio

$$\frac{I}{I_0} = \text{Transmittance } T$$

and $I/I_0 \times 100$ represents the percentage of light transmitted by the colored solution. For computations, it is preferable to have the extinction or optical density E which varies directly with concentration and thickness of the absorbing layer. This relationship is expressed in the well known Beer-Lambert Law:

$$E = K \cdot c \cdot d$$

where $E = \log_{10} I_0/I = \log_{10} I/T =$ extinction $=$ optical density $=$ absorbance; $K =$ specific constant $=$ specific extinction coefficient; $c =$ concentration of absorbing substance; and $d =$ thickness of absorbing layer.

It is evident that the application of the photometric analysis of the Feulgen dye–DNA complex is valid for the quantitative determination of DNA only if two criteria are fulfilled:

(1) The relation between DNA and Feulgen dye–DNA complex must be reproducible and stoichiometric or at least proportional, that is, increasing amounts of DNA must always combine with correspondingly increasing quantities of Feulgen dye.

(2) The Feulgen dye–DNA complex must follow Beer-Lambert's law.

The usefulness of the Feulgen reaction for the quantitative determination of DNA in solutions was first reported by Widström (1928) and Caspersson (1932). Using known concentrations of DNA, Caspersson found a linear relationship between DNA and the magenta-colored compound (Feulgen dye–DNA complex) in hydrolyzed solutions of DNA in concentrations from 0.1 mg. to 10 mg. DNA per cc. buffer solution. However he obtained unfavorable results if the DNA solutions contained proteins.

Similarly Sibatani (1950) and Naora *et al.* (1951) found that histones would interfere with the quantitative determinations of DNA in solution. Lessler (1951) who used the Feulgen reaction to determine DNA quantitatively in 20% gelatin preparations, found a linear increase of the Feulgen dye–DNA complex for concentrations of DNA from 0.2 mg. to 1 mg. per ml. gelatin solution. Thus it can be seen that, in solutions, the Feulgen dye–DNA complex usually follows Beer-Lambert's law and that photometric analysis of the Feulgen reaction can be used for the determination of DNA in solutions in a certain range of concentration.

However, that this is true for solutions does not necessarily make it applicable in the same sense to the Feulgen dye–DNA complex in cellular structures. It is obvious that a substance in a cell must be in a physical state completely different from what it would be in dilute solution. Furthermore, the distribution of the absorbing substances, their size and shape, their relatively high concentrations, and their relation to other chemical components in the cell are just a few of the pertinent factors which have to be taken into account when attempting photometry of chemical components in cells. A study of these factors is especially necessary since they present potential sources of error in Feulgen microspectrophotometry. They will be discussed in Section V (Critique of Method) of this chapter.

Caspersson who first introduced photometric analysis for the study of chemical components in cells, although he measured the natural absorption of chemical components in unstained cell structures in the ultraviolet range, was fully aware of these problems and has discussed them in great detail (1936, 1950). An excellent critical treatise of the problems which face the scientist who attempts to apply the photometric chemical analysis to stained cell structures has been published recently by Pollister and Ornstein (1955). They not only consider in detail all the factors relating to the fact that the photometric analysis is done in a cell structure and not in a solution, but they also discuss at great length the physcial problems arising from the necessity to introduce a microscope for microscopic photometric analysis.

It is evident that in order to provide a sound basis for measuring light absorption of stained substances in cellular structures, knowledge of the functioning of the absorption laws for these cellular substances is urgently needed. This is by no means a simple matter because, as Pollister and Ornstein (1955) state "There are no data on these intracellular physicochemical systems from which this can be predicted and hence there must be a large element of empiricism in the development of cytophotometry."

This lack of theoretical data is of course discouraging and has led some workers such as Glick (1953) to criticize severely microspectrophotometry of stained intracellular substances. On the other hand, the *empirical* investigation of such pertinent questions as to whether Beer-Lambert's law holds for the Feulgen dye–DNA complex in cells must certainly be recognized as a definite step forward toward the goal of obtaining a reliable foundation for light absorption measurements of colored substances in cells.

One of the empirical approaches which has been used for testing the validity of DNA determinations in cellular structures *in situ* by Feulgen microspectrophotometry is based on the comparison of the results obtained by this method with results obtained on the same material independently by other standard analytical methods such as biochemical analysis (Ris and Mirsky, 1949; Leuchtenberger *et al.*, 1951), chromosomal counts (Schrader and Leuchtenberger, 1950; Leuchtenberger, 1954a, 1954b), and ultraviolet microspectrophotometrical analysis (Leuchtenberger, 1954a). Since these data will be discussed in greater detail in Sections V and VI of this chapter, suffice it to say here that these comparative studies have shown a surprisingly good agreement between the Feulgen-DNA data and the other methods.

Another attempt to find out how the light absorption laws operate in cellular structures comes from studies in which Lambert's law was tested directly on Feulgen-stained nuclei. Measuring the light absorption of the same nuclei but in sections cut at different thicknesses, Swift (1950a) and Hoover and Thomas (1951) found a linear relationship between thickness and optical density indicating that $E = K \cdot d$ (Lambert's law) held in the Feulgen-stained nuclei on which the test was made.

Unfortunately, Beer's law ($E = K \cdot c$) cannot be tested directly in this manner because nuclei with known different concentrations of Feulgen dye–DNA complex would be needed and such experimentally prepared nuclei are not available at the present time. However, sufficient indirect evidence has been accumulated to make the assumption highly probable that the Feulgen dye–DNA complex does not show in practice major deviations from Beer's law in nuclei as one may perhaps surmise on theoretical grounds (Glick, 1953; Pollister and Ornstein, 1955). One of the most convincing illustrations for this concept is the finding that nuclei of greatly varying sizes but containing the same amount of DNA (as verified by biochemical analysis and/or chromosomal status) reveal on photometric analysis, the same amount of Feulgen dye–DNA complex. In other words there is a good relationship between extinction E and computed concentration c of the Feulgen dye–DNA complex in these nuclei [see Section V of this chapter, (Fig. 14)].

III. DETAILS OF INSTRUMENTATION

A. Description of the Instrument

Figure 5 is a photograph of one of the four instruments used in the Cytochemistry Laboratory at Western Reserve University for measuring the Feulgen dye–DNA complex in individual nuclei. With the exception of minor modifications these apparatuses are essentially the same as the ones described by Pollister and Moses (1949) and Pollister (1952).

Since most of the Feulgen microspectrophotometric work published so far has been done with such or similar types of instruments (Grundmann and Marquardt, 1953; Frazer and Davidson, 1953; Firket *et al.*, 1955; Pasteels and Lison, 1950; Vialli and Perugini, 1954) the description will be confined to the models used in this laboratory. Figure 6 is a diagram of the assembled parts of the apparatus. The major parts of the instruments are a light source, filters, or monochromator, a microscope and a photometric device for the light absorption measurements.

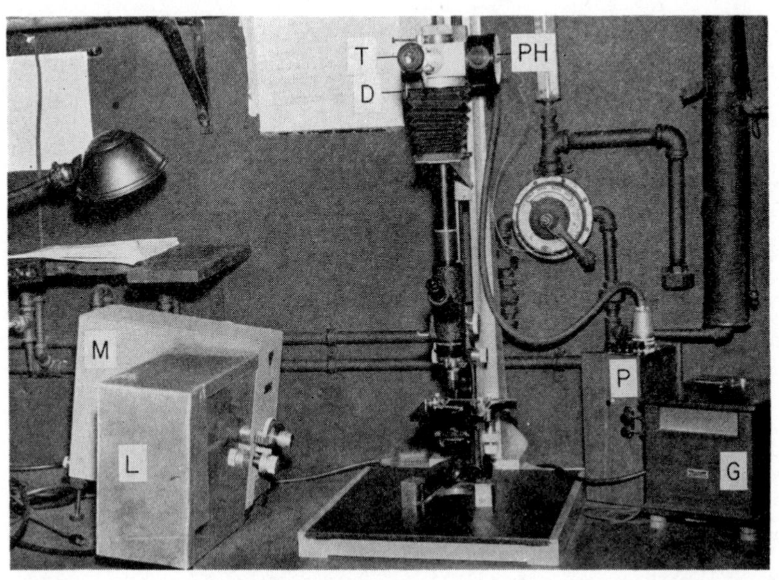

FIG. 5. Photograph of microspectrophotometric apparatus in combination with monochromator as used in the Cytochemistry Laboratory at the Institute of Pathology, Western Reserve University. (D) diaphragm lever, (G) galvanometer, (L) light source, (M) monochromator, (P) power supply, (PH) phototube housing, (T) telescope.

1. *The Light Source*

The light source consists of a 100 watt, 130 volt, AH_4 mercury vapor lamp (General Electric Co.). Using appropriate filters (see below) this line source permits good isolation of the mercury green line at 546 mμ used for Feulgen microspectrophotometry. Selection of this wavelength which is approximately 20% below that of the maximum absorption peak of the Feulgen dye–DNA complex is of advantage since lower extinctions reduce the error due to inhomogeneous distribution of the absorbing material in nuclei.

The lamp is mounted in a housing, or on an optical bench which contains a condensing lens and an iris diaphragm. The distance between lamp, lens and iris diaphragm can be changed so that Köhler illumination used for the measurements can be selected.

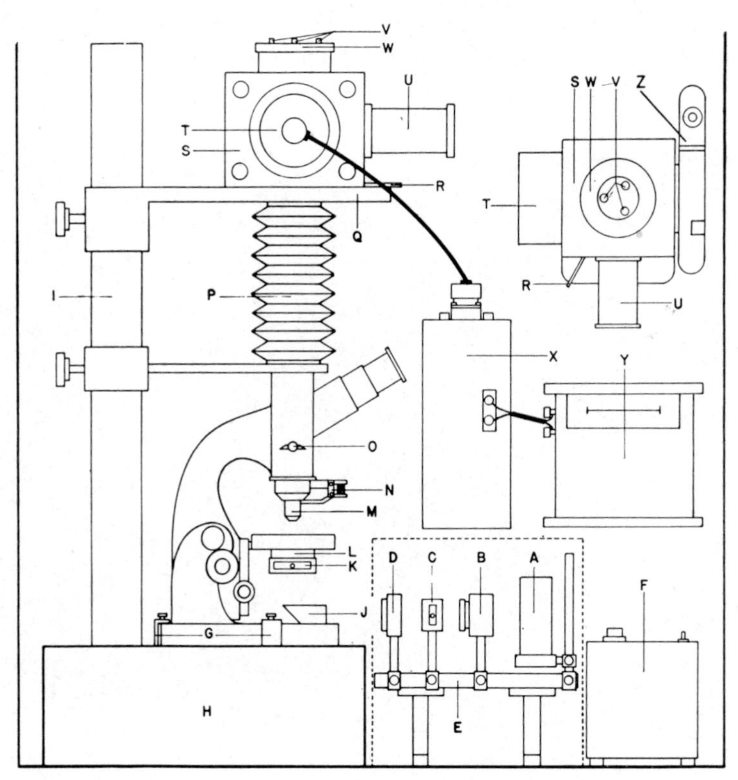

FIG. 6. Diagrammatic drawing of microspectrophotometric apparatus. (A) light source, (B) condenser lens, (C) iris diaphragm, (D) filter holder, (E) optical bench, (F) light source power supply, (G) base clamp for microscope, (H) base, (I) pillar (J) prism, (K) substage diaphragm, (L) substage condenser, (M) objective lens, (N) objective centering clutch, (O) prism, (P) bellows, (Q) photometer head baseplate, (R) diaphragm adjust, (S) head block, (T) phototube assembly, (U) telescope, (V) vertical mirror adjust and mirror knob lock screws, (W) mirror knob, (X) photometer power supply, (Y) galvanometer, (Z) photographic camera.

2. *Filters, Monochromator*

To isolate the desired wavelength, either a combination of two filters (Wratten 58 and 77A) or a grating monochromator (Bausch and Lomb) are placed in front of the light source.

3. *The Microscope*

The microscope is a research instrument (Leitz) with a side observation tube. It is tightly fixed to the base of a steel support to insure immobility. In order to avoid disturbing light reflection, the usual microscope mirror is replaced by a right angle prism or a first surface mirror. Above a substage diaphragm a condenser of a numerical aperture of 1.3 is mounted in such a way that precise centering is easily possible. The stage on which the microscopic slide is placed is equipped with a "research mechanical stage," double plate type, which permits fine movement in horizontal and vertical directions for centering the object. The objective is a 2 mm. apochromatic oil immersion with a numerical aperture (NA) of 1.3 and is mounted in a centrable objective clutch. Above this objective there is a prism in the microscope tube which permits the deflection of the light beam for side tube observation. This prism can be easily shifted out of the field so that the light beam continues vertically. The tube length of the microscope is 170 mm. and a compensating type of ocular is used to minimize residual lateral chromatic and spherical aberrations of the objective (Pollister and Ornstein, 1955).

4. *The Photometric Device*

The photometric device consists of the photometer head, the power supply, and a galvanometer. The photometer head is mounted on a steel pillar and is connected in a fixed position to the microscope by a bellows or a sleeve to form a light trap. The photometer head (Fig. 7) consists of a supporting base block to which a diaphragm is attached. Firmly fixed to the supporting base plate is a block with a central opening in which a first surface mirror mounted on a support is fitted. The mirror support consists of a bearing mount and a knurled knob which allows a 90° rotation of the mirror toward the telescope or the phototube. The mirror can be adjusted in the vertical direction by a cam which can be locked by a set screw. The two screws act as stops for positioning the mirror after it is aligned in respect to the telescope and phototube. The telescope is threaded into the front of the block exactly in a 90° angle to the phototube. The telescope, which is focused on the diaphragm, contains a calibrated micrometer disc which is adjustable through a 90° slot. The phototube (type 931A or IP21) is carried in a light-tight housing mounted on the block by four screws exactly 90° in line with the telescope. To measure the light intensity, the photomultiplier tube is run from a battery power supply (Farrand) and used with a Rubicon galvanometer.

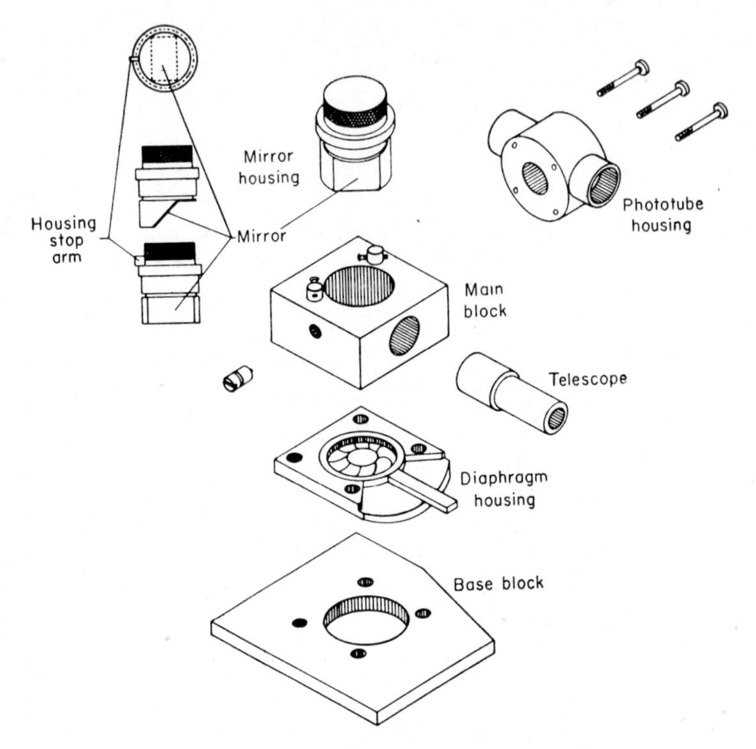

FIG. 7. Exploded view of photometer head assembly.

B. ALIGNMENT AND STANDARDIZATION OF THE INSTRUMENT

Before discussing the alignment of the microspectrophotometric apparatus, a few words should be said about the housing of the instrument. The instrument should be placed in a special light-tight room which is as dust free as possible, preferably air conditioned and sound proof. These precautions are necessary because even minute particles of dust and dirt on lenses or on slides produce light scatter leading to errors in photometric measurements. The room should be in a quiet location of the building since any kind of vibration (centrifuges, pumps, etc.) interferes with proper focusing and lead to instability of the galvanometer. These arrangements are also of advantage to the worker who does the microspectrophotometry. It is a rather tiresome and strenuous task to do meticulous cytological microscopic and photometric work in a light-tight room for many hours and therefore everything should be made as comfortable as possible. This includes a convenient placing of the instrument so that the worker can use it in a sitting position. Fatigue contributes definitely to errors in proper focusing, determination of nuclear size, and galvanometer readings. There should be ample working space available so that data can be easily recorded. If several instruments are used simultaneously in the same room, they should be sufficiently separated from each other.

The proper alignment of the microspectrophotometric apparatus is a specific requirement which must be fulfilled before microspectrophotometry can be carried out. This includes alignment of the light source, of the microscope, and of the photometric device, and also their position to each other. The microscope should be placed under the photometric device in such a way that the margin of the photometer diaphragm opening is concentric with the microscopic field.

1. *Alignment of the Lamp and the Microscope*

To produce an effective homogeneous illumination of the image, either Köhler (1893, 1931) or "critical" illumination (Nelson, 1910) must be established. A detailed description of the procedures is found in Shillaber's book (1944) and in an article by Dempster (1944). Figure 8 shows the path of the light rays if Köhler illumination is used between light source and microscope.

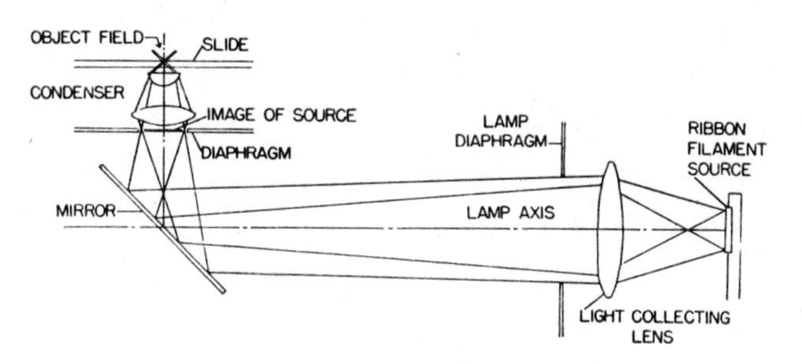

FIG. 8. Diagram of Köhler illumination.

A short practical outline for installing Köhler illumination is: A cardboard with a drawn target is placed about 18 inches in front of the lamp, the lamp diaphragm is cut down to its smallest aperture and the lens or lamp adjusted until the image of the lamp filament is in sharp focus on the target screen and all color fringes and shadows are symmetrical. After adjusting the mirror or prism of the microscope, the image of the properly aligned lamp is focused sharply and symmetrically on the surface of the condenser diaphragm so that it fills the diaphragm aperture which is used during the microspectrophotometric procedure. When the lamp is aligned, the condenser, objective, and ocular of the microscope should be brought into good alignment as described by Shillaber (1944).

The optical alignment can be done in a relatively short time with the help of a telescope used to align phase microscopes and by adding and centering the optical elements consecutively from the source out (Pollister and Ornstein, 1955).

In order to diminish flare (glare and stray light) in the microscope which may result in errors of large magnitude in the measured transmission value of a stained nucleus (Naora, 1951), the size of the illuminated area should be reduced as much as possible. Stopping down the lamp diaphragm to 1 mm. and closing the substage condenser diaphragm to 4 mm. so that the NA of the condenser is reduced to approximately 0.2, will eliminate a considerable portion of the flare (Uber, 1939; Grey, 1952). If a monochromator (instead of filters) is employed between the light source and the microscope, the vertical slits should be closed down also so that the illuminated area is only a few diameters larger than the nucleus to be measured. The alignment of commercially available monochromators when used for microspectrophotometry of cells is discussed by Pollister and Ornstein (1955) and by Swift and Rasch (1956).

2. *Alignment of the Photometric Device*

Before attempting any microspectrophotometric measurements of cells, the performance of the photometric device should be tested. The photosensitive surface of the phototube should show a uniform response to intensity and area of illumination. Consequently two requirements must be met: (1) the current output (galvanometer readings) should be directly proportional to the intensity of incident light; (2) the current output should show a linear response to the size of the area illuminated.

(1) To test the linear response of the phototube to light intensity, a Feulgen-stained slide is put under the properly illuminated and aligned microscope. A blank area on this slide is focused and centered in the microscopic field. A reticle with a cross-hair mounted in the sidetube ocular is convenient for centering the area. The prism is flipped out of the field and by turning the mirror in the photometer head, the enlarged image of the blank area can be observed through the telescope. The enlargement is usually adjusted in such a way that 1 unit of the micrometer disc mounted in the telescope corresponds to 1 mm. This calibration can be easily done with a stage micrometer slide. By means of the photometer head diaphragm and the calibrated disc, an area with a diameter of 2 or 3 mm. is delimited and the mirror is turned so that the image of the blank area falls directly on the cathode of the phototube. Then the phototube in the housing is adjusted by rotating it slightly until

the galvanometer shows the highest deflection; at this peak reading the phototube is locked into position. This is to insure that the photosensitive surface of the phototube is perpendicular to the light beam. By turning the sensitivity knobs on the power supply, the sensitivity is changed until the galvanometer shows its full deflection (10 microamperes). A series of neutral density filters with known transmissions are put in front of the lamp, their galvanometer readings recorded and the readings are checked with the known values. Comparison between the transmission data of neutral density filters obtained in a Beckmann spectrophotometer and in a microspectrophotometric apparatus show

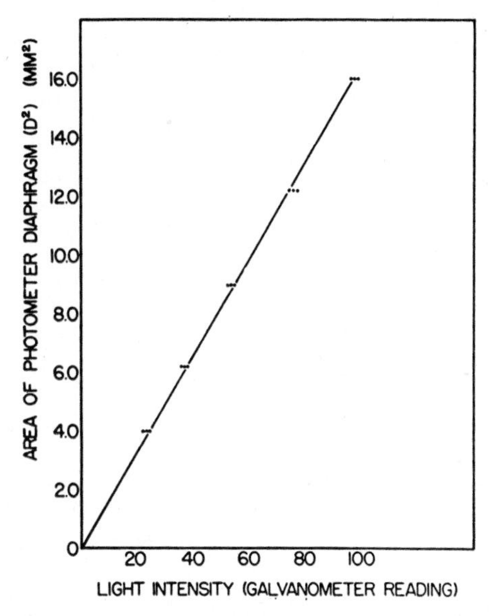

FIG. 9. Relationship between phototube response and area of image measured.

excellent agreement (Commoner, 1948; Kasten, 1956a; Ris and Mirsky, 1949) provided there is no light leakage and the power supply and phototube are in good condition.

(2) To test the uniformity of response of the phototube to size changes of the illuminated area of the image, galvanometer readings are taken at various openings of the photometer diaphragm. Since the maximum and uniform sensitivity of the 931A electron multiplier tubes is only in the exact center within a diameter of approximately 0–4 mm. (Marshall *et al.*, 1948), a linear response can be expected within a diameter of 4 mm. and is obtained as demonstrated in Fig. 9 where galvanometer readings are plotted as a function of the area of the image projected on the sensitivity cathode of the phototube. If areas with a

diameter larger than 4 mm. have to be measured, a focusing lens can be mounted in the photometer so that the back lens of the objective is imaged as a small spot in the center of the photosensitive area. This permits the measurement of the part of the microscopic image approaching the diameter of this lens (Pollister, 1952).

After the light source, microscope, and photometric device have all been aligned with respect to each other, the microspectrophotometric apparatus should be tested for the presence of flare light. To test for flare light in this system, the light transmittance should be measured in opaque objects with properties otherwise similar to those of the objects to be tested. Lison (1953) suggests measuring erythrocytes and lymphocytes heavily overstained with iron hematoxylin either in smear preparations or in paraffin sections. Smears of lampblack particles, 1–20 μ in diameter may also be used. If there is no glare in the microspectrophotometric apparatus, the transmittance value of opaque objects should obviously be zero, but under no circumstances should the extinction values be lower than 2.5 (Swift and Rasch, 1956). If these values are lower than 2.5, the apparatus should be checked again especially for light leaking into the apparatus, surfaces in the apparatus which might produce internal reflections, and the size of the numerical aperture of the condenser.

IV. PROCEDURE IN METHOD[1]

Since Feulgen microspectrophotometry attempts a quantitative analysis of the chemical constituent DNA in a biological material, all the steps involved in the procedure must be performed with the same rigid precautions and standardizations that are observed in any quantitative chemical analysis. The procedure actually consists of 3 main steps: (1) preparation of the biological material; (2) application of the Feulgen reaction on slides; (3) microspectrophotometric measurement of the amount of Feulgen dye contained in the individual nuclei.

A. Preparation of the Biological Material

Material from plants and animals can be prepared in various ways: (a) squash, touch, or smear preparations of tissues and fluids, (b) smears of isolated nuclei and cells, and (c) sections of embedded tissues. The choice among these methods depends on the type of material to be examined. For fluids such as blood, seminal, and ascitic fluids, smear preparations are advisable since, in the cells to be measured, the nuclei usually do not overlap and are whole, factors which greatly

[1] Composition and preparation of the fixatives and reagents concerned with the Feulgen reaction are given in Section VII.

facilitate the subsequent photometric measurements. These points also hold true for smears of nuclei isolated from tissues. A great advantage of doing Feulgen microspectrophotometry on smears made from suspensions of isolated nuclei is that a biochemical analysis of DNA can be made on the same suspension (Leuchtenberger *et al.*, 1952a; Leuchtenberger *et al.*, 1952b; Leuchtenberger *et al.*, 1951; Ris and Mirsky, 1949). Reference standards gained from biochemical analyses on the same material which is used for photometric measurements are quite important because the two types of analyses serve as a check on each other. Such comparisons have been made in this laboratory whenever possible, especially when unfamiliar material was to be examined by Feulgen microspectrophotometry. The advantages and limitations of biochemical analyses for DNA when compared with Feulgen microspectrophotometric analysis will be discussed in Section V of this chapter. An efficient method for isolating nuclei from tissues for DNA determinations has been given by Vendrely and Vendrely (1948).

Since some tissues contain different cell types which can be recognized only when the architecture of the whole tissue is preserved (such as the spermatogenic and oogenic cells in the testis and ovaries, respectively, or cells in different zones of the adrenal, pancreas, and skin), examination of tissue sections is of the utmost importance.

After removal, the tissues should be fixed as soon as possible. Methods which preserve tissues especially well are nonchemical fixation by the freeze-drying technique (Caspersson, 1950; Danielli, 1953), and rapid fixation in buffered osmium tetroxide according to Palade (Leuchtenberger *et al.*, 1955). Satisfactory fixation fluids most commonly used are 10% formalin, acetic-alcohol (Carnoy), and formalin-acetic-alcohol (Lavdowsky). Some investigators such as Swift (1950a, b) also recommend 50% formalin. However, the enormous shrinkage of tissues and cells after this fixative is a definite disadvantage for microspectrophotometry. In addition, aldehydes from this strong fixative which may remain in the tissues would interfere with the specificity of the Feulgen reaction.

The pieces used for fixation should not be more than 1–2 mm. in thickness to insure rapid fixation. Time of fixation should not usually exceed 5 hours because certain chemical constituents may be extracted and the tissue may get too brittle for satisfactory sectioning after the paraffin embedding. If, for some reason, dehydration cannot be carried out immediately following fixation, the tissues may be temporarily stored in 70% alcohol. It is advisable to fix a tissue with a known DNA content every time with the one to be investigated so that a standard of reference will be available. A schedule which has been found satisfactory in this laboratory for fixation, dehydration, and embedding is as follows:

From the fixative, the tissues are rinsed in several changes of 70% alcohol about 10–15 minutes each. They can be left in 70% alcohol overnight. From 70%, they are changed several times in 85% and 95% alcohol, about 15 minutes each; then 4 changes of absolute alcohol, 15 minutes each; equal parts of benzene and absolute alcohol, 15–30 minutes; 4 changes of benzene, 15 minutes each; 3 changes of paraffin (Tissuemat, 56–58° C.), 30–60 minutes each.

Fixing solutions and all alcohols are discarded after each use. The benzenes may be used a few times before discarding. The first paraffin is discarded after using it a few times; the subsequent changes of paraffin are moved up and a fresh jar of paraffin is prepared for the last change. This fresh jar of paraffin is filtered through lintless paper or cloth. Miracloth, a viscose product made by The Visking Corporation, 400 W. Madison St., Chicago 6, Ill., is very good for this purpose.

The sections are cut at various thicknesses depending on the size of the nuclei. The approximate average diameter of the nuclei can be determined in sections cut at 6–8 μ, either by camera lucida drawings or filar micrometers. This will permit the selection of a section thickness in which a considerable number of the nuclei will remain whole. Since thick sections may often lead to overlapping, it is advisable to cut at several thicknesses; this is also advantageous for tissues containing nuclei of different sizes such as polyploid liver. These sections should be mounted on the same slide together with a section from a standard tissue with a known DNA content. Adjacent slides are prepared for Feulgen controls and enzyme digestion.

B. Application of Feulgen Reaction on Slides

With minor modifications, the Feulgen reaction is essentially the same as that described by Stowell (1945). The reagents should all be of the highest chemical purity and the same brands should always be used. This is especially important to insure qualitative and quantitative reproducibility of the Feulgen reaction. The Feulgen reaction itself should be carried out according to a rigidly kept time schedule. This includes exposure to the Schiff reagent and bleaching solutions.

For each new tissue, different hydrolysis times should be run to determine at what period the maximum amount of Feulgen dye will be bound (Figs. 10 and 11). Once this time is determined, strict adherence to it is required. The temperature at which the hydrolysis is run should be kept at exactly 60° C. This implies of course that the hydrolysis solutions are preheated to the proper temperature before the slides are exposed to them.

Whenever possible, slides made for comparative DNA studies should

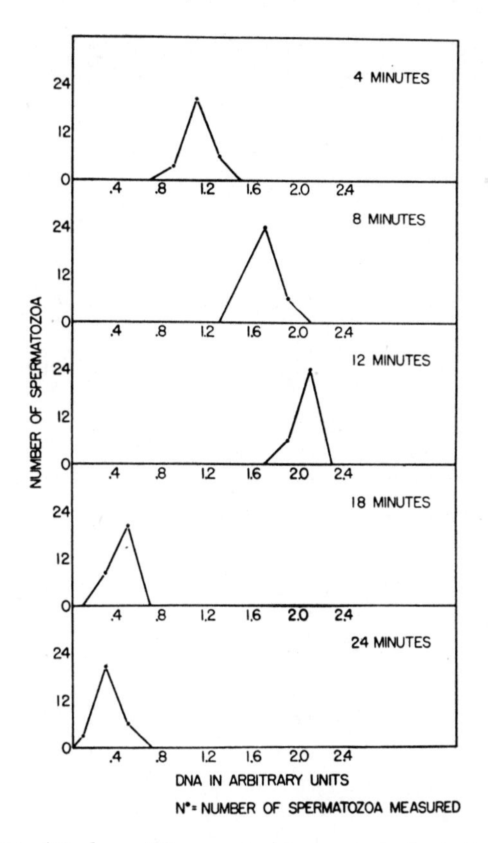

FIG. 10. DNA (Feulgen Microspectrophotometry) in individual bull sperm nuclei after various times of hydrolysis at 60°C. N° = 150.

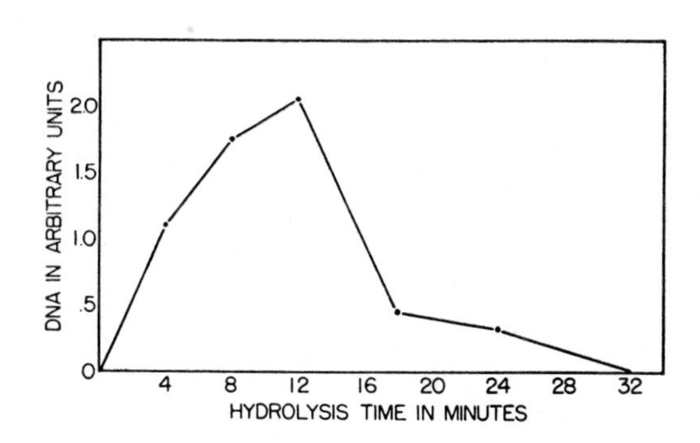

FIG. 11. Mean amount of DNA (Feulgen Microspectrophotometry) in 150 individual bull sperm nuclei after different times of hydrolysis.

all be placed in the same slide racks (glass) so that they will all be exposed to the same solutions simultaneously and for the same length of time. A Feulgen control (unhydrolyzed adjacent section) should be run simultaneously with each Feulgen slide.

The technique for the Feulgen reaction is as follows:

(1) Carry paraffin sections down to water through xylene, 95, 80, 70, 50, and 30% alcohols.

(2) Rinse in 1 *N* HCl at room temperature.

(3) Hydrolyze in 1 *N* HCl at 60° C. for the optimum time period previously determined for the particular tissue. In our experience, 12 minutes has been found to be the optimal time for most of the biological material examined here.

(4) Rinse carefully in distilled water.

(5) Place in Schiff's solution in a glass-covered dish for 2 hours (in the dark).

(6) Rinse rapidly but thoroughly in several changes of distilled water.

(7) Place in 3 changes of bleaching solution, 10 minutes each (to remove unspecifically absorbed dye).

(8) Wash in running tap water, 5 minutes.

(9) Rinse in distilled water.

(10) Dehydrate through 30, 50, 70, 80, 95%, and 3 absolute alcohols.

(11) Clear through 3 changes of xylene and mount in a medium having a refractive index close to that of the tissue. Clarite mounting medium is especially suitable for animal tissues fixed in Carnoy (acetic-alcohol) (Pollister and Ornstein, 1955).

If a counterstain is desired, after dehydrating up to 95% alcohol, stain 30–60 seconds in a highly dilute solution of Fast Green. Follow with 3 changes of absolute alcohol, clear, and mount. Do not use the same jars of absolute alcohol contaminated with Fast Green for other staining procedures.

In case there is any question whether Feulgen positive material in a section is DNA, enzyme experiments with deoxyribonuclease (DNase) should be run. Carnoy is the fixative of choice for digestion by DNase because, as far as is known at present, this enzyme is not effective after formalin-containing fixatives. Since citric acid will also inhibit the enzyme, all traces of it must be thoroughly washed out if smears of nuclei isolated in this reagent are used. Three consecutive slides should be prepared: (a) one for Feulgen, (b) one for Feulgen after the buffer, and (c) one for Feulgen after the enzyme. The technique is as follows:

(1) All 3 slides are brought to 70% alcohol where slide (a) is held until treatment of (b) and (c) is finished.

(2) From 70% alcohol, slides (b) and (c) are carried down to distilled water.

(3) Slide (b) is placed in the buffer solution which must be from the same lot of buffer which was used to dissolve the enzyme. Slide (c) is placed in the enzyme solution. Both slides are treated at the same temperature and for the same length of time. A suggested time and temperature is 1 hour at room temperature.

(4) After treatment, both slides are rinsed thoroughly in many changes of distilled water (especially the one from the enzyme solution), but to avoid contamination, they should be rinsed separately and slide (b) should be rinsed first.

(5) After rinsing, carry slide (a) down to water and place all 3 slides in the same rack so that they can now be subjected to the Feulgen procedure simultaneously.

C. Microspectrophotometric Measurements

The procedure which will be described in this section is relatively simple and can serve as an excellent foundation for the investigator who wants to attempt chemical analysis of cells by more elaborate and complicated microspectrophotometric techniques such as scanning techniques (Caspersson *et al.*, 1955; Davies and Walker, 1953; Pollister and Ornstein, 1955; Walker and Deeley, 1956), two-wavelength methods (Ornstein, 1952; Patau, 1952), and absorption measurements in the ultraviolet range (Caspersson, 1936, 1950, 1954).

The technique, which has been used in this laboratory for a number of years, involves photometric measurements of the Feulgen dye–DNA complex at one particular wavelength, 546 mμ, and through a central region of the nucleus. This "plug method" was first introduced by Swift (1950a) who showed empirically that measurements of light transmittance through a central cylindrical region give more accurate results than do measurements of light transmittance through a whole nucleus (DiStefano, 1948; Leuchtenberger, 1950; Pollister and Ris, 1947; Ris and Mirsky, 1949). Further discussion of this method is given below.

Before starting the microspectrophotometric measurements, the Feulgen-stained slides should be examined microscopically. It should be ascertained that the unhydrolyzed Feulgen controls and DNase digested slides are Feulgen-negative. Furthermore, it is of great importance to study thoroughly the histological and cytological features of the specimen in which the amount of DNA is to be determined. The various cell types in different zones of such tissues as the adrenal, pancreas, and reproductive organs, or in pathological and mitotic areas, should be specially designated either by a field finder or by preparing a photographic record so that the Feulgen measurements for DNA can be done in the same cell. Without these records, it is difficult to find and recognize the same cells at the high magnification used in microspectrophotometry. Particular attention should also be paid to the distribution of the Feulgen-positive material in the nuclei since strikingly inhomogeneous distribution may lead to serious errors in the measurements (see Section V of this chapter). Nuclei in which the Feulgen-positive material appears in large clumps located only at the nuclear membrane and where the central portion is practically empty, such as are frequently observed in tissues showing degenerative changes, or late prophases and diakinesis, are not suitable for Feulgen microspectrophotometry.

Attention should also be paid to the shape of the nuclei. Most desirable are the ones which resemble a sphere. Nuclei whose longer diameters are more than twice the size of the shorter diameter are still usable if special formulae are employed in the computation (see below). Spindle-shaped or odd shaped nuclei are not suitable for this type of Feulgen microspectrophotometry in which the measurements are made at one wavelength (546 mμ), but can be measured with "the two-wavelength method" (Ornstein, 1952; Patau, 1952).

A brief outline of the actual microspectrophotometric measurement of the Feulgen dye in a nucleus is as follows:

(1) Place a drop of high viscosity Shillaber immersion oil ($N_D^{25° C.} = 1.5150$) on the condenser lens and bring it into contact with the under surface of the slide carefully avoiding air bubbles.

(2) Put a drop of low viscosity Shillaber immersion oil (same refractive index as the high) on the coverslip and lower the oil immersion objective into it.

(3) Bring the image into the center of the field and into sharp focus.

(4) Select the nucleus to be measured.

(5) Ascertain that it is uncut by focusing on material below and above the entire nucleus and be sure that it is not overlapped.

(6) Flip out the prism so that the nucleus can be observed through the telescope in the photometer head.

(7) Rotate the mirror so that the image is projected toward the telescope and bring the enlarged nucleus into the center of the field and into sharp focus.

(8) Measure the major and minor axes of the nucleus by means of the ocular micrometer in the telescope.

(9) Place the nucleus in the exact center of the field diaphragm opening by a slight movement of the mechanical stage.

(10) Delimit a plug corresponding to approximately 75% of the

nucleus by gradually narrowing the opening of the field diaphragm and measure the diameter of the plug.

(11) Open the phototube shutter and rotate the mirror so that the nuclear image is projected onto the photosensitive surface of the aligned tube.

(12) Take the galvanometer reading of the transmission of light through the nuclear plug.

(13) Without changing the diaphragm opening, move the slide to a blank area as close as possible to the nucleus. Check this through the side tube of the microscope.

(14) Take the galvanometer reading of the transmission of light through this blank area.

(15) Close the phototube shutter and select the next nucleus.

1. *Additional Useful Instructions Concerned with the Actual Measurements*

(1) Turn on the lamp and photometric measuring device at least 15–20 minutes before starting to measure.

(2) Be sure that the apparatus, lenses and slides are clean and free of dust.

(3) Recheck the alignment of the apparatus (see Section III, B).

(4) Be sure the dark current of the galvanometer is stable and set at zero.

(5) Set the sensitivity of the galvanometer in such a way that the reading for a blank area of the same plug size selected for the nucleus is approximately 100. Under no circumstances, should this reading be above 10 microamperes.

(6) Make at least 2 or 3 measurements of the size and light transmission for each nucleus.

(7) In addition to these readings, keep exact records of the location of the measured nuclei and any special cytological features, especially distribution of absorbing material. It is very useful to have prepared mimeographed record sheets on which these raw data can be easily recorded with carbon copies.

(8) Measure at least 50 nuclei at random for each tissue with an unknown DNA content. If the statistical analysis (see below) of the DNA content shows a nonuniform population, then measure as many more nuclei as the statistical analysis requires. This will hold especially for tissues where there is mitosis, ploidy, growth (normal and abnormal), and pathological change.

(9) Always measure at least 20 nuclei in the standard material (known DNA content) mounted on the same slide.

2. *Computation of the Amount of DNA in a Nucleus*

On the basis of the raw data gathered as described above, namely, sizes of nuclear and plug diameters and transmittance through the nuclei and blanks, the amount of Feulgen dye–DNA complex in arbitrary units can be computed for each nucleus by the use of the following equation (Swift, 1950a).

$$A = \frac{E}{F} C^2$$

Where $A =$ amount of DNA in arbitrary units, $E =$ extinction, $C =$ radius of the central plug, and $F =$ fraction of the total nuclear volume occupied by the cylindrical region.

E can be easily computed from the transmittance data since $E = \log_{10} 1/T$ or it can be obtained from conversion tables such as are given in Brode (1949, p. 615).

In order to compute A, it must be kept in mind that measuring the transmittance through a central region of a spherical nucleus gives a value only for a cylinder cut through this nucleus (see Fig. 12). Consequently, the extinction E must be divided by this fraction F. The main steps in computing F are given in Fig. 12.

F = The fraction of the total nuclear volume occupied by the cylindrical region.

V = Volume of a central cylindrical region cut from the center of a sphere.

R = Radius of the sphere.

H = Half the height of the cylinder.

C = Radius of the central plug.

$$F = \frac{V}{4/3\,\pi R^3}$$

$$V = 4/3\,\pi (R^3 - H^3)$$

$$F = \frac{R^3 - H^3}{R^3}$$

$$H = \sqrt{R^2 - C^2}$$

$$F = \frac{R^3 - (R^2 - C^2)^{3/2}}{R^3}$$

or

$$F = 1 - \left(1 - \frac{C^2}{R^2}\right)^{3/2}$$

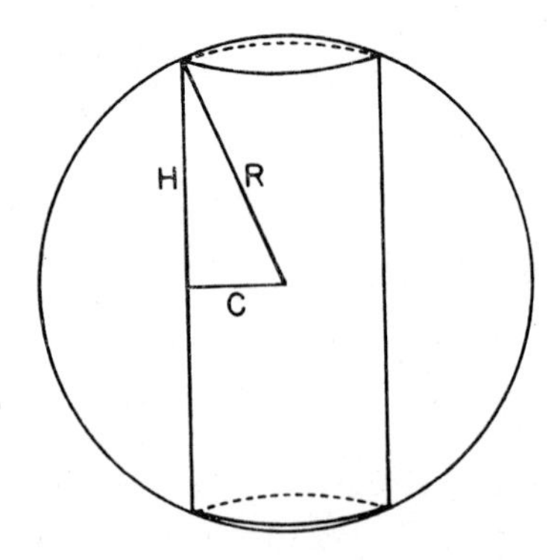

FIG. 12. Computation of the fraction "*F*" of the total nuclear volume occupied by a cylindrical region.

To avoid repetition in the computation for F, it is advisable to prepare tables from which F can be read directly. An example of this table is given in Section VII (Table II). The value of F as derived above applies to nuclei which are nearly spherical and where the nuclear volumes are treated as spheres. In such nuclei the major and minor diameters are averaged and one-half of this mean is taken as the radius of the nucleus. This introduces only a small error; but for nuclei which are prolate spheriod or ellipsoid with a major axis $2A$ and a minor axis $2R$, Kasten (1956a) gives the following formula for F:

$$F = \frac{R}{A}\left\{1 - \left[1 - \left(\frac{C}{R}\right)^2\right]^{3/2}\right\}$$

An example of raw data obtained on 60 individual nuclei from 3 human tissues and their calculations are given in Table I. These data were taken at random from our files on human material which comprise several thousand measurements. Although the number of nuclei measured for each tissue is much too small to permit any definite conclusions, still these sample data do indicate the trend of the results. For example, the spermatid values are approximately one-half the kidney values and in each of these tissues there is only a moderate variation from nucleus to nucleus. In the liver the values seem to fall into 3 classes: the lowest class is similar to that of the kidney; the next class seems to be an approximate double; and the highest class seems to be a quadruple. Here again, within each class, the variation from nucleus to nucleus is only moderate. It can also be seen that one of the liver values (1.6) is as low as that of the spermatids. Whether this low value may be due to an error in technique (cut nucleus) or may be inherent in the nucleus itself, cannot be answered with certainty. As pointed out before, many more nuclei should be measured for each one of the tissues but an especially large number is needed for the liver where the population does not seem to be homogeneous. Needless to say, the values should be submitted to statistical analysis (Dahlberg, 1948; Fisher, 1950) and tested for their reproducibility in similar specimens.

For the interpretation of the data, it is advisable to give not only the calculated mean value with its standard error (S.E.) per nucleus for each tissue, but the individual data obtained in a tissue should be plotted (histograms) so that the variation from nucleus to nucleus can be assessed.

3. Errors in the Microspectrophotometric Procedure

There are certain errors in Feulgen microspectrophotometry which are caused by factors inherent in the investigator, the biological material, and the apparatus.

The use of the human eye introduces certain physical errors which

TABLE I
SAMPLE RAW DATA ON NUCLEI (FEULGEN MICROSPECTROPHOTOMETRY) AND COMPUTATIONS OF THE AMOUNT OF DNA IN ARBITRARY UNITS

Type of Material	No. of Nucleus	Diameters in Arbitrary Units 1 Unit = 1.6 μ		I	I_0	$T\%$	E	F	Amount of DNA in Arbitrary Units $\frac{E\,C^2}{F}$
		Nucleus	Plug						
Human Kidney	1	3.2 × 3.0	2	29	99	29	0.538	0.558	2.4
Fixed in 10% Formalin	2	2.9 × 2.6	2	26	100	26	0.585	0.665	2.2
	3	3.0 × 2.8	2	29	99	29	0.538	0.620	2.2
Female Age 60	4	3.0 × 2.6	2	27	97	28	0.553	0.634	2.2
	5	3.0 × 2.9	2	26	97	27	0.569	0.603	2.4
	6	3.0 × 3.0	2	28	98	29	0.538	0.586	2.4
	7	3.0 × 2.9	2	25	98	26	0.585	0.603	2.5
	8	3.0 × 2.7	2	31	98	32	0.495	0.627	2.0
	9	3.0 × 3.0	2	31	98	32	0.495	0.586	2.2
	10	3.0 × 2.7	2	31	97	32	0.495	0.627	2.0
	11	3.7 × 3.4	3	38	99	38	0.420	0.848	2.9
	12	3.9 × 3.6	3	41	100	41	0.387	0.784	2.8
	13	3.8 × 3.5	3	39	98	40	0.398	0.816	2.8
	14	4.0 × 4.0	3	47	99	48	0.319	0.711	2.5
	15	3.7 × 3.5	3	34	97	35	0.456	0.831	3.2
	16	3.9 × 3.3	3	31	95	33	0.481	0.831	3.3
	17	3.5 × 3.5	3	30	99	30	0.528	0.865	3.5
	18	3.7 × 3.5	3	40	100	40	0.398	0.831	2.8
	19	4.0 × 4.0	3	40	98	41	0.387	0.711	3.1
	20	4.0 × 3.7	3	46	99	46	0.337	0.752	2.5
Human Liver	1	4.1 × 4.0	3	62	96	65	0.187	0.697	1.6
Fixed in 10% Formalin	2	4.0 × 3.9	3	51	97	53	0.277	0.724	2.2
	3	4.1 × 3.8	3	47	95	49	0.310	0.724	2.5
Female Age 11	4	4.2 × 3.5	3	38	97	39	0.409	0.753	3.1
	5	5.2 × 4.3	3	32	98	33	0.481	0.535	5.2
	6	4.9 × 4.4	4	32	96	33	0.481	0.868	5.6
	7	5.1 × 4.6	4	39	96	41	0.387	0.819	4.8
	8	4.2 × 4.0	4	44	92	48	0.319	0.986	3.2
	9	4.2 × 4.1	4	44	92	48	0.319	0.979	3.4
	10	4.7 × 4.7	4	43	96	45	0.347	0.857	4.2
	11	4.0 × 4.0	3	48	97	50	0.301	0.711	2.4
	12	4.1 × 3.7	3	48	97	50	0.301	0.737	2.4
	13	4.1 × 3.8	3	53	97	55	0.260	0.724	2.1
	14	4.0 × 3.8	3	48	98	49	0.310	0.737	2.4
	15	4.4 × 3.9	3	49	98	50	0.301	0.669	2.6

TABLE I (*Continued*)

	16	4.2 × 3.7	3	45	96	47	0.328	0.724	2.6
	17	3.6 × 3.5	3	39	99	39	0.409	0.848	2.8
	18	3.9 × 3.8	3	43	97	44	0.357	0.753	2.7
	19	5.4 × 4.4	4	36	98	37	0.432	0.507	8.7
	20	5.1 × 4.6	4	38	98	39	0.409	0.516	8.1
Human Spermatids	1	2.2 × 2.0	2	45	100	45	0.347	0.975	1.3
	2	2.4 × 2.4	2	48	100	48	0.319	0.833	1.4
Fixed in 10% Formalin	3	2.4 × 2.2	2	46	100	46	0.337	0.875	1.4
Male Age 22	4	2.4 × 2.3	2	47	100	47	0.328	0.854	1.4
	5	2.5 × 2.3	2	45	99	45	0.347	0.833	1.5
	6	2.1 × 2.0	2	44	99	44	0.357	0.988	1.3
	7	2.1 × 2.0	2	37	96	39	0.409	0.988	1.5
	8	2.4 × 2.1	2	48	98	49	0.310	0.900	1.2
	9	2.3 × 2.3	2	44	99	44	0.357	0.875	1.5
	10	2.2 × 2.2	2	42	98	43	0.367	0.925	1.4
	11	2.4 × 2.3	2	45	98	46	0.337	0.854	1.4
	12	2.3 × 2.2	2	41	100	41	0.387	0.900	1.6
	13	2.2 × 2.1	2	43	100	43	0.367	0.950	1.4
	14	2.3 × 2.3	2	47	100	47	0.328	0.875	1.3
	15	2.4 × 2.1	2	48	100	48	0.319	0.900	1.3
	16	2.4 × 2.1	2	48	100	48	0.319	0.900	1.3
	17	2.0 × 2.0	2	40	99	40	0.398	1.000	1.4
	18	2.3 × 2.3	2	49	99	50	0.301	0.875	1.2
	19	2.5 × 2.2	2	47	99	48	0.319	0.854	1.3
	20	2.3 × 2.3	2	46	100	46	0.337	0.875	1.4

must be considered. Measurements of nuclear sizes with the ocular micrometer and estimations of completeness of the nuclei carry, of necessity, subjective elements which are dependent on the individual ability of the investigator. Training can of course mitigate some of these elements, but there will always be certain individual differences among investigators such as accuracy of observation and threshold of fatigue.

Possible errors caused by inherent properties of the biological materials are mainly the inhomogeneous distribution of the absorbing material (Feulgen dye–DNA complex), unspecific light loss, and the shape of the nuclei. The problem of inhomogeneous distribution of the absorbing molecules is perhaps the most definite defect and will be discussed in detail in Section V. Suffice it to say here that inhomogeneous distribution will tend to give values somewhat too low, although the exact magnitude of error for each nucleus cannot be assessed at present.

Unspecific light loss due to scatter of light in the biological material must also be considered as a possible error. Nuclei can easily be tested for unspecific light loss by determining the transmittance in nuclei of unstained Feulgen controls and in the Feulgen-negative cytoplasm. Usually the transmittance is about the same or is only a little lower than the transmittance of a real blank on the slide; that is, unspecific light loss must be negligible. However, if the Feulgen dye–DNA complex itself shows a very high transmittance, then unspecific light loss has to be considered and the extinction value obtained has to be deducted from the extinction value of the Feulgen-stained nucleus.

Since nuclei are regarded in the computations as ideal mathematical objects such as spheres, prolate spheroids, or ellipsoids, but are in reality only approximations of such structures, a small but unavoidable error is thereby introduced.

Variables in the microspectrophotometric apparatus which may lead to errors are mainly: instrument alignment, fluctuations in the electric current, light source, and phototube (noise) and flare light. The errors due to faulty alignment of the instrument can be completely avoided and *must* be avoided (see Section III, B). Fluctuations in the electric current and light source can be brought to a minimum with a voltage stabilizer. Noise in the phototube is usually very small and can be checked by taking several readings on each object. Flare light in the apparatus can be reduced to a minimum by taking the precautions discussed in Section III, B.

V. CRITIQUE OF THE METHOD

It is indeed not surprising that Feulgen microspectrophotometry for DNA in nuclei has been criticized on theoretical grounds by several authors (Glick, 1953; Glick *et al.*, 1951; Naora, 1951). Even to those microscopists who have had little training in theoretical physics, it is obvious that there must be a striking difference in the behavior of light when absorbed by molecules in a colored solution or by molecules in a complex system such as a colored nucleus. A priori, one would hardly expect that the absorption laws operating in solutions could be applied to structures so intricate as nuclei. In addition, since unfortunately no data are available on the physicochemical properties of nuclei, there is no way at present to lay a theoretical foundation on which the absorption laws could be developed for such material.

A further difficulty is the fact that, in contrast to solutions which can be diluted or concentrated at will, biological materials are inflexible to a large degree and must be handled as such. This is especially serious because, even if the arrangement of the absorbing molecules is uniform in living nuclei, the distribution becomes inhomogeneous after biological materials have been subjected to the various procedures needed to prepare them for Feulgen microspectrophotometry (fixation, dehydration, embedding, and hydrolysis) (Leuchtenberger, 1954a).

A. Distributional Error

Actually, the nonrandom distribution of the absorbing molecules in nuclei is one of the strongest objections brought against the use of Feulgen microspectrophotometry for DNA determinations (Glick, 1953). The error which can be expected on theoretical grounds from a nonhomogeneous distribution is called the distributional error. Ornstein (1952) who has discussed the problem in detail and has proposed several methods to minimize the error defines distributional error in the following way: "Distributional error is the error introduced into a photometric calculation by using the total measured transmittance of a photometric field as if that field contained a uniform and random distribution of chromophores, when in fact, the field contains a nonuniform distribution. This gives rise to variations of local transmittance within the field."

The magnitude of the error which can be caused by different degrees of nonhomogeneous distribution is exemplified by the diagrams in Fig. 13. It can be seen that absorbing molecules which are homogeneously

Distribution of Absorbing Molecules in the Two Halves of a Sphere

I		II		III		IV	
1/2	1/2	0	ALL	1/4	3/4	3/8	5/8

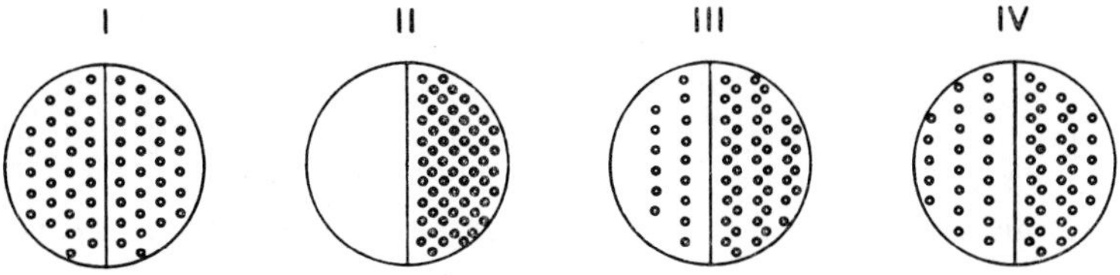

I II III IV

I		II		III		IV	
E=1.0	E=1.0	E=0	E=2.0	E=.5	E=1.5	E=.75	E=1.25
T=10%	T=10%	T=100%	T=1%	T=31.5%	T=3.15%	T=17.7%	T=5.6%
TM=10%		TM=50.5%		TM=17.3%		TM=11.6%	
EM=1.0		EM=.297		EM=.750		EM=.934	

EM = Mean Extinction
TM = Mean Transmittance

Fig. 13. Effect of inhomogeneous distribution of absorbing molecules on extinction values.

distributed in a sphere (Fig. 13, I) give a considerably higher extinction than if the same number of absorbing molecules are inhomogeneously distributed in this sphere (Fig. 13, II, III, IV). In case I, Liebhafsky and Pfeiffer's extension of Beer's law (1953) applies, namely that the mean extinction obtained in a photometric field containing a number of absorbing centers is the same as the extinction which would be obtained by measuring each of the absorbing centers separately, provided that the number of absorbing centers is distributed uniformly. This law obviously does not apply in cases such as II, III, and IV. Here the true extinction could be obtained only by summating the separate extinctions obtained by measuring each absorbing center individually (Caspersson *et al.*, 1955; Deeley, 1955).

The discrepancy between the true extinction and the computed mean extinction is especially striking when all the absorbing molecules are concentrated in one area causing the presence of a large nonabsorbing area. In example II (Fig. 13), the extinction value is less than 30% of the true value. But the error decreases markedly when the difference in concentration of absorbing molecules decreases between two adjacent areas (Fig. 13, III and IV). This implies that nuclei having the same pattern of nonuniform distribution of absorbing molecules, but having different *concentrations* of these, will have different distributional errors, that is, the smaller the extinction, the smaller the distributional error and vice versa. In other words, if measurements are made on a nucleus containing a nonabsorbing area adjacent to an area with *high* absorption, the distributional error is much greater than if the measurements are made on a nucleus containing the same nonabsorbing area but adjacent to an area with *low* absorption.

Fortunately enough there are only very few nuclei in which densely staining large clumps adjacent to blank spaces are observed after the Feulgen reaction. Such nuclei are not suitable for the type of microspectrophotometry discussed in this chapter, but must be examined by scanning procedures (Caspersson *et al.*, 1955; Deeley, 1955) or by the two-wavelength method (Ornstein, 1952; Patau, 1952) in order to determine the amount of the Feulgen dye–DNA complex more accurately. In most nuclei, the Feulgen dye–DNA complex aggregates are only of moderate size and absorbance. To decrease their optical density further, the measurements can be carried out off the absorption peak. At very high transmittances, the nonrandom distribution becomes negligible (Ornstein, 1952). However, measurements too far off the absorption peak may introduce other errors. Factors which usually play a minor role such as unspecific light loss in the unstained nuclei, or noise in the phototube leading to light disturbances, become greatly magnified since the absorption differences between absorbing nuclei and blanks become smaller and smaller (Pollister and Ornstein, 1955).

Another technique which decreases the absorption of the Feulgen dye–DNA complex has been suggested by Davies *et al.* (1954). By using a crushing condenser, "the cells are compressed and flattened so that the whole absorbing region is converted into a thin sheet of low optical density (~ 0.1) lying in the plane of the slide." While this method is certainly an elegant means of reducing the distributional error in cell structures such as chromosomes in squash preparations, more work is needed to see whether the crushing technique is applicable to nuclei in tissue sections.

An illustration that nonrandom distribution in nuclei probably does not introduce a serious error of large magnitude in Feulgen microspectrophotometry as has been suggested (Glick *et al.*, 1951) is presented in Fig. 14. Nuclei from 9 different human tissues each with a

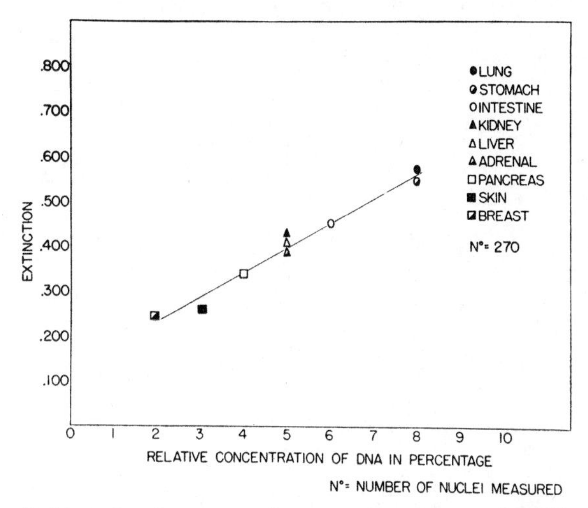

FIG. 14. Relationship between relative concentration of DNA and mean extinction (Feulgen Microspectrophotometry) in diploid nuclei of various human tissues.

presumably diploid chromosomal and diploid DNA content but with varying relative concentrations of Feulgen dye–DNA complex ranging from 2 to 8% show a nearly linear relationship between extinction and concentration, indicating that the Feulgen dye–DNA complex follows Beer's law. This finding, which is just one characteristic example of a large series of measurements made in this laboratory on nuclei from human tissues (Harkin, 1956; Leuchtenberger, 1954b, c; Leuchtenberger *et al.*, 1955; Leuchtenberger *et al.*, 1954b; Leuchtenberger *et al.*, 1956a; Leuchtenberger and Lund, 1952; Leuchtenberger *et al.*, 1953; Moritz and Leuchtenberger, 1955; Swartz, 1956), seems all the more significant since the relationship between extinction and concentration holds in a sample even as small as that shown in this curve where each point represents the mean for only 30 nuclei from a tissue. Swift (1950a) who examined a variety of mouse tissues for their DNA content reported that the relationship holds in nuclei in a concentration range even between 1 to 20.

B. Schwarzschild-Villiger Effect

The second major objection against the use of Feulgen microspectrophotometry for DNA determinations in nuclei is the occurrence of the "Schwarzschild-Villiger effect" (Naora, 1951, 1952, 1955). This effect, which was first described by Schwarzschild and Villiger (1906) for microdensitometry of photographic images of stars, may lead to erroneously high light transmittance values in an object whenever the illuminating light beam is not confined solely to the test object, but is also spread over the surrounding area. In such cases, this light passing through the surrounding area may lead to the formation of flare light which falsely adds to the true transmittance determined in the object. Flare light is caused to a great extent by internal light reflections and scattering at glass–air interfaces in the optical system of the microspectrophotometric apparatus.

Naora (1955) stresses the seriousness of this error especially in cases of nuclei with low transmittance surrounded by an area with high transmittance. He has calculated that a Feulgen-stained rat liver nucleus with a "true transmittance" of 1% will show a faulty transmittance of 20% if measured in the presence of the Schwarzschild-Villiger effect. In order to reduce this effect, Naora urges that the illuminated area be less than the diameter of the nucleus to be measured for light transmittance (preferably, one-third). To achieve illumination limited to a minute portion of the specimen, he described modifications of the microspectrophotometric apparatus, one main feature of which is the replacement of the microscope condenser with a high-power oil immersion objective with a numerical aperture of 1.25–1.30.

Ornstein and Pollister (1952) have expressed serious doubts as to the validity of the theoretical foundation of Naora's calculation for the actual situation in microspectrophotometry. They are of the opinion that the error must be much smaller than that given by Naora. They further justifiably stress the fact that in Feulgen microspectrophotometry, precautions (such as using the condenser at a low numerical aperture) have always been taken to reduce flare as much as possible.

Lison (1953), on the basis of his own experimental evidence, also criticizes Naora's conclusions as to the magnitude of the error. Using opaque objects such as heavily overstained iron-hematoxylin lymphocytes or lampblack particles, Lison measured their transmittance with his microspectrophotometric apparatus and found the flare light to be 2%. By measuring objects of different transmittances, he calculated that for objects with a transmittance between 90–45%, the relative error is less than 1%, while for objects with transmittances between 40–10% the error increases up to 5%.

Swift and Rasch (1956) report a flare light error of 0.3% for opaque objects in their microspectrophotometric apparatus, which is essentially the same as the one described in this chapter. They state that this error is negligible in objects with extinctions below 0.5 but increases to 1.3% at an extinction of 1.0.

Although it appears as if Naora's calculation of the flare light error may indeed be greatly overestimated still, all precautions should be taken to eliminate flare light from the microspectrophotometric system as much as possible.

C. Spatial Orientation

Another possible source of error in microspectrophotometry of biological materials was pointed out by Commoner (1949) and Commoner and Lipkin (1949). They raised the question whether the absorbing molecules in cell structures may not show a spatial orientation. The presence of spatial orientation would lead to errors in absorption measurements since light polarized in one plane shows higher absorption than in another plane (dichroism) (Schmidt, 1947) and measurements with nonpolarized light would give erroneously low extinctions. Although spatial orientation of the DNA in grasshopper sperms has been observed (Caspersson, 1940), orientation in nuclei and chromosomes has been shown to be of insufficient degree to produce a significant error (Mellors *et al.*, 1950; Thorell and Ruch, 1951; Wilkins, 1950). Pollister and Swift (1950) who examined a variety of Feulgen-stained materials did not find dichroism in any of them. Nevertheless, whenever new biological materials, especially thin, elongated cell structures are to be used for microspectrophotometry, they should be investigated for spatial orientation.

D. Proportionality Factor

Another criticism voiced against Feulgen microspectrophotometry is concerned with one particular aspect of the Feulgen reaction itself, namely the relation between the actual amount of DNA and the amount of Feulgen dye–DNA complex formed in a nucleus. As pointed out previously in this chapter (Section II) this relationship must be at least proportional, if not stoichiometric, that is, increasing amounts of DNA in nuclei must always bind correspondingly increasing amounts of Feulgen dye before the Feulgen reaction can be used for quantitative purposes. Since there is so little theoretical knowledge about the quantitation of the steps involved in the reaction between the aldehydes and the Schiff reagent, the applicability of the Feulgen reaction for the quantitative estimation of DNA has been questioned (Glick, 1953). The doubts have been raised particularly since some workers have reported loss of Feulgen-positive material after hydrolysis of tissues (Ely and Ross, 1949) and incompleteness of the Feulgen reaction (Li and Stacey, 1949). There are also indications that proteins may interfere with the quantitative aspects of the Feulgen reaction (Caspersson, 1932; Sibatani, 1950) although recent microspectrophotometric studies by Kasten (personal communication) show that the amounts of Feulgen dye–DNA complex are essentially the same before and after protein digestion or histone extraction.

There is sufficient indirect evidence to show that the nonproportionality factor cannot play a major role in Feulgen microspectrophotometry if the technique is always carried out under the same rigidly controlled conditions with a known standard as a control. There are many and various data which refute this criticism. A few examples are given here:

(1) Nuclei with a chromosomal complement of $4n$, $2n$, and $1n$ such as primary spermatocytes, secondary spermatocytes, and spermatids, also show 4 DNA, 2 DNA and 1 DNA with Feulgen microspectrophotometry (Leuchtenberger, 1954b; see also Table II).

(2) Nuclei with the same chromosomal status show the same amount of DNA in spite of varying amounts of protein (Schrader and Leuchtenberger, 1950; Ris and Mirsky, 1949).

(3) DNA ratios found by biochemical analyses on uniform cell populations agree very well with the DNA ratios found by Feulgen microspectrophotometric analyses of the same material (Leuchtenberger *et al.*, 1951; Ris and Mirsky, 1949).

E. Evaluation from Actual Data

After having discussed these major objections raised mainly on the basis of theoretical considerations, an appraisal of Feulgen microspectrophotometry on the basis of actual data seems warranted. A special clarification of the problem might be derived from a comparison between the DNA data obtained by Feulgen microspectrophotometry and the DNA data gathered by other independent methods on the same material. As a baseline for such a survey, Feulgen microspectrophotometric DNA data obtained in the Cytochemistry Laboratory at Western Reserve University on over 36,000 individual nuclei are presented in Table II. Such an approach seems justified because of the considerable number of determinations done under strictly standardized and controlled conditions on a variety of tissues with the microspectrophotometric method and apparatus just described. Furthermore, whenever possible the Feulgen data were checked by other procedures. Ultraviolet microspectrophotometry, chromosomal studies, as well as

DNA determinations by biochemical analysis were carried out simultaneously on the same material. The DNA data are given in Table II as mean DNA values per nucleus expressed in arbitrary units.

When comparing the mean DNA values per nucleus derived from biochemical and microspectrophotometric analyses, it must be kept in mind that not only is there a striking difference in the type of method employed for the determination of DNA, but the way in which the DNA is *computed* is also completely different. In the biochemical analysis a large number of isolated nuclei are analyzed for their total DNA content. This value is then divided by the number of nuclei in the aliquot sample in order to obtain the DNA value for an individual nucleus. In the microspectrophotometric analysis, individual nuclei are analyzed each for their DNA content and the mean DNA value per nucleus is obtained by adding up all the individual DNA values and then this sum is divided by the number of nuclei measured. Therefore, the microspectrophotometric mean is based on a *number* of analyses done on *individual* nuclei while the biochemical mean is based on *one* analysis done on a *pooled number* of nuclei. Consequently the microspectrophotometric DNA analysis at once reveals whether a cell population in a tissue is uniform with respect to its DNA content since variations from nucleus to nucleus can be easily assessed, while obviously biochemical analysis cannot give this type of information. This does not by any means lessen the importance of the chemical analysis for DNA in nuclei, particularly if we keep in mind that if the chemist presents a value of DNA for a nucleus it is based on much larger samples of nuclei (billions), while the cytochemist—even if he were able to study 1000 nuclei of each tissue—always analyzes only a comparatively small population of nuclei with respect to the whole tissue. There seems to be no doubt as to the desirability of employing both methods simultaneously on the same material.

The data presented in Table II may be considered mainly from 2 points of view which seem pertinent in examining the validity of a method: (1) How do the DNA results obtained by Feulgen microspectrophotometry compare with DNA results by other analytic methods? (2) What is the degree of accuracy with which amounts of DNA can be determined?

Comparing first the DNA data by Feulgen microspectrophotometry with the results of biochemical analysis it can be seen that there is a striking agreement between both methods whenever tissues with a uniform cell population are examined. On the other hand, it can also be seen that if a tissue such as rat liver which has a nonuniform cell population with nuclei containing 2 DNA, 4 DNA, and 8 DNA is

TABLE II

MEAN DNA VALUES [a] IN NUCLEI OF VARIOUS BIOLOGICAL MATERIALS OBTAINED BY FEULGEN MICROSPECTROPHOTOMETRY, BIOCHEMICAL ANALYSES, AND ULTRAVIOLET MICROSPECTROPHOTOMETRY

Type of material	Type of tissue or cell	Assumed chromosomal status	No. of nuclei measured by Feulgen microspectrophotometry	Mean amount of DNA per nucleus in basic arbitrary units		
				Feulgen microspectrophotometry	Biochemical analysis	Ultraviolet microspectrophotometry
Clam	Sperm	1n	200	1 DNA		
	Epithelium	2n	50	2 DNA		
Mouse	Bronchus	2n	4400	2 DNA		
	Kidney	2n	160	2 DNA		
	Liver	2n, 4n, 8n	320	2 DNA, 4 DNA, 8 DNA		
	Lung	2n	500	2 DNA		
	Pancreas	2n, 4n	320	2 DNA, 4 DNA, 8 DNA		
	Spermatid	1n	160	1 DNA		
	Ascites lymphoma tumor	2n [c]	350	2 DNA [b]	2 DNA [b]	2 DNA [b]
	Ascites Ehrlich tumor	4n [c]	350	4 DNA [b]	4 DNA [b]	4 DNA [b]
Rat	Liver	2n, 4n, 8n	100	2 DNA, 4 DNA, 8 DNA [b]	Average value of all DNA [b] classes	2 DNA, 4 DNA, 8 DNA [b]
	Thyroid	2n	215	2 DNA		
	Kidney	2n	38	2 DNA [b]	2 DNA [b]	
Hamster	Kidney	2n	50	2 DNA		
	Liver	2n, 4n, 8n	400	2 DNA, 4 DNA, 8 DNA		
	Pouch (normal)	2n	30	2 DNA		
	Pouch (tumor)		230	2 DNA to 4 DNA to 8 DNA		

Dog	Adrenal	$2n$	550	2 DNA		
	Kidney	$2n$	60	2 DNA		
	Liver	$2n$	1460	2 DNA	2 DNA (Vendrely and Vendrely, 1948)	
	Pancreas	$2n$	110	2 DNA	2 DNA (Vendrely and Vendrely, 1948)	
	Skin	$2n$	500	2 DNA		
	Thyroid	$2n$	180	2 DNA		
Beef	Liver (young animal)	$2n$	80	2 DNA[b]	2 DNA[b]	2 DNA[b]
	Liver (mature animal)	$2n, 4n, 8n$	880	2 DNA, 4 DNA, 8 DNA		
	Kidney	$2n$	50	2 DNA[b]	2 DNA[b]	
	Spleen	$2n$	75	2 DNA[b]	2 DNA[b]	2 DNA[b]
	Sperm	$1n$	8000	1 DNA[b]	1 DNA[b]	1 DNA[b]
	Spermatid	$1n$	400	1 DNA		
	Secondary spermatocyte	$2n$	320	2 DNA		
	Primary spermatocyte	$4n$	340	4 DNA		
	Spermatogonia	$2n$-$4n$	320	2 DNA to 4 DNA		
Human	Adrenal	$2n$	70	2 DNA		
	Breast (normal)	$2n$	40	2 DNA		
	Breast (tumor)		150	2 DNA to 4 DNA		
	Intestine (normal)	$2n$	150	2 DNA		
	Intestine (tumor)		160	2 DNA to 4 DNA		
	Kidney (normal)	$2n$	60	2 DNA		
	Kidney (tumor)		100	2 DNA to 4 DNA		
	Liver (child)	$2n$	900	2 DNA[b]	2 DNA[b]	
	Liver (adult)	$2n, 4n, 8n$	2000	2 DNA, 4 DNA, 8 DNA		
	Lung (normal)	$2n$	30	2 DNA		
	Lung (tumor)		30	2 DNA to 4 DNA		
	Lymph nodes, (normal)	$2n$	60	2 DNA	2 DNA (Métais and Mandel, 1950)	

TABLE II (*Continued*)

Type of material	Type of tissue or cell	Assumed chromosomal status	No. of nuclei measured by Feulgen micro-spectrophotometry	Mean amount of DNA per nucleus in basic arbitrary units		
				Feulgen microspectro-photometry	Biochemical analysis	Ultraviolet microspectro-photometry
Human	Lymph nodes, (tumor)		60	2 DNA to 4 DNA		
	Pancreas (normal)	$2n$, $4n$	30	2 DNA, 4 DNA		
	Pancreas (tumor)		70	2 DNA to 4 DNA to 8 DNA		
	Prostate (normal)	$2n$	500	2 DNA		
	Prostate (hypoplasia)		500	2 DNA		
	Prostate (hypertrophy)		500	2 DNA		
	Prostate (tumor)		300	2 DNA to 4 DNA to 8 DNA		
	Skin (normal)	$2n$	800	2 DNA		
	Skin (tumor)		100	2 DNA to 4 DNA to 8 DNA		
	Stomach (normal)	$2n$	30	2 DNA		
	Stomach (tumor)		70	2 DNA to 4 DNA to 8 DNA		
	Urinary bladder (normal)	$2n$, $4n$	60	2 DNA, 4 DNA		
	Urinary bladder (tumor)		120	2 DNA to 4 DNA to 8 DNA		
	Sperm	$1n$	7000	1 DNA [b]	1 DNA [b]	

	Spermatid	1n	340	1 DNA
	Secondary spermatocyte	2n	330	2 DNA
Human	Primary spermatocyte	4n	300	4 DNA
	Spermatogonia	2n-4n	370	2 DNA to 4 DNA

[a] All the DNA data unless otherwise indicated were obtained in this laboratory.
[b] Simultaneous analyses on the same material.
[c] Actual counts on the same material.

examined, the microspectrophotometric analysis of individual nuclei is capable of revealing the individual DNA classes whereas the biochemical analysis that is done on a mass of nuclei gives only an average value. This average value is not a representative mean for any one of the nuclear classes.

The second standard method which can also be used as a criterion for the examination of the validity of Feulgen microspectrophotometry is the chromosomal status of the nuclei in different tissues. It is evident from the table that there is good accordance between the DNA results from Feulgen microspectrophotometry and the chromosomal numbers, a comparison that is of special significance in tissues showing multiple DNA values.

The third method used for comparison concerns DNA determinations by *ultraviolet* microspectrophotometry. Although it may be argued that another microspectrophotometric method is not too suitable for comparison, nevertheless, the different principles of the two methods and accordance of the ultraviolet data with the Feulgen data are of sufficient interest to justify its inclusion in this table.

Coming now to the second point—namely, the sensitivity and accuracy of the microspectrophotometric DNA determinations—the excellent agreement of the multiple DNA values obtained by Feulgen microspectrophotometry with chromosome numbers, as well as with biochemical analysis, illustrates the order of magnitude with which the method can be safely used. Even if the actual DNA values obtained by microspectrophotometry of the Feulgen stain may sometimes show variations up to 15% from the theoretical ratios of 1:2:4:8 DNA, such a difference would hardly be capable of obscuring the multiplicity of the DNA values.

On the other hand, the question of the significance of *intermediate* DNA values occurring in nuclei of tissues such as tumors cannot be so readily answered. Although there does not seem to be any doubt about the constant occurrence of such intermediate DNA values in nuclei of proliferating tissues (Leuchtenberger *et al.*, 1954b; Lison and Pasteels, 1950; Swift, 1953), it is felt that much more work and probably more exact methods are needed to interpret the degree of validity of the intermediate values.

On the basis of the data presented in Table II, it seems justifiable to say that microspectrophotometry of Feulgen stain can be utilized as a reliable tool for the comparison of relative mean amounts of DNA in nuclei of different cells, provided that the differences to be detected are sufficiently large.

In support of the validity of the method, it may be added that microspectrophotometry of Feulgen stain also permits an excellent reproducibility of the DNA results, if the preparation and measurements of nuclei are done under carefully standardized and controlled conditions. In this laboratory, a group of 23 trained workers, using 4 different microspectrophotometric setups within a period of more than 6 years, obtained nearly identical DNA results in the same tissues, the largest variation ever encountered being 10%. Furthermore, results from other laboratories are essentially in accordance with the data presented in this table (Leuchtenberger, 1954a; Pollister, 1950; Swift, 1953).

Taking into consideration the reproducibility of the DNA data obtained by Feulgen microspectrophotometry, their good correlation with chromosomal counts, with ultraviolet microspectrophotometric, and biochemical data, one can hardly escape the conclusion that the method has its merits and can be used safely for DNA determination at this level of investigation. This, of course, does not exclude the necessity of cautious interpretation, of checking the DNA data by other methods, and of the possible interference of other factors such as those discussed at the beginning of this section, if the method is used for the detection of smaller differences in DNA content.

VI. BRIEF ASSESSMENT OF RESULTS

A. Normal Processes

The value of a new method such as Feulgen microspectrophotometry is gauged not only by its comparison with other standard methods but after its validity is established, the question should be raised, what has this method contributed to biological problems? Has Feulgen microspectrophotometry really opened new pathways for the study of intracellular components such as DNA and has knowledge been gained which could not have been obtained by previously used methods?

One of the important contributions of the method is the reenforcement of the concept (Caspersson, 1936) that chemical studies of intracellular constituents can be made at the level of a single cell in which the morphological architecture is still preserved, thus permitting a direct correlation between morphology and chemical composition. By this method only, can a specific change in a specific *cell structure* be expressed in terms of a change in the quantity of a specific *chemical component*, such as DNA, of that cell. This possibility and the possibility of detecting individual DNA differences from cell to cell within a tissue offers obvious advantages over a method where the DNA analysis is based on masses of cells and after DNA is separated from its normal cell environment.

A great variety of biological problems have been attacked by Feulgen microspectrophotometry and valuable information has been gathered on the behavior of DNA in cells. Due to lack of space, only a few problems will be briefly assessed from which obvious fundamental contributions have been derived. For further details, the reader is referred to the reviews published by Davidson (1953), Leuchtenberger (1954a), Milovidov (1954), Pollister (1950), Pollister and Ornstein (1955), Swift (1953), Vendrely (1952), and Vendrely (1955).

1. *Growth*

In cell growth, it has been shown that there are at least two types of activity with regard to DNA: (a) Changes in nuclear size may occur without a change in DNA content and may be explained by a change in the content of protein and other constituents (Bern and Alfert, 1954; Leuchtenberger, 1950; Leuchtenberger and Schrader, 1951; Schrader and Leuchtenberger, 1950). (b) Changes in nuclear sizes may be associated with a corresponding increase in DNA content as evidenced by multiple DNA classes (Leuchtenberger *et al.*, 1951; Leuchtenberger *et al.*, 1954b; Lison and Pasteels, 1950; Ris and Mirsky, 1949; Swartz, 1956; Swift, 1950a). Furthermore, there are indications that the formation of multiple DNA classes may be controlled by genetic factors associated with the activity of a pituitary growth hormone (Leuchtenberger *et al.*, 1954a).

2. *Evolution*

Another interesting finding obtained by Feulgen microspectrophotometry of DNA in cells pertains to evolutionary processes. Hughes-Schrader (1951, 1953) on studying the DNA content in several species of Mantidae found that the DNA complement can be used as "a cytotaxonomic tool in evaluating evolutionary relationship among species whose karyotypes are not analysable by the method of comparative cytology." Similarly the DNA differences among certain invertebrate and vertebrate species have been interpreted to signify evolutionary trends (Mirsky and Ris, 1951; Moses and Yerganian, 1952).

3. *Differentiation and Development*

For embryological problems such as differentiation of tissues and blood cells, Feulgen microspectrophotometry has also given interesting information. The main consensus among the workers in this field seems to be that differentiation is associated with variation in the DNA content in nuclei (Lison and Pasteels, 1951; Marinone, 1951; Moore, 1952; Reisner and Korson, 1951).

4. *DNA Constancy Concept*

One of the problems perhaps most intensively examined by Feulgen microspectrophotometry and which has led to controversy in the literature (see below), concerns the DNA content in the normal "resting" nucleus. This great interest arose from the work of Boivin *et al.* (1948) who, on the basis of their biochemical analysis, stated that the DNA content in diploid somatic nondividing nuclei is constant for various tissues in the same species and is twice that of the haploid DNA value found in the sperms from the same species. Feulgen microspectrophotometric studies of DNA on the whole are in accord with this constancy concept. Regardless of the function of the tissues, *mean* DNA values per diploid nucleus are very similar from tissue to tissue, correlate with chromosomal status (see Table II), and usually do not depend on the state of the physiological activity of the cell.

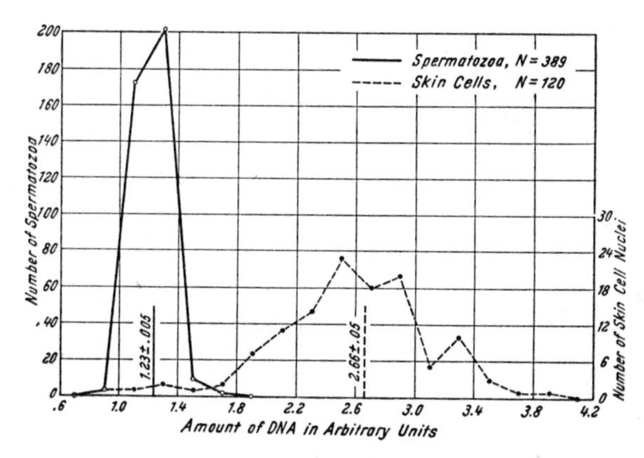

FIG. 15. Comparison of the amount of DNA in individual spermatozoa of fertile males and of individual normal human skin cells. Mean DNA values at vertical lines.

On the other hand, Feulgen microspectrophotometric studies have also demonstrated that there are certain variations from nucleus to nucleus in a somatic tissue. Figure 15 (see also raw data in Table I) is a typical example of the variability of DNA values in individual nuclei which may be encountered in many tissues. While the mean DNA value for the skin nuclei corresponds very nicely to the 2 DNA value which would be expected on the basis of the 1 DNA value found in the haploid sperms, this *mean* value actually masks the variation in the DNA values of the *individual* nuclei. As repeatedly pointed out by various investigators, a constant mean DNA value for a tissue does not of necessity imply that each nucleus contains the same constant amount of DNA

(Leuchtenberger *et al.*, 1951; Leuchtenberger *et al.*, 1953; Lison and Valeri, 1956).

Whether the differences in DNA content from nucleus to nucleus are due to errors inherent in the Feulgen microspectrophotometric technique or whether they are due to true biological differences cannot be decided at the present time. Some support for the view that this variability may be truly biological comes from chromosomal studies which have shown that deviations form the normal number of chromosomes is quite common among somatic tissues (Hsu, 1952; Tanaka, 1951). Perhaps, as Roels suggests (1954), the constant mean DNA value for the tissues in a given species may represent an "equilibrium value" which infers the possibility that the DNA content of a nucleus may vary sometimes during early differentiation or during the catabolic and anabolic activity of its cell life.

5. *Exceptions to the DNA Constancy*

Although, as demonstrated in Table II, most tissues do not reflect metabolic activity in their *mean* DNA values, but maintain the DNA content expected on the basis of their chromosomal complement, there is an increasing number of observations showing that in some species and some tissues, the DNA content may be affected by physiological and metabolic activity. The first evidence suggesting that the DNA content in nuclei may change with metabolic activity was presented by Leuchtenberger and Schrader in the salivary gland nuclei of the snail *Helix* (1952) and by Schrader and Leuchtenberger in the nurse cells of the bug *Acanthocephala* (1952). Striking changes in the DNA content of *Helix* salivary gland nuclei were observed during the production of secretory granules containing polysaccharides and considerable variation in the DNA content was found in *Acanthocephala* nurse cells during the process of providing nutritive substances for the egg. Similarly Fautrez and Moerman (1954) described changes in the DNA content correlated with physiological activity in the liver nuclei of a fish, *Lebistes reticulatus*. Govaert (1953) found greater DNA variations in the cells of *Fasciola hepatica* during intense metabolic activity than during the resting phase. Roels (1954) reported changes in the DNA content of rat thyroid nuclei under the influence of thyroxine and thiouracil which respectively inhibit and stimulate the activity of thyroid cells. An increase in the DNA content of rat liver nuclei after alloxan administration was found by Diermeier *et al.* (1951). Recently La Cour *et al.* (1956) observed that plants grown at low temperatures showed a striking reduction in the DNA content of their nuclei which returned to normality when the plants were returned to their normal temperature.

The DNA variability observed in the nuclei of some tissues during changes in metabolic and physiological cell activity or in processes of differentiation (Moore, 1952) should evoke caution against a sweeping generalization of the DNA constancy concept. However, in view of the overwhelming data in which DNA has been found constant, some workers have vigorously denied the validity of these exceptions to the constancy hypothesis (Alfert and Swift, 1953; Patau and Swift, 1953; Swift, 1953, 1955) and have interpreted them in terms of technical errors. This skepticism is perhaps best reflected in their own words "Simple numerical relationships in a sufficiently large number of data are, therefore, inherent evidence of reliability. As it happens, the simple relations that have been found agree with our working hypothesis. Those published data which seem to contradict it lack at the same time that inherent evidence of reliability" (Patau and Swift, 1953).

While of course, the possibility cannot be entirely excluded that heretofore unknown and unpredictable errors may be responsible for the examples in which deviations from the constancy rule were reported, this explanation seems unlikely. The author is especially reluctant to accept this interpretation since the data which have been gathered in this laboratory have all been obtained with the same standardized Feulgen microspectrophotometric technique. Therefore it would seem strange indeed that technical errors should become apparent only in those cases which are regarded as exceptions to the constancy concept. Furthermore, when these few exceptions were encountered in this laboratory, special precautions were taken to recheck the findings by examination of more specimens and by repeat measurements made by different workers with different instruments.

6. *Mitosis*

Feulgen microspectrophotometric studies of DNA synthesis during the mitotic cycle have also yielded relevant findings. Contrary to what may have been expected on the basis of visual inspection of the mitotic process, the period in which DNA is synthesized seems to be relatively brief and does not parallel the period of the *complete* mitotic cycle (Bryan, 1951). While there are many instances in which DNA doubling is already completed in interphase as was first described by Swift for plant nuclei (1950b), increasing evidence has been presented that the stage at which DNA synthesis occurs may vary for various tissues (for discussion, see Leuchtenberger, 1954a; Swift, 1953).

B. Pathological Processes

The examples presented so far have been concerned with the assessment of DNA data pertaining mainly to *normal* cellular processes. In the following, some results will be discussed which demonstrate the fascinating potentialities of Feulgen microspectrophotometry for the study of pathological cell processes.

In recent years, some pathologists have become increasingly aware of the necessity to extend their light microscope studies of tissues and cells. Although these structural studies have yielded a wealth of information on the *morphological* alterations in diseases, nevertheless pathologists have recognized the limitations imposed by purely structural studies even if performed at the level of the electron microscope.

After all, alterations in structure are actually the expression of alterations in the chemical constituents which make up that structure. Since the cell can be considered as occupying a fundamental position in disease and probably is the primary target hit by injurious agents, investigation of the chemical cellular constituents seems of special significance. The investigation of disease at this level has been of great interest to our own laboratory because it was felt that such studies may not only throw light on etiology and pathogenesis, but may help greatly in the early diagnosis of disease. Since DNA is perhaps the most important building stone in all cells, quantitative studies of DNA by Feulgen microspectrophotometry in individual cells *in situ* and their correlation with structural alterations in disease afford obviously intriguing possibilities. A variety of pathological conditions such as tumors, dwarfism, sterility, and virus infections have been explored by this technique and significant information has been gained.

1. *Tumors*

In an extensive study of a great diversity of malignant human tumors, Leuchtenberger *et al.* (1954b) have found that interphase nuclei from precancerous and malignant tissues carry higher amounts of DNA and reveal much greater DNA scatter from cell to cell than do the nuclei from the normal homologous tissues. However, the deviating DNA data cannot be considered a specific criterion for malignant transformation of cells, but may be explained mainly on the basis of growth and mitotic processes present in most tumor tissues. Bader (1953), after studying a few animal tumors, came to approximately the same conclusion. As pointed out before, DNA synthesis occurs also in interphase nuclei of normal tissues which are undergoing mitoses and growth. On the other hand since most normal tissues in adult humans do not exhibit mitoses, or in other words, show usually the characteristic DNA value with relatively little scattering, an increase in and a large scatter of DNA in such a tissue must be looked upon with suspicion in regard to malignancy, unless regeneration is to be expected. Feulgen microspectrophotometric studies may be particularly helpful in diagnosing tumor cases in which mitotic figures are scanty or absent and which the pathologist designates as borderline cases, such as carcinoma *in situ.*

2. *Dwarfism*

Another question in pathology for which Feulgen microspectrophotometry has provided new and helpful information concerns the problem of dwarfism. On the basis of studies by this technique it became evident that dwarfism in animals may be associated with deficiency of DNA and disturbances in DNA synthesis. For example, certain dwarf mice reveal the absence of the usual multiple DNA classes in liver nuclei, but these classes may be restored by administration of a pituitary growth hormone (Leuchtenberger *et al.,* 1954a). In a sterile dwarf bull, Leuchtenberger and Schrader (1955) found extreme DNA deficiency in both somatic and spermatogenic cells. Similarly, in a recent DNA analysis of somatic and spermatogenic tissues from several dwarf bulls and dwarf carriers, deficiency and/or variability in the DNA content was observed (Leuchtenberger *et al.,* 1956c).

3. *Infertility*

One abnormality in which DNA studies by Feulgen microspectrophotometry have proved to be of special diagnostic value concerns the problem of male infertility. In contrast to the remarkably constant and uniform haploid DNA content found in the spermatozoa of fertile males (both human and bulls), the DNA content in the spermatozoa from infertile males is variable and significantly lower than that from the fertile ones (Leuchtenberger *et al.,* 1953, 1956a; Leuchtenberger *et al.,* 1954c; Leuchtenberger *et al.,* 1955). This finding is especially pertinent since the cytological appearance and the dry weight of the spermatozoa containing the normal and deficient amounts of DNA are identical (Leuchtenberger *et al.,* 1956b). The same holds true for the spermatogenic cells; primary and secondary spermatocytes and spermatids each have a deficient DNA content in infertile males, as compared with the 4 DNA, 2 DNA, and 1 DNA found in the fertile males, although the histological and cytological features of the testes and germ cells may be completely normal (Leuchtenberger *et al.,* 1954c, 1956a). It thus appears that the DNA deficiency found by Feulgen microspectrophotometry in spermatogenic cells of normal cytological appearance can be considered as a criterion for gauging at least one type of male infertility and may have a bearing on the fertilization process (Ito and Leuchtenberger, 1955).

4. *Virus Diseases*

Another pathological process where the potentialities of Feulgen microspectrophotometric studies have just begun to be realized are the virus diseases. Because of the chemical composition of the viruses (most of them contain DNA) and their peculiar relationship to cells, quantitation of DNA by Feulgen microspectrophotometry may be a valuable tool for disclosing the presence of viruses in cells. In virus diseases such as molluscum contagiosum or verruca vulgaris, multiplication of the viruses within the infected cells has been demonstrated (Caspersson, 1950). If Feulgen microspectrophotometric DNA studies are done on such virus-infected cells, unusually large quantities of DNA can be

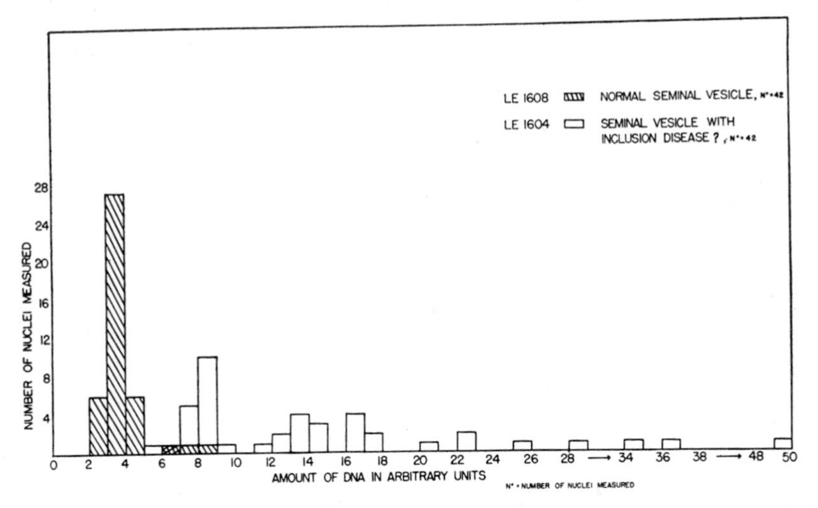

FIG. 16. Amount of DNA (Microspectrophotometry of Feulgen reaction) in individual nuclei of human seminal vesicle.

found in individual cells. These DNA values are far beyond those observed in malignant tumors and are usually found in the absence of mitotic processes. With the increasing recognition of the importance of viruses for etiology in disease, the pathologist when confronted with a peculiar kind of abnormality in tissues, has become aware of the possible virus etiology. Recently, Moritz and Leuchtenberger (1955) observed in a young man who died of an unexplained respiratory disease, an enormous increase in nuclear size (karyomegaly) of cells from nearly all the tissues. Feulgen microspectrophotometric studies disclosed enormous amounts of DNA in these nuclei (Fig. 16). This finding correlated with the cytological and histological appearance of the tissues was quite similar to that obtained in known virus diseases and consequently suggested the possibility of a virus disease.

It is hoped that the rather brief assessment of a few results will suffice to demonstrate the applicability of Feulgen microspectrophotometry for DNA determinations in various normal and abnormal biological conditions. There are many unexplored fields in the biological and medical sciences where Feulgen microspectrophotometry may prove of great value. [For example, we are just beginning to apply this technique to elucidate the mechanisms operating in stress (Cole and Leuchtenberger, 1956).] This is especially true when the data are correlated closely with cytological and histological studies and when this technique is used in combination with methods which yield additional information about other intercellular substances and their relation to DNA.

ACKNOWLEDGMENTS

This chapter is dedicated to Professor Alan R. Moritz, Director of the Institute of Pathology, Western Reserve University, whose vision and continuous help made possible the development of a Cytochemistry Laboratory at the Institute of Pathology.

The author wishes to express her sincere thanks to the entire staff of the Cytochemistry Laboratory, to Professor T. Caspersson, Karolinska Institute, Stockholm, Sweden, to Dr. Rudolf Leuchtenberger, Doctors Hospital, Cleveland, Ohio and to all the visiting scientists for their many stimulating discussions about the problems considered in this chapter.

This work was aided by grants from the U. S. Public Health Service A-787, RG 4268 and C 1814; from the Brush Foundation, from the Elsa U. Pardee Foundation, and from the Franchester Fertility Fund of Cleveland, Ohio.

VII. APPENDICES

Appendix 1: Fixatives

All fixatives are prepared fresh each time just before use.

a. Carnoy

30 cc. Alcohol, 95% or absolute
10 cc. Glacial Acetic Acid

b. 10% Formalin

45 cc. Formaldehyde, C.P.
5 cc. Distilled water

c. Lavdowsky

25 cc. Alcohol, 95%
20 cc. Distilled water
5 cc. Formaldehyde, C.P.
1 cc. Glacial Acetic Acid

Appendix 2: Feulgen Reagents

All chemicals are of reagent grade. All dyes are certified and obtained from the National Aniline Co., 40 Rector Street, New York 6, N. Y. All glassware including slides is chemically cleaned.

a. 1 N Hydrochloric Acid

In a volumetric flask, dilute 80.7 cc. concentrated HCl (sp. gr. 1.19) to 1000 cc. This solution can be stored in the refrigerator, but an adequate preheated supply should always be kept in a glass-stoppered bottle in the 60° C. incubator. Also keep some preheated glass dishes in this incubator for the hydrolysis. Use a fresh portion of acid each time and for each rack of dishes.

b. Schiff Solution (Leucofuchsin)

(1) Add 1 gm. certified basic fuchsin to 200 cc. boiling water measured after it has come to a boil). (Caution—may boil over!) Continue boiling for 5 minutes while constantly agitating the flask.

(2) Cool to 50° C. (may be cooled by running cold water over the flask).

(3) Filter (Whatman no. 1 grade of filter paper).

(4) Add 20 cc. 1 *N* HCl.

(5) Cool to 25° C. (room temperature).

(6) Add 2 gm. Na or K metabisulfite. Shake well. Let solution stand overnight in the dark in a tightly closed (glass-stoppered) bottle. Color will change to a light yellow or orange.

(7) The next morning add 0.5 gm. neutral activated charcoal (Merck 18351) and immediately shake for *exactly* 1 minute.

(8) Filter rapidly through double layer of coarse filter paper. (Arthur H. Thomas no. 5160 grade of filter paper.)

Store in well-stoppered (glass stopper) bottle and keep in refrigerator. Return used portion of Schiff solution to the stock bottle. The solution may be used for a month as long as it remains clear, colorless and free from precipitate. Always exercise great caution to avoid contamination with any oxidizing agent; do not leave solution exposed to the air any longer than necessary.

c. Bleaching Solution

60 cc. 1 *N* HCl
60 cc. 10% aqueous Na or K metabisulfite
1080 cc. Distilled water

Mix fresh each time. The metabisulfite solution may be kept as a stock solution. The 1 *N* HCl is the same as that used for the hydrolysis.

d. Fast Green Counterstain

Dissolve approximately 100 mg. of Fast Green in about 10 cc. of distilled water. Add to 400 cc. absolute alcohol.

Appendix 3: Deoxyribonuclease Reagents

All glassware must be chemically cleaned as for the Feulgen reagents. All chemicals are of reagent grade. The gelatin is the bacteriological grade made by Coleman and Bell Co., Norwood, Ohio. For the deoxyribonuclease, use only the highly purified crystalline product made by the Worthington Laboratories in Freehold, New Jersey. It should be kept in the refrigerator because it is inactivated by heat.

(1) Dissolve 0.4 gm. NaOH in 150 cc. distilled water. (A)
Dissolve 1.3 gm. KH_2PO_4 in 150 cc. distilled water. (B)
These solutions may be temporarily stored in the refrigerator.

(2) To make the buffer solution, mix 20 cc. of (A) and 40 cc. of (B). Dilute with equal quantity of water. The pH should be 7 and should be checked with a pH meter.

(3) To 100 cc. of this mixture, add 100 mg. of gelatin. Heat very slightly with constant shaking until the gelatin is in solution.

(4) Add 36 mg. anhydrous $MgSO_4$ per 100 cc. of above solution (0.003 *M*). Dissolve.

(5) Cool to room temperature and add 7–8 mg. of DNase per 50 cc. of Mg-gelatin buffer solution. DNase goes into solution with difficulty. Be sure it is all dissolved before using the solution. This solution may be used once or twice more over a course of 2 weeks if it is kept refrigerated at all times. The same is true for the buffer.

Appendix 4: Computation of Cylindrical Region "F"

TABLE III
Computation of the Fraction "*F*" of the Total Nuclear Volume Occupied by a Cylindrical Region

Diameter (2*C*) of central plug =							
3				4			
Nuclear diameter	*F*	Nuclear diameter	*F*	Nuclear diameter	*F*	Nuclear diameter	*F*
3.0	100	5.0	49.0	4.0	100	6.0	58.7
3.05	99.3	5.05	48.1	4.05	99.3	6.05	57.8
3.1	98.5	5.1	47.2	4.1	98.6	6.1	57.0
3.15	97.2	5.15	46.3	4.15	97.9	6.15	56.2
3.2	95.9	5.2	45.5	4.2	97.2	6.2	55.4
3.25	94.4	5.25	44.7	4.25	96.1	6.25	54.7

TABLE III (Continued)

Diameter (2C) of central plug =							
3				4			
Nuclear diameter	*F*	Nuclear diameter	*F*	Nuclear diameter	*F*	*F*	Nuclear diameter
3.3	92.8	5.3	44.0	4.3	95.0	6.3	53.9
3.35	91.3	5.35	43.3	4.35	93.9	6.35	53.2
3.4	89.7	5.4	42.5	4.4	92.8	6.4	52.5
3.45	88.1	5.45	41.8	4.45	91.5	6.45	51.8
3.5	86.5	5.5	41.1	4.5	90.3	6.5	51.1
3.55	84.8	5.6	39.7	4.55	89.1	6.55	50.4
		5.7	38.4				
3.6	83.1	5.8	37.2	4.6	87.9	6.6	49.8
3.65	81.6	5.9	36.1	4.65	86.8	6.65	49.1
3.7	80.0			4.7	85.7	6.7	48.5
3.75	78.4	6.0	35.0	4.75	84.4	6.75	47.9
3.8	76.8	6.1	34.0	4.8	83.1	6.8	47.3
3.85	75.3	6.2	33.0	4.85	81.9	6.85	46.7
3.9	73.7	6.3	32.0	4.9	80.7	6.9	46.1
3.95	72.4			4.95	79.6	6.95	45.6
		6.4	31.1				
4.0	71.1	6.5	30.2	5.0	78.4	7.0	45.0
4.05	69.7	6.6	29.2	5.05	77.3	7.1	43.7
4.1	68.3	6.7	28.7	5.1	76.1	7.15	43.1
4.15	66.9	6.8	27.7	5.15	75.0	7.2	42.5
4.2	65.5	6.9	27.0	5.2	73.8	7.25	41.9
4.25	64.4			5.25	72.7	7.3	41.4
4.3	63.3	7.0	26.2	5.3	71.7		
4.35	62.2	7.1	25.5	5.35	70.6	7.4	40.4
4.4	61.0	7.2	24.8	5.4	69.6	7.5	39.4
4.45	59.9	7.3	24.2	5.45	68.6	7.6	38.5
4.5	58.7	7.4	23.6	5.5	67.6	7.7	37.5
4.55	57.6	7.5	23.1	5.55	66.5	7.8	36.6
		7.6	22.5			7.9	35.5
4.6	56.5	7.7	22.0	5.6	65.7		
4.65	55.5			5.65	64.8	8.0	35.0
4.7	54.5			5.7	63.8	8.1	34.2
4.75	53.5			5.75	62.9	8.2	33.5
4.8	52.5			5.8	62.0	8.3	32.8
4.85	51.6			5.85	61.1	8.4	32.1
4.9	50.7			5.9	60.3	8.5	31.3
4.95	49.8			5.95	59.5	8.6	30.6
						8.7	

REFERENCES

Alfert, M., and Swift, H. (1953). *Exptl. Cell Research* **5,** 455.
Bader, S. (1953). *Proc. Soc. Exptl. Biol. Med.* **82,** 312.

Baker, J. (1942). *In* "Cytology and Cell Physiology" (G. Bourne, ed.). Oxford Univ. Press, London and New York.
Bauer, H. (1933a). *Z. Zellforsch. u. mikroskop. Anat.* **15,** 225.
Bauer, H. (1933b). *Z. mikroskop.-anat. Forsch.* **33,** 143.
Bern, H. A., and Alfert, M. (1954). *Rev. brasil. biol.* **14,** 25.
Boivin, A., Vendrely, R., and Vendrely, C. (1948). *Compt. rend.* **226,** 1061.
Brachet, J. (1947). *Compt. rend. soc. biol.* **142,** 1280.
Brode, W. R. (1949). "Chemical Spectroscopy." Wiley, New York.
Bryan, J. H. D. (1951). *Chromosoma* **4,** 369.
Carr, J. S. (1945). *Nature* **156,** 143.
Caspersson, T. (1932). *Biochem. Z.* **253,** 97.
Caspersson, T. (1936). *Skand. Arch. Physiol.* **73** suppl. 8, 1.
Caspersson, T. (1940). *Chromosoma* **1,** 605.
Caspersson, T. (1950). "Cell Growth and Cell Function." Norton, New York.
Caspersson, T. (1954). *Anais acad. brasil. cienc.* **26,** 199.
Caspersson, T., Lomakka, G., Svensson, G., and Säfström, R. (1955). *Exptl. Cell Research suppl. 3,* 40.
Catcheside, D. S., and Holmes, B. E. (1947). *Symp. Soc. Exptl. Biol.* **No. 1,** 225.
Chargaff, E., and Davidson, J. N., (eds.) (1955). "The Nucleic Acids," Vols. I and II. Academic Press, New York.
Cole, J. W., and Leuchtenberger, C. (1956). *Surgery* **40,** 113.
Commoner, B. (1948). *Ann. Missouri Botan. Garden* **35,** 239.
Commoner, B. (1949). *Science* **110,** 31.
Commoner, B., and Lipkin, D. (1949). *Science* **110,** 41.
Dahlberg, G. (1948). "Statistical Methods for Medical and Biological Students." George Allen & Unwin, London.
Danielli, J. F. (1953). "Cytochemistry. A Critical Approach." Wiley, New York.
Davidson, J. N. (1953). "The Biochemistry of the Nucleic Acids." Methuen, London.
Davies, H. G., and Walker, P. M. B. (1953). *Progr. in Biophys. and Biophys. Chem.* **3,** 195.
Davies, H. G., Wilkins, M. H. F., and Boddy, R. G. H. B. (1954). *Exptl. Cell Research* **6,** 550.
Deeley, E. M. (1955). *J. Sci. Instr.* **32,** 263.
Delamater, E. D., Mescon, H., and Barger, J. D. (1950). *J. Invest. Dermatol.* **14,** 133.
Dempster, W. T. (1944). *J. Opt. Soc. Am.* **34,** 695.
Diermeier, H. F., DiStefano, H. S., Tepperman, J., and Bass, H. D. (1951). *Proc. Soc. Exptl. Biol. Med.* **77,** 769.
DiStefano, H. (1948). *Chromosoma* **3,** 282.
DiStefano, H. (1952). *Stain Technol.* **27,** 171.
Ely, J. O., and Ross, M. H. (1949). *Anat. Rec.* **104,** 103.
Fautrez, J., and Moerman, J. (1954). *Compt. rend.* **80,** 554.
Feulgen, R. (1932). *In* "Handbuch der biologischen Arbeitsmethoden," Abt. V, Teil 2, II, 1055. Urban and Schwarzenberg, Berlin.
Feulgen, R., and Rossenbeck, H. (1924). *Z. physiol. Chem.* **135,** 203.
Feulgen, R., and Voit, K. (1924a). *Z. physiol. chem.* **135,** 249.
Feulgen, R., and Voit, K. (1924b). *Arch. ges. Physiol. Pflüger's* **206,** 289.
Firket, H., Chèvremont-Comhaire, S., and Chèvremont, M. (1955). *Nature* **176,** 1075.
Fisher, R. A. (1950). "Statistical Methods for Research Workers." Oliver and Boyd, London.
Frazer, S. C., and Davidson, J. N. (1953). *Exptl. Cell Research* **4,** 316.

Gibb, T. R., Jr. (1942). "Optical Methods of Chemical Analysis." McGraw-Hill, New York.
Glick, D. (1953). *Intern. Rev. Cytol.* **2,** 447.
Glick, D., Engström, A., and Malmström, B. G. (1951). *Science* **114,** 253.
Gomori, G. (1952). "Microscopic Histochemistry." Univ. Chicago Press, Chicago, Illinois.
Govaert, J. (1953). *Compt. rend.* **147,** 1494.
Grey, D. S. (1952). *Lab. Invest.* **1,** 85.
Grundmann, E., and Marquardt, S. (1953). *Chromosoma* **6,** 115.
Hammarsten, E. (1924). *Biochem. Z.* **144,** 383.
Harkin, J. C. (1956). A. M. A. *Arch Pathol.* **61,** 24.
Hoover, C. R., and Thomas, L. E. (1951). *J. Natl. Cancer Inst* **12,** 219.
Hotchkiss, R. D. (1955). *In* "The Nucleic Acids" (E. Chargaff and J. N. Davidson, eds.), Vol. II, p. 435. Academic Press, New York.
Hsu, T. C. (1952). *J. Heredity* **43,** 167.
Hughes-Schrader, S. (1951). *Biol. Bull* **100,** 178.
Hughes-Schrader, S. (1953). *Chromosoma* **6,** 79.
Ito, S., and Leuchtenberger, C. (1955). *Chromosoma* **7,** 328.
Kasten, F. H. (1956a). *Physiol. Zoöl.* **29,** 1.
Kasten, F. H. (1956b). *J. Histochem. and Cytochem.* **4,** 462.
Kaufmann, B. P., McDonald, M. R., and Gay, H. (1951). *J. Cellular Comp. Physiol.* **38,** suppl. 1, 71.
Köhler, A. (1893). *Z. wiss. Mikroskop.* **10,** 433.
Köhler, A. (1931). "Handbuch der biologischen Arbeitsmethoden," Abt. 2, Teil 2, No. 2, 1691. Urban and Schwarzenberg, Berlin.
Kossel, A. (1891). *Arch. Anat. u. Physiol., Physiol. Abt.,* p. 181.
La Cour, L. F., Deeley, E. M., and Chayen, J. (1956). *Nature* **177,** 272.
Lessler, M. A. (1951). *Arch. Biochem. Biophys.* **32,** 42.
Lessler, M. A. (1953). *Intern. Rev. Cytol.* **2,** 231.
Leuchtenberger, C. (1950). *Chromosoma* **3,** 449.
Leuchtenberger, C. (1954a). *In* "Statistics and Mathematics in Biology" (O. Kempthorne, T. A. Bancroft, J. W. Gowen and J. L. Lush, eds.), p. 557. Iowa State Coll. Press, Ames, Iowa.
Leuchtenberger, C. (1954b). *Science* **120,** 1022.
Leuchtenberger, C. (1954c). *Lab. Invest.* **3,** 132.
Leuchtenberger, C., and Lund, H. Z. (1952). *Cancer Research* **12,** 278.
Leuchtenberger, C., and Schrader, F. (1951). *Biol. Bull.* **101,** 95.
Leuchtenberger, C., and Schrader, F. (1952). *Proc. Natl. Acad. Sci. U. S.* **38,** 99.
Leuchtenberger, C., and Schrader, F. (1955). *J. Biophys. Biochem. Cytol.* **1,** 615
Leuchtenberger, C., Doolin, P. F., and Kutsakis, A. H. (1955). *J. Biophys. Biochem. Cytol.* **1,** 385.
Leuchtenberger, C., Helweg-Larsen, H. F., and Murmanis, L. (1954a). *Lab. Invest.* **3,** 245.
Leuchtenberger, C., Klein, G., and Klein, E. (1952a). *Cancer Research* **12,** 480.
Leuchtenberger, C., Leuchtenberger, R., Vendrely, C., and Vendrely, R. (1952b). *Exptl. Cell Research* **3,** 240.
Leuchtenberger, C., Leuchtenberger, R., and Davis, A. M. (1954b). *Am. J. Pathol.* **30,** 65.
Leuchtenberger, C., Leuchtenberger, R., Schrader, F., and Weir, D. R. (1956a). *Lab. Invest.* **5,** 422.

Leuchtenberger, C., Murmanis, I., Murmanis, L., Ito, S., and Weir, D. R. (1956b). *Chromosoma* **8,** 73.
Leuchtenberger, C., Schrader, F., Hughes-Schrader, S., and Gregory, P. W. (1956c). *J. Morphol.* **99,** 481.
Leuchtenberger, C., Schrader, F., Weir, D. R., and Gentile, D. P. (1953). *Chromosoma* **6,** 61.
Leuchtenberger, C., Vendrely, R., and Vendrely, C. (1951). *Proc. Natl. Acad. Sci. U. S.* **37,** 33.
Leuchtenberger, C., Weir, D. R., Schrader, F., and Leuchtenberger, R. (1954c). *Excerpta Med. Sect. I* **8,** 418.
Leuchtenberger, C., Weir, D. R., Schrader, F., and Murmanis, L. (1955). *J. Lab. Clin. Med.* **45,** 851.
Levene, P., and Bass, L. (1931) "Nucleic Acids." Chemical Catalog, New York.
Li, C., and Stacey, M. (1949). *Nature* **163,** 538.
Liebhafsky, H. A., and Pfeiffer, H. G. (1953). *J. Chem. Educ.* **30,** 450.
Lison, L. (1953). *Science* **118,** 382.
Lison, L., and Pasteels, J. (1950). *Bull. acad. roy. méd. Belg.* **36,** 348.
Lison, L., and Pasteels, J. (1951). *Arch. biol. (Liège)* **62,** 1.
Lison, L., and Valeri, V. (1956). *Chromosoma* **7,** 497.
McCarty, M. (1946). *Bacteriol. Revs.* **10,** 63.
Marinone, G. (1951). *Sang Le* **22,** 89.
Marshall, C., Coltman, J. W., and Bennett, A. J. (1948). *Rev. Sci. Instr.* **19,** 744.
Mellors, R. C., Berger, R. E., and Streim, H. G. (1950). *Science* **111,** 627.
Métais, P., and Mandel, P. (1950). *Compt. rend.* **144,** 277.
Miescher, F. (1871). *H.-S. Mediz.-chem. Unters. Labor. angew. Chem.,* **4,** 441.
Miescher, F. (1897). "Histochemischen und Physiologischen Arbeiten." Vogel, Leipzig.
Miller, E. S. (1939). "Quantitative Biological Spectroscopy." Burgess, Minneapolis, Minnesota.
Milovidov, P. F. (1954). *Protoplasma Monograph.* **20, 21.**
Mirsky, A. E., and Ris, H. (1951). *J. Gen. Physiol.* **34,** 451.
Moore, B. C. (1952). *Chromosoma* **4,** 563.
Moritz, A. R., and Leuchtenberger, C. (1955). *Am. J. Pathol.* **31,** 564.
Moses, M. J. (1951). *J. Natl. Cancer Inst.* **12,** 205.
Moses, M. J. (1952). *Exptl. Cell Research suppl.* **2,** 75.
Moses, M. J., and Yerganian, C. (1952). *Records. Genet. Soc. Am.* **No. 21,** 51.
Naora, H. (1951). *Science* **114,** 279.
Naora, H. (1952). *Science* **115,** 248.
Naora, H. (1955). *Exptl. Cell Research* **8,** 259.
Naora, H., Matsuda, H., Fukuda, M., and Sibatani, A. (1951). *J. Japan. Chem.* **5,** 729.
Nelson, E. M. (1910). *J. Roy. Microscop. Soc.* **00,** 282.
Ornstein, L. (1952). *Lab. Invest.* **1,** 250.
Ornstein, L., and Pollister, A. W. (1952). *Science* **116,** 203.
Overend, W., and Stacey, M. (1949). *Nature* **163,** 538.
Pasteels, H., and Lison, L. (1950). *Arch. biol. (Liège)* **61,** 445.
Patau, K. (1952). *Chromosoma* **5,** 341.
Patau, K., and Swift, H. (1953). *Chromosoma* **6,** 149.
Pollister, A. W. (1950). *Rev. hématol.* **5,** 527.
Pollister, A. W. (1952). *Lab. Invest.* **1,** 106.

Pollister, A. W., and Moses, M. J. (1949). *J. Gen. Physiol.* **32,** 567.
Pollister, A. W., and Ornstein, L. (1955). *In* "Analytical Cytology" (R. C. Mellors, ed.). McGraw-Hill, New York.
Pollister, A. W., and Ris, H. (1947). *Cold Spring Harbor Symposia Quant. Biol.* **12,** 147.
Pollister, A. W., and Swift, H. (1950). *Science* **111,** 68.
Reisner, E. H., and Korson, R. (1951). *Blood* **6,** 344.
Ris, H., and Mirsky, A. E. (1949). *J. Gen. Physiol.* **33,** 125.
Roels, H. (1954). *Nature* **174,** 514.
Schiff, H. (1866). *Compt. rend.* **64,** 182.
Schmidt, W. J. (1947). *Naturwissenschaften* **26,** 413.
Schneider, W. (1945). *J. Biol. Chem.* **161,** 293.
Schrader, F., and Leuchtenberger, C. (1950). *Exptl. Cell Research* **1,** 421.
Schrader, F., and Leuchtenberger, C. (1952). *Exptl. Cell Research* **3,** 136.
Schwarzschild, K., and Villiger, W. (1906) *Astrophys. J.* **23,** 286.
Shillaber, C. P. (1944). "Photomicrography in Theory and Practice." Wiley, New York.
Sibatani, A. (1950). *Nature* **166,** 365.
Stowell, R. (1945). *Stain Technol.* **20,** 45.
Stowell, R., and Albers, V. (1943). *Stain Technol.* **18,** 57.
Swartz, F. J. (1956). *Chromosoma* **8,** 53.
Swift, H. (1950a). *Physiol. Zoöl.* **23,** 169.
Swift, H. (1950b). *Proc. Natl. Acad. Sci. U. S.* **36,** 643.
Swift, H. (1953). *Intern. Rev. Cytol.* **2,** 1.
Swift, H. (1955). *In* "The Nucleic Acids" (E. Chargaff and J. N. Davidson, eds.), Vol. II, p. 51. Academic Press, New York.
Swift, H., and Rasch, E. (1956). *In* "Physical Techniques in Biological Research" (G. Oster and A. W. Pollister, eds.), Vol. III, p. 354. Academic Press, New York.
Tanaka, T. (1951). *Kromosomo* **2,** 39.
Thorell, B., and Ruch, F. (1951). *Nature* **167,** 815.
Uber, F. M. (1939). *Am. J. Botany* **26,** 799.
Vendrely, C. (1952). *Bull. biol. France et Belg.* **86,** 1.
Vendrely, R. (1955). *In* "The Nucleic Acids" (E. Chargaff and J. N. Davidson, eds.), Vol. II, p. 155. Academic Press, New York.
Vendrely, R., and Vendrely, C. (1948). *Experientia* **4,** 434.
Vendrely, R., and Vendrely, C. (1949). *Experientia* **5,** 327.
Vialli, M., and Perugini, S. (1954). *Riv. istochim. norm. e patol.* **1,** 149.
Walker, P. M. B., and Deeley, E. M. (1956). *Exptl. Cell Research* **10,** 155.
Widström, G. (1928). *Biochem. Z.* **199,** 298.
Wieland, H., and Scheuing, G. (1921). *Ber. deut. chem. Ges.* **54,** 2527.
Wilkins, M. H. F. (1950). *Discussions Faraday Soc.* **9,** 363.

AUTORADIOGRAPHY AS A CYTOCHEMICAL METHOD, WITH SPECIAL REFERENCE TO C^{14} AND S^{35}

S. R. Pelc

Medical Research Council, Biophysics Research Unit, King's College, London, England

I. INTRODUCTION

Autoradiography makes it possible to locate radioactive material in a specimen, and since it is possible to obtain a resolution of 1–2 μ, autoradiography can be regarded as a cytochemical technique. Animal and plant cells normally contain only negligible traces of radioactive elements, and the method is highly specific for the tracer. Radioactive isotopes of stable elements offered in a given chemical form to an organism may be incorporated into some of the compounds in the tissues;

whether or not they are taken up will depend on the metabolic state of the tissue. An autoradiograph therefore can reveal the location of a compound which has been metabolically active during the time of application of labeled material and fixation; it will not show a compound which has been inert during this period of time. Biosynthesis of a compound normally involves many steps which may or may not be performed in the same part of a cell or tissue, and fully formed compounds may move subsequently by secretion or movement of the cells in which they are located. Frequently these movements can be followed by preparing autoradiographs of specimen fixed at various times after labeling.

Autoradiography thus makes it possible to investigate directly the dynamics of the metabolic processes within cells and tissues and considerably extends the scope of cytochemistry, whose task until recently was regarded as exploring the location and amounts of chemical compounds in cells.

It is the purpose of this article to deal with autoradiography as a cytochemical method, and various aspects of the subject, such as technical details and applications outside the scope of cytochemistry, may be only mentioned or entirely omitted. Experience has shown that the actual preparation of autoradiographs is relatively easy; the main difficulties are found in choosing a suitable labeled precursor and in interpreting results. For these reasons I wish to emphasize the methodology rather than the techniques of autoradiography.

II. PHYSICAL PRINCIPLES OF AUTORADIOGRAPHY

A. Introduction

If a specimen containing radioactive material is placed in contact with a photographic emulsion (Fig. 1), the ionizing radiations emitted during radioactive decay will change the emulsion in such a way that a blackening is produced in the film after development. The number of

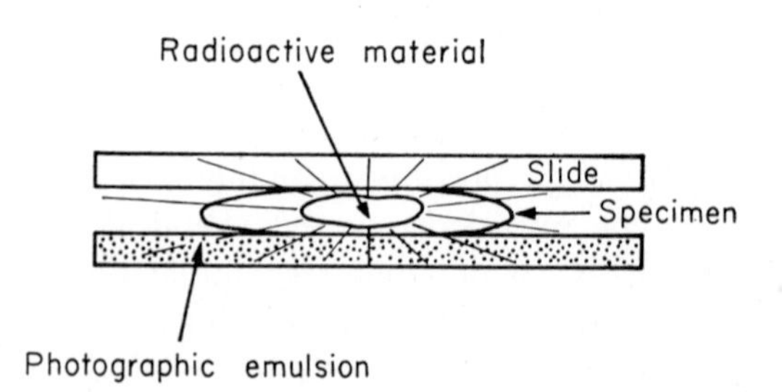

FIG. 1. Basic scheme of autoradiograph.

developed grains in the film depends on the characteristics of the emulsion, the physical properties, and number of ionizing particles which have hit the film. The size and shape of the photographic image should ideally be identical with that of the structure containing the tracer. This ideal could be realized if all ionizing particles were emitted perpendicular to the surface of the specimen; since they are emitted in all directions, some spread is inevitable.

B. Photographic Action of Ionizing Radiations

Photographic emulsions are not genuine emulsions, but suspensions of silver bromide crystals in gelatin; the size, and to some extent the chemical properties, of the crystals or grains depend on the manufacture.

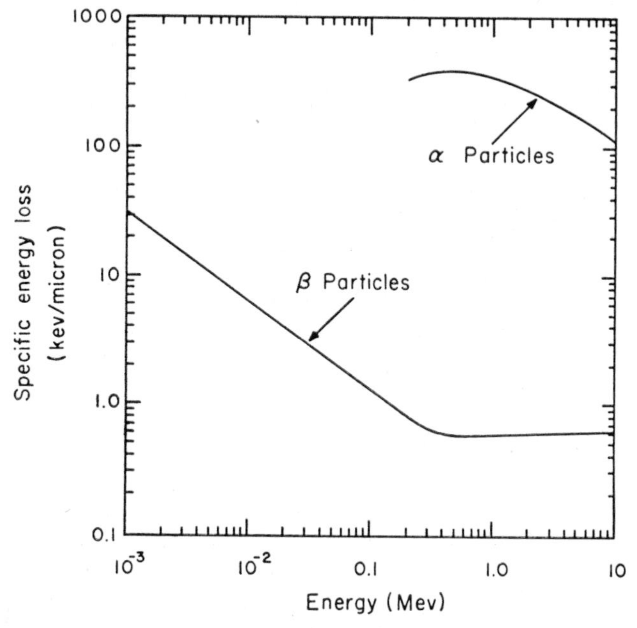

Fig. 2. Specific energy loss in nuclear emulsions as a function of particle energy (from Norris and Woodruff, 1955).

During an exposure to light, or to ionizing radiations, some of the grains are changed in such a way that the developer reduces them to silver, while only a very small number is reduced without previous exposure. Subsequent fixation dissolves the unused silver bromide and hardens the soft gelatin.

The theory of the photographic process was proposed by Gurney and Mott (1938) and later modified by Mitchell (1949). According to this theory dissipation of energy in an AgBr crystal results in the liberation of electrons by a photoelectric effect. These electrons are mobile in a perfect silver bromide crystal, but are trapped by impurities. It is assumed that small accumulations of silver sulfide, the so-called sensitivity specks, act as traps. The additional electrons give these specks a negative charge, and positive silver ions are attracted and form a small accumulation of metallic silver, which catalyzes reduction of the silver bromide crystal to silver during development. A minimal number of silver atoms is necessary to form an effective catalyst.

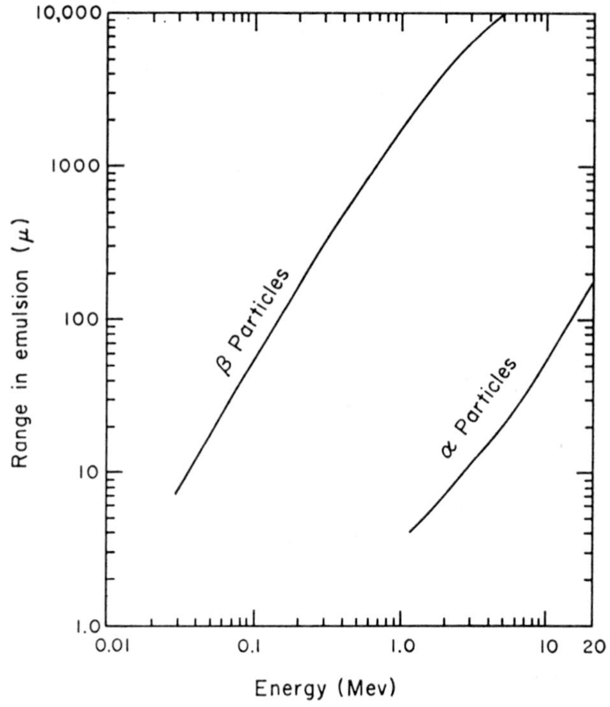

FIG. 3. Range in nuclear emulsion as a function of particle energy (from Norris and Woodruff, 1955).

Ionizing particles lose energy at a different rate, usually expressed in keV per micron. On this specific energy loss (Fig. 2) and the detailed characteristics of the grains depends the number of developed grains per unit length of path. Larger grains will absorb more energy and will therefore tend to be more sensitive, but the mode of manufacture of a given emulsion exerts a great influence on the intrinsic sensitivity of grains. The range in nuclear track emulsions of α- and β-particles of various energies is shown in Fig. 3.

It would be surprising if all grains were perfectly manufactured; some will carry a silver speck instead of a sensitivity speck, and be developed without previous exposure; others will be relatively insensitive. Therefore a few grains are developed in a completely unexposed emulsion, to which number have to be added the grains which were accidentally made developable by dark-room light, cosmic rays, and other sources of ionizing radiations. When judging autoradiographs, the observer has to evaluate the concentration of grains above the background value, which necessitates greatest care in avoiding unnecessary background, especially since autoradiographs of low grain density offer the advantage of better resolving power, less obscuring of cytological detail, and economy in the use of radioactive material.

The liquid emulsion technique (Bélanger and Leblond, 1946), especially when electron track emulsions are used, or the stripping film technique (Pelc, 1947; Doniach and Pelc, 1950) can be used for cytochemical purposes. The experimenter has a wide choice of emulsions which can be poured in varying thicknesses to give autoradiographs by the original technique of Bélanger and Leblond, for which it is therefore impossible to quote any figures of general validity for the response of the emulsion.

C. Quantitative Autoradiography

In track emulsions each particle hitting the film will produce one track, and the content of radioactive material can be calculated if the number of particles per decaying atom is known for a given tracer (usually one). For stripping film (Kodak Ltd.) the relevant figures are given in Table I (Cormack, 1955). No recent determinations are available for S^{35} and C^{14}; the values given in Table I are estimations.

TABLE I
Grain Yields in Stripping Film (Kodak Ltd.)

Isotope	Maximum energy (MeV.)	Grains per β-particle
P^{32}	1.7	0.78 ± 0.10
I^{131}	0.25 – 0.82	1.8 ± 0.2
Fe^{59}	0.26 – 0.46	1.6 ± 0.3
C^{14}	0.15	approx. 2.0
S^{35}	0.17	approx. 2.0

Using P^{32}, an exposure of 5.6×10^8 β-particles per cm.^{2} is necessary to obtain an optical density (extinction) of 0.5 above fog with this film (Herz, 1950, 1951). The lowest densities which might be conveniently measured by densitometry can be assumed to be approximately 0.1, needing an exposure of the order of 1.1×10^8 β-particles per cm.^{2} or 110 β-particles per 100 μ^2, the approximate area of a small cell. Experience has shown that such a heavy autoradiograph obscures all cytological detail, while autoradiographs of 10–30 grains per cell are easily observable without interfering with the microscopic image of the specimen. For this reason counting of developed grains or tracks is preferable to densitometry in cytochemical applications of autoradiography, though it can be of value in histochemical work (Dudley and Dobyns, 1949; Odeblad, 1952).

The number of grains per unit area is proportional to the exposure up to approximately 100 grains per 100 μ^2 (Dudley, 1954; Andresen *et al.*, 1952; Lamerton and Harriss, 1954; Taylor *et al.*, 1955), if fading of the latent image can be neglected (Ray and Stevens, 1953), which is the case for exposure times of up to 3 weeks under the usual conditions of exposure, or for longer periods if the preparations are kept under N_2 or CO_2.

Radioactive decay and the grain yield have to be taken into account to calculate the total incorporation of tracer in a specimen. For emitters of very soft β-rays, such as tritium, self-absorption in the specimen has to be allowed for. One-half of the particles are emitted in the opposite direction from the photographic emulsion, and therefore the true number of particles emitted by the material in the specimen will be twice the number indicated by the emulsion. Details regarding the calculations for evaluating the raw data will be found in Appendix 1.

Counting of photographic grains by eye is very laborious and tiring, though consistent results can be obtained after some experience. An automatic grain counter based on the principle of the flying-spot-microscope has been designed (Dudley and Pelc, 1953), but further development of the electronic system is necessary before the instrument can be put to routine uses. Odeblad (1952), and Mazia *et al.* (1955) have developed techniques of measuring the density of small areas in autoradiographs.

D. Resolving Power

Highest accuracy of localization of radioactive material is essential if autoradiography is to be useful as a tool in cytochemistry. Particles are emitted in all directions and will therefore tend to cause blackening of the photographic emulsion over an area of the film which is wider than the area in which the tracer is concentrated. The spread of the image obviously cannot be more than the range of the particles; in fact the resolving power of autoradiographs is much better than could be achieved if the range of particles were of primary importance. The resolving power is, however, somewhat better for tracers emitting soft β-particles than hard ones. It is fortunate for cytochemistry that H^3, C^{14}, and S^{35} are very favorable in this respect.

The main factors in determining resolving power are the geometrical arrangement and the grain size of the photographic emulsion. Obviously, the resolving power of an autoradiograph cannot be better than the diameter of the grains. Since resolving powers of 1–2 μ can be achieved with the best techniques available at present, photographic emulsions with grains of 0.2–0.3 μ diameter can be regarded as a reasonable compromise. The gain in resolving power through using finer-grained emulsions would not be appreciable, and could only be obtained at the expense of sensitivity and bad visibility of the grains at low power.

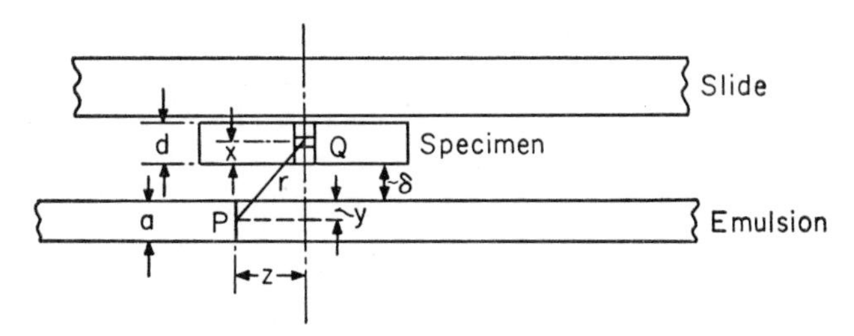

FIG. 4. Schematic representation of source and emulsion in autoradiography for calculating resolving power.

Doniach and Pelc (1950) calculated the dependence of resolving power on the geometrical arrangement, assuming that the particles proceed in straight lines and therefore the intensity of radiation diminishes with the square of the distance. If a point source of radioactive material is located in the specimen at point Q, the intensity of radiation at point P in the emulsion will be inversely proportional to the square of the distance r (see Fig. 4). The resolving power will be good when the ratio of intensities at P to P_0 a point on the perpendicular is as small as possible. Numerical values for the number of grains at a cross-section of the film at various distances from the perpendicular were found by forming the integral of the intensity at P over the thickness of emulsion a. Since the radioactive material might be at any depth in the specimen, the numerical values given in Table II were calculated for a line source (instead of a point source) of radioactive material subtended in the specimen which is equivalent to calculating an average value at different depths. Twice the distance at which the integral falls to one-half of its maximal value was taken as a measure of the resolving power.

TABLE II
RESOLVING POWER CALCULATED BY THE METHOD OF DONIACH AND PELC
(All data in microns)

Thickness of		Gap	Resolving power
Specimen	Emulsion		
d	a		
5	15	0.1	3
5	15	1.0	9
5	15	3.0	17
2	2	0.1	2
2	2	1.0	5
2	5	0.1	2
2	5	1.0	5.5
2	5	3.0	10.0

It can be seen from Table II that the gap between specimen and photographic emulsion is of great influence on the resolving power. This point has to be considered when the specimen has to be coated with a layer of impermeable material such as celloidin, etc. Layers thicker than 0.1 μ will seriously affect resolving power. Table II also shows that the thickness of both the specimen and emulsion influence the resolving power of autoradiographs. It is clear that better resolving powers could be obtained by the use of much thinner sections and emulsions, which is technically feasible. The sensitivity of such preparations would be lower because, for a given specimen, the amount of tracer available is proportional to the thickness; thinner emulsions contain fewer grains per unit area than thicker ones, and smaller-grained emulsions, which are usually less sensitive, would have to be used.

Observations on biological materials and experimental work by

TABLE III
RESOLVING POWER FOR DISC SOURCES (NADLER) OF 40 μ DIAMETER
(All data in microns)

Thickness of		Gap	Resolving power
Specimen	Emulsion		
2	5	0.1	5.0
5	5	0.1	6.0
5	5	0.5	6.4
5	10	0.1	8.0
5	20	0.1	9.0
5	35	0.1	10.2

Stevens (1950) showed that this theory provides a reasonably good guide, in spite of the somewhat artificial assumption of a line source.

Nadler (1951) calculated the resolving power for cylindrical discs of 40 μ diameter, taking as a measure the distance from the edge of the source at which the density falls to one-half. The results shown in Table III indicate that the resolving power is definitely worse for such large sources. Nadler's calculations should also apply when relatively small organelles are found in a relatively large labeled source, e.g., cell nuclei in cells with considerable amounts of labeled cytoplasm.

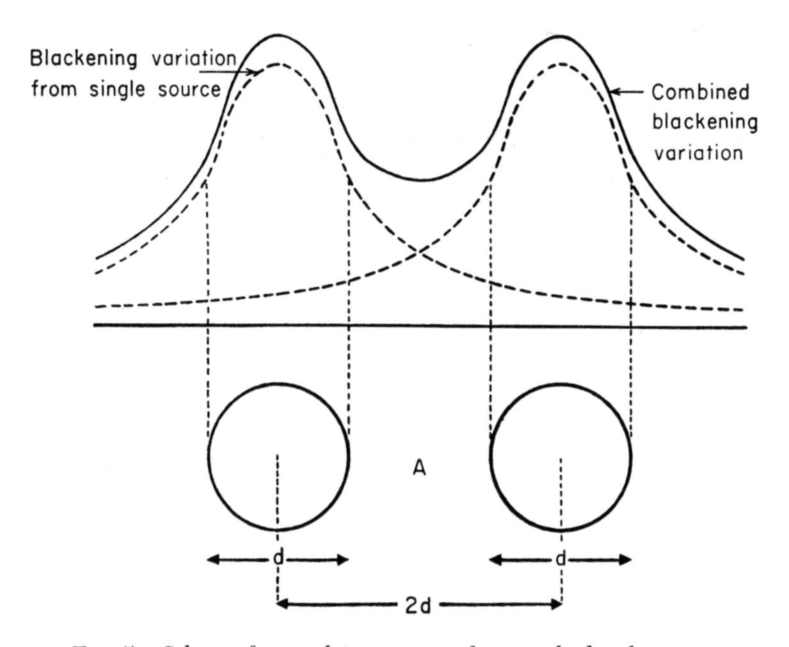

FIG. 5. Scheme for resolving power of two cylindrical sources.

To avoid the objections which can be raised against these calculations, Lamerton and Harriss (1954) suggested the following definition of resolving power: "The resolution of a given autoradiographic technique shall be defined as the distance d, if the images of two uniformly active cylindrical sources of diameter d can just be resolved when the centres are separated by a distance d" (Fig. 5). For thin preparations, the results of their calculations listed in Table IV do not differ significantly from those in Table II. These authors also found that relatively low concentrations of tracer evenly distributed in a specimen impair the observability of the labeling of enclosed small sources.

TABLE IV
RESOLVING POWER ACCORDING TO LAMERTON AND HARRISS
(All data in microns)

Thickness of		Gap	Resolving power
Specimen	Emulsion		
2	2	0	2.1
2	2	0.5	3.4
5	5	0	5.1
5	5	0.5	6.4
2	5	0	3.3
2	5	0.5	5.0
5	20	0	9.3
5	20	0.5	20.6

All authors are unanimous in finding that heavy exposures lead to bad resolving power.

III. CONCENTRATION OF RADIOACTIVE MATERIAL, PHYSIOLOGICAL DOSE, AND RADIATION DOSE

A certain minimal concentration of tracer in the specimen is necessary to produce an observable autoradiograph. This can only be obtained if sufficient tracer is applied to the organism, whose metabolic processes might, however, be affected by the amount of the chemical applied or by the radiation emitted by the tracer. These factors can represent a severe limitation of experiments involving autoradiography, and have to be considered in planning experiments. A method of calculation has been described by the present author (Pelc, 1951).

A. MINIMUM CONCENTRATION OF TRACER TO OBTAIN AUTORADIOGRAPHS

The calculations are based on the following considerations. In contrast autoradiography approximately 10 grains per 100 μ^2 of emulsion can be regarded as just sufficient, necessitating an exposure to $10/\delta$ β-particles, (where δ is the yield of developed grains per particle hitting the film), i.e., which are obtained from the decay of twice that number of radioactive atoms during the time of exposure. Exposure times of two half-lives for short-lived isotopes result in the decay of three-quarters of the atoms present, and longer times would therefore not add substantially to the density of an autoradiograph. For some isotopes exposure to two half-lives would be too long, and a given time of exposure has to be assumed for the calculations. Only a part of the specimen will usually be labeled, (e.g., nuclei, but not cytoplasm in experiments on DNA synthesis); consequently the average concentration of isotope per unit volume of tissue can be lower the smaller the proportion of actually labeled tissue (Fig. 6). Compilation of the various factors gives the following equations for the minimum concentration of tracer at the time autoradiographs are started.

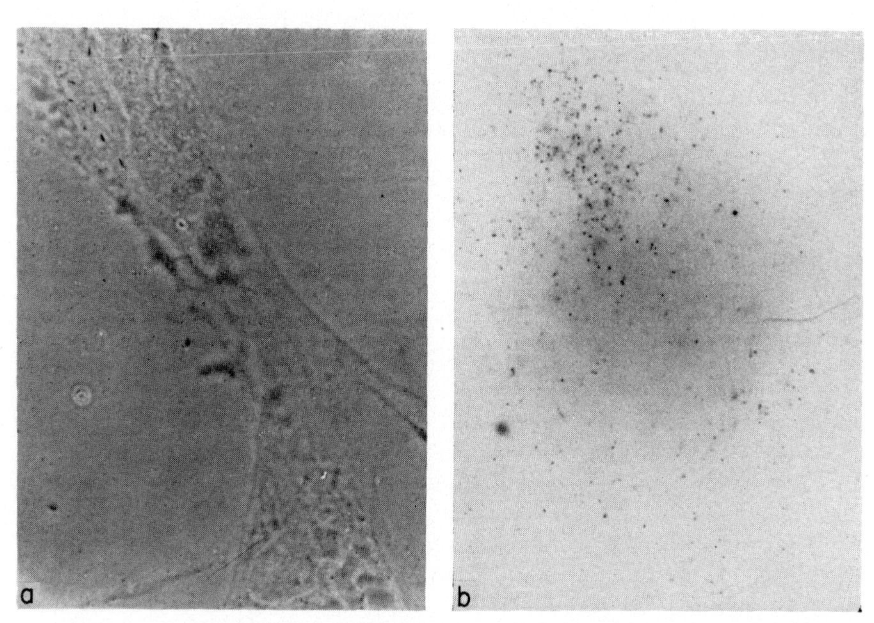

FIG. 6. Incorporation of P^{32} as phosphate in cell nuclei of avian fibroblasts grown in tissue culture. A: Phase contrast; B: same field in conventional illumination.

For short-lived isotopes of half-life H in days, assuming 5 μ sections and emulsions and a proportion f of labeled to unlabeled tissue, the minimum concentration is:

$$C = \frac{11 \cdot 5f}{H\delta} \mu\text{C. per ml.} \qquad (1)$$

For long-lived isotopes, when radioactive decay during the time (d in days) of exposure can be neglected:

$$C = \frac{12 \cdot 5f}{d\delta} \mu\text{C. per ml.} \qquad (2)$$

For specimens of different thickness the values have to be changed accordingly; for smears and squashes, the cell volume before drying can be assumed to be the thickness of the specimen. Such calculations can only give very approximate results, since the knowledge of the basic data will usually be incomplete. For preliminary calculations of work with fast X-ray film or fine-grained stripping film δ can be taken as 20 or 1, respectively.

B. Amount of Tracer to be Applied to Obtain Autoradiographs

To arrive at concentrations as calculated by Eqs. 1 or 2 the following factors have to be taken into account: radioactive decay between the time of application of tracer and setting up of autoradiographs; the fate of the applied chemical compound in the organism, i.e., incorporation, secretion, and excretion; extraction of labeled compounds during preparation.

Radioactive decay is easily allowed for by the use of the methods outlined in Appendix 1; it is of importance when short-lived isotopes are used.

In most instances our knowledge of the fate of a given chemical in an organism is not sufficiently well known to allow of an accurate calculation, and the investigator will have to be satisfied with approximations along the following lines.

After application, the tracer is diluted by the pool of metabolites present in the organism, e.g., P^{32} injected into a mammal will mix with the free phosphate in the organism, and the labeled fraction has to be high enough to give an autoradiograph when a certain number of the mixture of labeled and stable atoms are incorporated. If the labeling of a given chemical compound is to be investigated, it will clearly be of advantage to use a precursor which is not present in large amounts in the organism; thus, to label the DNA in dividing cells in mice, concentrations of the order of 1 mc. per mouse of P^{32} compared with 25 μc. of C^{14}-labeled adenine had to be used.

It is important to use precursors which are as specific as possible for the chemical compound under investigation, to avoid incorporation of the tracer into other compounds which cannot be easily removed from the specimen. Thus C^{14} derived from labeled glycine is incorporated into proteins and nucleic acids, which would result in an autoradiograph indicating little else besides metabolic activity in general. More significant results can be obtained if specific precursors such as [8-C^{14}]-adenine for incorporation into nucleic acids and S^{35}-methionine for proteins are utilized.

The timing of incorporation of tracers, i.e., the time which elapses between application and fixation, is of paramount importance in experiments involving autoradiography. It can neither be taken for granted that compounds are synthesized where they are located in quantity by the more conventional cytochemical methods, nor that incorporation is always very quick. Thus Leblond *et al.* (1948) have found that P^{32} is rapidly incorporated into deoxyribonucleic acid in the small intestine of mice. The cells move rapidly up the villi and many labeled cells are already shed into the intestinal tract after 3 days (Fig. 7). The process of DNA synthesis is much slower in the testis, where additional

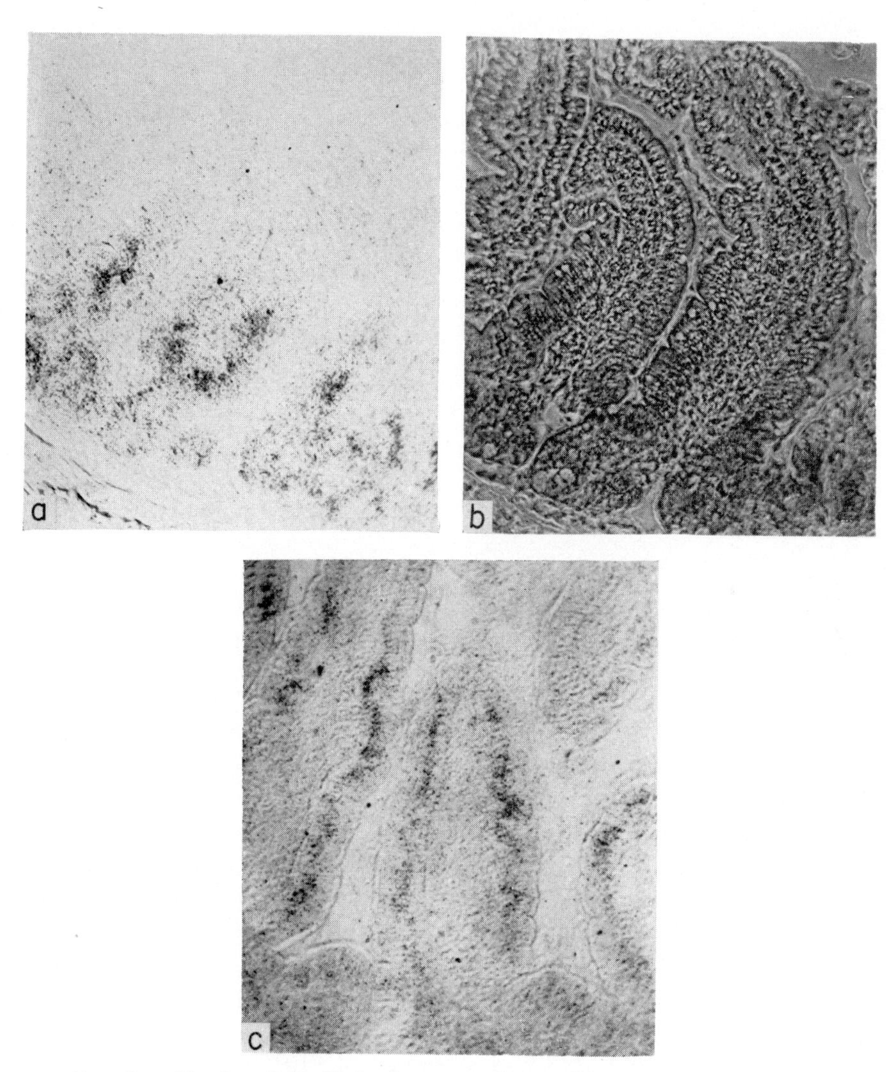

FIG. 7. Uptake of [8–C^{14}] adenine in the small intestine of a mouse. Cold perchloric acid extraction. A: Autoradiograph 24 hours after a single injection; B: same field as A, but phase contrast; C: autoradiograph 48 hours after a single injection.

spermatogonia still incorporate [8-C^{14}]-adenine into DNA 3–6 days after a single injection, while removal of labeled DNA begins after the full time of spermiogenesis (25–32 days after injection) when labeled sperm passes into the epididymis (Pelc and Howard, 1956b). After a single injection of labeled sulfate ($Na_2S^{35}O_4$) the tracer in the tracheal cartilage is at first incorporated into the cells which do not stain metachromatically (Fig. 8), then moves slowly into the strongly metachromatic matrix (Pelc and Glücksmann, 1955), where it is usually located.

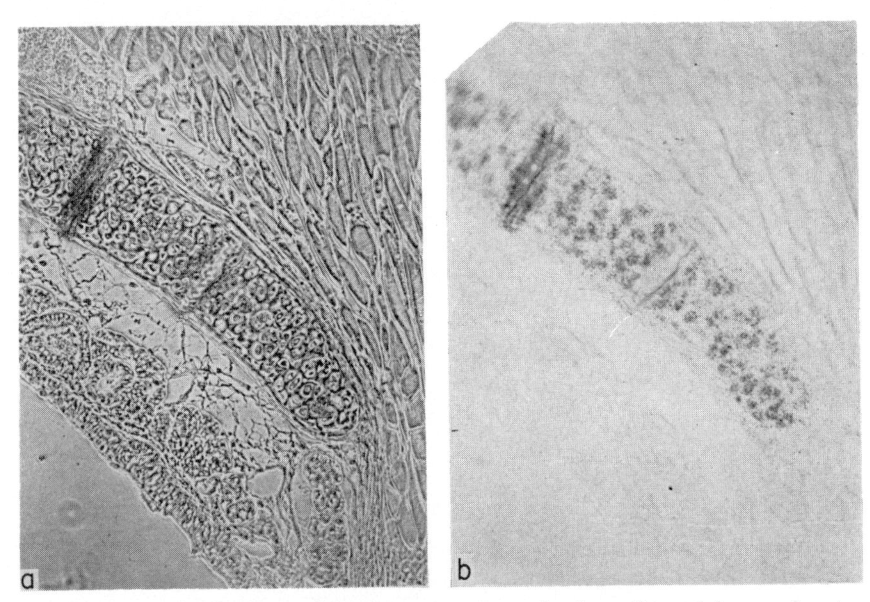

FIG. 8. Incorporation of $S^{35}O_4^-$ in mouse tracheal cartilage, 4 hours after injection. A: Phase contrast; B: autoradiograph.

Removal of tracer during preparation of the specimen is usually approached from two sides. As little as possible of the labeled compounds under investigation should be leached out, and ideally all of the unwanted labeled material should be removed. Compounds soluble in water, alcohol, and xylene are dissolved by the usual methods of fixation and preparation of sections. If such compounds are to be retained, good results can be obtained by freeze-drying followed by covering the specimen with a thin film of nylon before applying the photographic emulsion (Chapman-Andresen, 1953a), or by the use of Carbowax (Blank, 1949; Blank and McCarthy, 1950. See also Canny, 1955). A number of cytochemical procedures for the removal of certain compounds are well known, and many of these can be used if care is taken to check their efficiency for the purpose in hand. Loss of stainability or its retention do not necessarily imply that all components of one compound have been removed or that none of another have been lost. It is, for example, difficult to extract all labeled adenine with ribonuclease after incorporation into RNA (Fig. 9), possibly because some polynucleotides in the specimen withstand washing in water; treatment with cold perchloric acid has been found preferable. Conversely, hydrolysis of a specimen in normal hydrochloric acid at 60° C., which removes ribonucleic acid, does not affect P^{32} incorporated into DNA, but can remove C^{14}-adenine from it although the compound can be stained by the Feulgen reaction. Considerable improvement in the quality of the staining of autoradiographs can be obtained if the specimen is stained before the film is applied, but it has been found that occasionally tracer may be removed during the procedures involved.

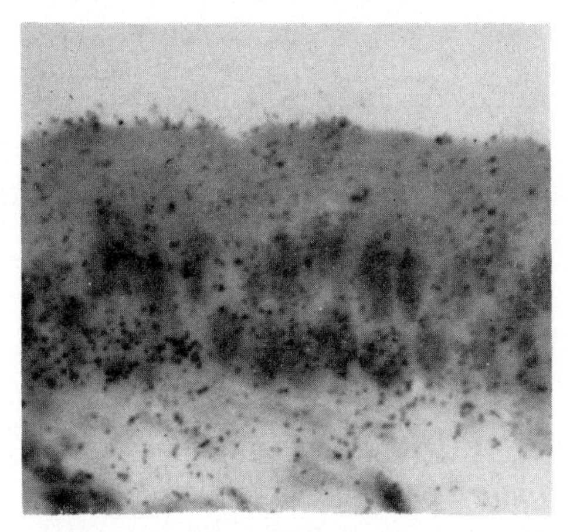

FIG. 9. Incorporation of [8–C^{14}]-adenine in the epithelium of the mouse trachea, 2 days after injection. Section treated with ribonuclease before autoradiography (hematoxylin and eosin).

It can be seen that considerable losses of tracer are frequently encountered before autoradiographs can be started, and the amount applied has to be large enough to result in a sufficient concentration of tracer after preparation. A method of calculating these amounts is given in Appendix 2; values of the amounts to be applied found suitable in experimental work can be found in the last section.

C. Physiological Dose and Radiation Dose

A given dose of a labeled compound, like an unlabeled compound, applied to an organism, can affect its normal metabolic processes. The total weight of tracer used is usually very small, and the experimenter is more frequently faced with the problem of subdividing small quantities or of the stability of extremely dilute solutions than with the danger of overdosing. It is of advantage to use compounds of high specific activity which make it possible to use smaller quantities for the same activity. To obtain an autoradiograph at least 30–40 radioactive atoms must decay during the time of exposure in a small cell or a cell nucleus, and occasionally it may be impossible to incorporate the requisite number of labeled molecules if the specific activity is low and only a small concentration of the compound in question is incorporated during the time of the experiment.

After its application, radioactive material present will decay and emit radiations which can affect the organism. The wide range of sensitivity found in living materials makes it impossible to quote a radiation dose which could be regarded as absolutely harmless. Doses of less than approximately 10 reps can, in nearly all cases, be regarded as safe; above this level the possibility of radiation damage has to be considered. The relation between concentration of evenly distributed radioactive material in a volume which is large compared with the range of particles emitted and the radiation dose due to β-rays emitted is approximately:

$$\text{d.r.} = 61\alpha\bar{E}_\beta = 20\alpha E_{\beta\max} \text{ reps per day} \tag{3}$$

where α is the concentration in μc. per ml, and $\bar{E}_\beta$ the average energy of the β-particles, and $E_{\beta\max}$ the usually quoted maximum energy in MeV. In small animals the γ-ray dose is very much smaller than that due to β-rays. If the tracer is unevenly distributed, Eq. 3 can only be used as a very rough approximation; a better approximation is obtained by carrying out the calculation under the assumption that the highest local concentration pertains. Such a result can be taken as valid for the volume which contains this high concentration and for adjacent parts. The total amount of radioactive material present in the organism will contribute to the radiation dose, and for this reason it is again advisable to use specific precursors because a higher relative amount of tracer which is finally utilized to that discarded can be obtained.

The mathematical treatment outlined in Appendix 2 shows that in experimental work it is preferable to use (a) isotopes with reasonably long half-life, since exposure times can be used which are long compared with the time the tracer is in the organism; (b) isotopes emitting radiations of low energy; (c) specific precursors. Radiation dose is now rarely a limiting factor in experiments on plants and animals, but severely limits successful autoradiography of human material.

IV. TECHNIQUES OF AUTORADIOGRAPHY

The following techniques of preparing autoradiographs suitable for cytochemical purposes are available: the coating technique, the inversion technique, the floating technique and the stripping film technique, technical details of which can be found in the original and various later publications (see References).

Bélanger and Leblond (1946) coated sections with liquid photographic emulsion. After cooling, the preparations are stored and processed after suitable exposure. In Bélanger's (1950) inversion technique, specimen and autoradiograph are inverted after processing, and the section stained. Emulsions of varying grain size can be applied in different thickness, which makes these techniques very flexible; however, for cytochemical work the very thin layers of emulsion of even thickness in contrast autoradiography are hard to achieve. The coating technique is usually used when α- or β-tracks are to be produced.

Evans (1947) and Endicott and Jagoda (1947) floated sections on a photographic plate or film which was then dried and left for exposure; the paraffin wax was removed before photographic processing. The thickness of the usual photographic materials does not allow of a sufficient resolving power and this technique is now rarely used in cytochemistry.

Stripping films are photographic emulsions which can be removed from their support. Autoradiographs are prepared by floating a piece on water (Pelc, 1947; Doniach and Pelc, 1950; Pelc, 1956a) until it is soaked through, and then lifted out in such a way that the film covers the specimen. The preparation is then dried, left for exposure, and processed. Resolving powers of 2–3 μ for emitters of hard β-rays and 1–2 μ for C^{14} and S^{35} are obtained with special stripping films for autoradiography, which are now produced by various makers. This technique is now widely used in cytochemical applications of contrast autoradiography.

All techniques are sensitive to artifacts, and careful work in the dark room is absolutely essential. Dust and precipitates, frequently due to impure solutions, have to be avoided.

V. OBSERVATION AND INTERPRETATION OF AUTORADIOGRAPHS

A. Observation

The techniques described in Section IV, result in autoradiographs which are superimposed on the specimen, an advantage as far as localization is concerned and a disadvantage with regard to staining.

The observation of unstained preparations is greatly facilitated if phase-contrast microscopy is used. The specimen can be viewed by phase-contrast; after removal of the condenser annulus the autoradiograph is visible. It has been found that localization is reasonably accurate by this procedure, which has the advantage that differences in intensity of the autoradiograph over a section are more easily spotted than in stained preparations.

The preparations can be stained before or after the photographic emulsion is applied. In the former case the specimen must usually be coated with a layer of some impermeable material, e.g., celloidin, to prevent bleaching of the stain during photographic processing. Such a layer should be thin, to avoid loss of resolving power; the calculations quoted in Section II, D indicate that a thickness of 0.1 μ or less can be tolerated. When this method is used, the possibility of removing tracer during prestaining has to be considered and control experiments are necessary.

The choice of stains is restricted to those which can penetrate the film without causing damage or unduly heavy staining of the film if preparations are stained after photographic processing. The gelatin of the film is easily damaged by strong chemicals and heat, which precludes the use of certain stains, some of which may also attack the silver grains or cause artefacts. The following stains can be used with special stripping film for autoradiography produced by Kodak Ltd., but have not been tested on other makes: celestine blue and hemalum, neutral red and carbol fuchsin, Ehrlich's hematoxylin and eosin, carmalum, Leishmann-Giemsa, toluidine blue and methyl green-pyronin (Pelc, 1956a). In all instances the gelatin of the film is stained so that the color contrast is not as good as in ordinary stained sections. There is, however, no difficulty in observing detail unless the autoradiograph is too heavy.

B. Interpretation

The first step in interpretation is to make certain that an autoradiograph has been produced, background and artifacts being the main causes of mistakes.

It is unfortunately impossible to obtain preparations which are completely free from background, i.e., a certain number of grains or tracks are always developed. The background can be kept to a minimum by working with the smallest possible amount of light during the preparation and storage of the emulsions, before and during exposure, in places which are as far as possible from sources of X-rays and other penetrating radiations. When electron-track emulsions are used, the preparations can be shielded with lead to reduce the background due to natural sources of radiations.

The most frequent artifacts are due to the action on the photographic emulsion of substances which diffuse from the specimen into the emulsion, making grains developable, and to dark granules, such as pigments, in the specimen which might be mistaken for grains. They can be distinguished by examination under high power, especially of unstained preparations whose film has been swollen with a few drops of water before a coverslip is placed on top. In such preparations, if present, dark granules can be seen at the lowest focal level in the specimen, chemical artifacts form a single layer of grains nearest to the section, while ionizing radiations produce grains throughout the thickness of the emulsion, except when the softest emitters such as H^3 are used. The preparation of controls without radioactive tracer will always reveal the same artifacts as the active specimen, and is by far the best check. The β-rays emitted by C^{14} and S^{35} frequently produce short tracks, which fact can be used to verify an autoradiograph, since visible light, chemical background, and artifacts consist of randomly spaced single grains.

Only concentrations of grains or tracks which are significantly above background can be regarded as an autoradiograph. With an average background of, say, 4 grains per 100 μ^2, none or six grains over such an area will frequently be observed, while reduction of the background to 2 grains per 100 μ^2 will make six grains a rare event over unlabeled cells. Due to the randomness of radioactive decay, the uncertainty attached to a given number of grains above background can easily be calculated. Thus, if in a specimen 10 grains are found over 1 cell nucleus and none over an adjacent one, it can be concluded that the positive nucleus is labeled and the negative one either unlabeled or very weakly labeled. If an electron-track emulsion necessitating shorter exposures has been used for the same specimen, and one or two tracks are found over one nucleus and none over another, it can only be said that, if the background is sufficiently low, the first nucleus can be regarded as labeled, while the second one may or may not be unlabeled. To obtain significant results the number of tracks over the nuclei might be counted, and if this number is significantly higher than over the same total area of cytoplasm the conclusion can be drawn that more tracer has been incorporated into cell nuclei than cytoplasm, but, at best, only very limited conclusions can be drawn regarding single nuclei, and in any case the advantage of autoradiography, the possibility of observing the metabolic activity of single cells, is lost.

A minimum number of grains or tracks per single structure is thus necessary to obtain definite information; this number, however, must not be too high in order to avoid loss of resolving power and visibility of cell structure. Experience has shown that in contrast autoradiography with fine-grained emulsions, 10–30 grains per 100 μ^2 fulfill these conditions if it is borne in mind that structures much smaller than 100 μ^2 still need a significant number of grains.

The limits of resolving power, dependent on the technique used, have to be borne in mind throughout the investigation. The borders of a labeled area frequently show these limits, e.g., the autoradiograph in Fig. 10 does not extend far beyond the labeled cytoplasm toward the

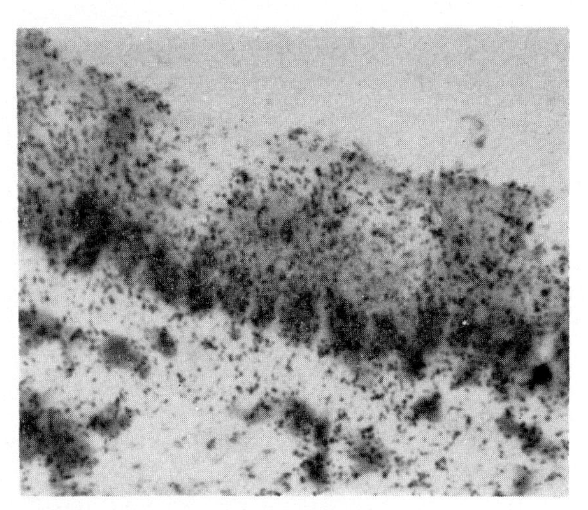

FIG. 10. S^{35}-DL-Methionine in the tracheal epithelium of a mouse, 4 hours after injection (hematoxylin and eosin).

lumen of the trachea, and therefore the autoradiograph over the nuclei can neither be due to scattered radiation from the cytoplasm, nor due to cytoplasm above or below the nuclei, since the spaces between them are conspicuously free from grains. The limitations due to crossfire from labeled adjacent structures (Lamerton and Harriss, 1954) are of great importance for a tightly packed specimen. Smears or squashes with well-separated cells can be of great advantage, especially in conjunction with sections, which preserve the spatial relationship of the cells.

An autoradiograph can only be specific for the applied isotope, not for a chemical compound or group of compounds. The importance of using specific precursors has already been stressed in previous sections. Even then the question of adsorption of the unmodified compound and the possibility of incomplete extraction has to be considered. The localization of labeled compounds can be of assistance; e.g., in experiments on DNA synthesis small remnants of labeled compounds in the cytoplasm can frequently be neglected. It is always advisable to prepare autoradiographs before and after extraction to estimate the effect of the methods used (Fig. 11).

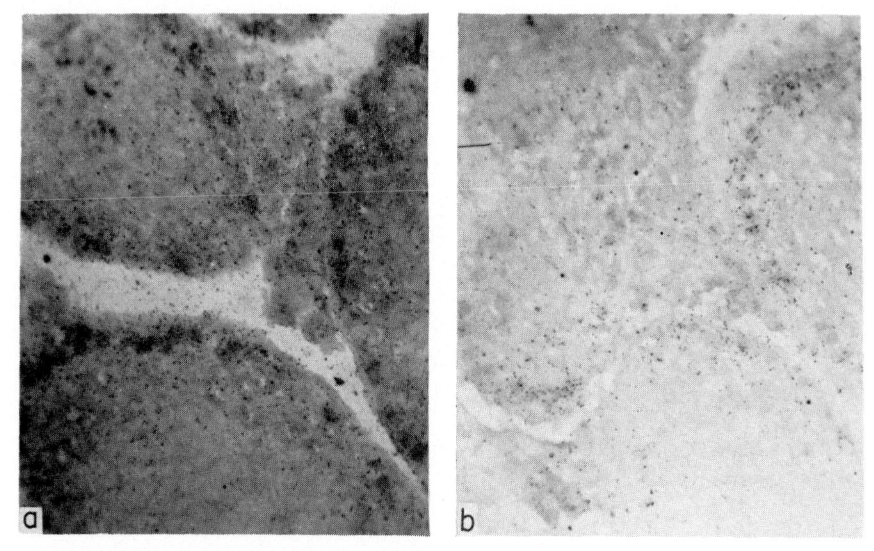

FIG. 11. [8–C^{14}]-Adenine in mouse testis, 2 days after injection (methyl green-pyronin). A: Unextracted; B: 18 hours cold perchloric acid.

VI. A SURVEY OF PREVIOUS CYTOCHEMICAL APPLICATIONS OF AUTORADIOGRAPHY

This survey is not intended as a complete review of previous work; it is to be a selection indicating details such as the amounts of tracer applied, exposure times, and extraction methods used. Omission of published work thus does not imply adverse criticism, availability of experimental detail rather than the importance of results having been the criteria for inclusion.

A. Nucleic Acids in Plants

(1) Seedlings of *Vicia faba* were cultured in tap water containing 0.2–2.4 μc. P^{32} (as ortho phosphate) per ml. Roots were fixed at times from 2 hours to 32 days in acetic acid-alcohol, hydrolyzed in 1 *N* HCl, and autoradiographs of squashes or sections prepared by the stripping film technique. After 28 days' exposure, autoradiographs of nuclei of interphase cells and of groups of chromosomes in dividing cells were observed (Howard and Pelc, 1951a). Seedlings were also grown in water containing 0.06 μc. [8-C^{14}]-adenine. Since acid hydrolysis removes adenine from DNA, ribonuclease was used to remove RNA. Exposure time was 28 days (Howard and Pelc, 1951a). Seedlings of *Zea mays* were grown in water containing 200 μc. per liter of [8-C^{14}]-adenine, and autoradiographs prepared by the stripping film technique (Clowes,

1956). The resolving power is much improved by the use of adenine instead of P^{32} (Fig. 12).

(2) Inflorescences of *Lilium henryi* were labeled in White's medium containing P^{32} in a concentration of 1–4 μc. per ml. (Plaut, 1953)

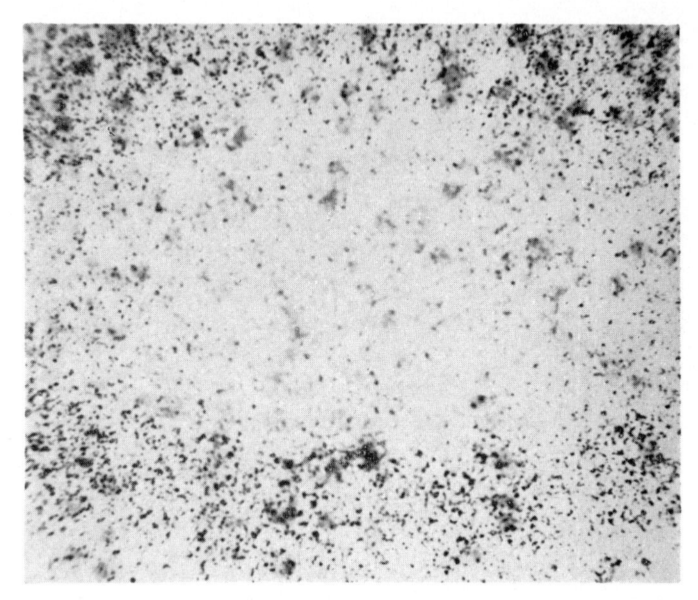

FIG. 12. Incorporation of [8–C^{14}]-adenine in the root meristem of *Zea mays* (courtesy F. A. L. Clowes).

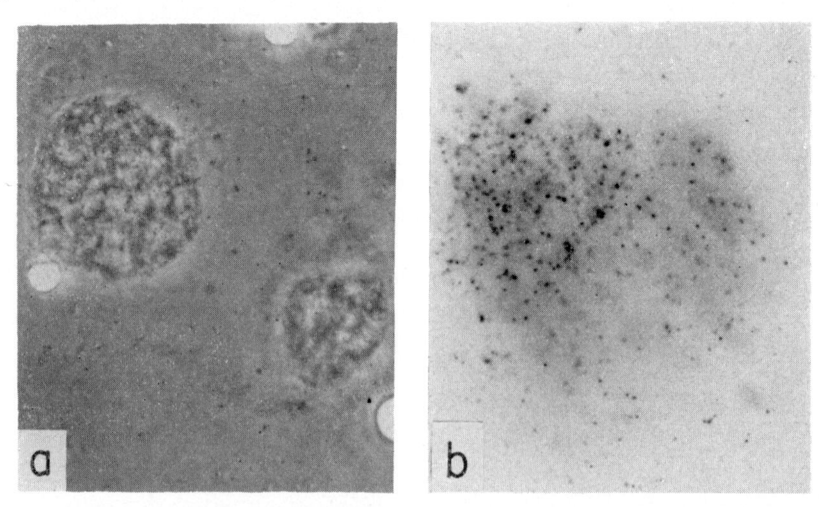

FIG. 13. A tapetal and a preleptotene sporogenous nucleus in *Lilium henryi*. A, by phase contrast; B, P^{32} autoradiograph, after extraction with hot HCl (courtesy Dr. W. S. Plaut).

(Fig. 13), and buds of *Tradescantia paludosa* or *Lilium longiflorum* in water containing 20–75 μc. per ml. of P^{32} (Taylor, 1953; Moses and Taylor, 1955). Other technical details as in *(1)*; in addition, perchloric acid was used for extraction of RNA and acid-soluble phosphorus.

B. Nucleic Acids in Mammals

(1) Intraperitoneal injections of 20–40 μc. P^{32} per gram have to be used to obtain autoradiographs due to P^{32} incorporated in nucleic acids by rodents. RNA can be extracted with ribonuclease or by hydrolysis with 1 *N* HCl. Autoradiographs of various organs were prepared, by the coating technique (Leblond *et al.*, 1948), of ascites tumor cells by

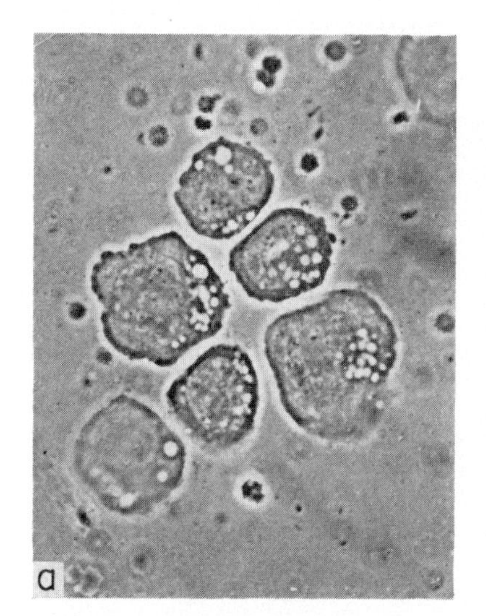

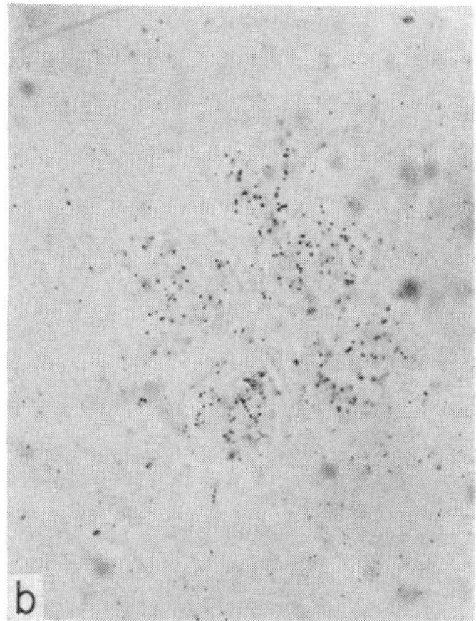

Fig. 14. Incorporation of [8–C^{14}]-adenine in mouse ascites tumor 32 hours after injection (ribonuclease). A: Phase contrast; B: same field, normal illumination (courtesy S. Hornsey and A. Howard).

the stripping film technique (Hornsey and Pelc, 1954). Exposure time 14–28 days. The radiation dose, due to the high concentration combined with the high energy of the β-particles emitted by P^{32}, delays cell division and interferes with DNA synthesis.

(2) Nucleic acids can be labeled without appreciable radiation dose if labeled adenine, usually [8–C^{14}]-adenine, is used. Rats were injected (i.p.) with 0.3 μc. per gram of this compound and killed 24 hours after injection or later. After fixation in acetic acid-alcohol followed by formol-saline, stripping film autoradiographs of smears or sections were prepared with or without extraction with ribonuclease (Pelc and Scott, 1952). Mice were given 1.0 μc. per gram and autoradiographs prepared.

Extraction of RNA with ribonuclease (Fig. 14) or cold perchloric acid (Fig. 11), of DNA with deoxyribonuclease, and of RNA and DNA with both enzymes or hot perchloric acid have been used (Hornsey and Howard, 1956; Pelc and Howard, 1956b; Pelc, 1956b). Nuclear (DNA) autoradiographs and separate autoradiographs of the daughter nuclei in telophase can be observed. RNA labeling of nucleoli could be distinguished from nuclear labeling in preparations of mouse prostates.

C. Nucleic Acids in Tissue Culture

(1) Human bone marrow was grown in medium containing 0.5 μc. per ml. of [8-C^{14}]-adenine, or of 1 μc. per ml. of P^{32} (Lajtha, 1952). Stripping film autoradiographs of methanol-fixed smears were obtained

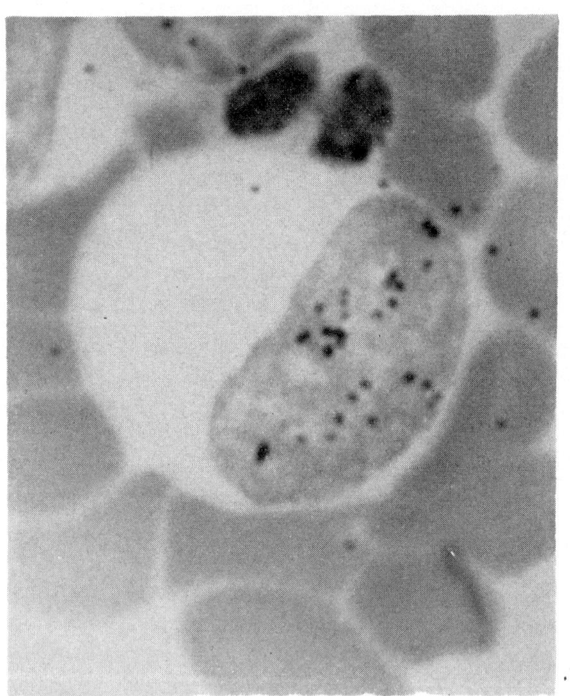

Fig. 15. Incorporation of [8–C^{14}]-adenine in a human myelocyte grown in *vitro* 20 hours after injection. Extraction with 1 *N* HCl for 6 minutes at 60°C. (Leischmann-Giemsa) (courtesy Dr. L. G. Lajtha).

after exposure times of 14 or 28 days, respectively. Ribonucleic acid was extracted with ribonuclease, cold perchloric acid, or hot 1 *N* HCl. The time of hydrolysis by the last technique is very critical, since the acid soon begins to remove adenine from the DNA (Fig. 15).

(2) Chick heart fibroblasts were grown by a modified hanging drop

technique. One drop of a solution containing 33 μc. of P^{32} per ml. was pipetted on the explants which were fixed 5–24 hours later. Differences in the concentration of P^{32} in cytoplasm and nuclei was observed; after treatment with hot 1 *N* HCl only nuclear autoradiographs remained (Fig. 6). Exposure times up to 28 days were used (Pelc and Spear, 1950).

(3) Mouse prostates were grown in tissue culture for various periods of time. One drop of a solution containing 50 μc. of [8–C^{14}]-adenine per ml. of Tyrode's solution was placed on the explants; the remaining liquid was pipetted off 30 seconds later. After 5–48 hours, the explants were fixed in acetic acid-alcohol followed by formol-saline. Stripping film autoradiographs were prepared with or without extraction with cold perchloric acid, and developed after 20 days' exposure (Lasnitzky and Pelc, 1956). For methods of extraction and resolving power, see Section VI, B *(2)*.

D. Nucleic Acids in Various Species

(1) Frogs were given intraperitoneal injections of 4 μc. of [8–C^{14}]-adenine. Animals were killed 1–3 days later and electron-track autoradiographs of sections prepared by the coating technique (Fig. 16). Intracellular localization, such as autoradiographs of nucleoli, was obtained (Ficq, 1955).

(2) Early third instar larvae of *Drosophila repleta* were fed P^{32} in

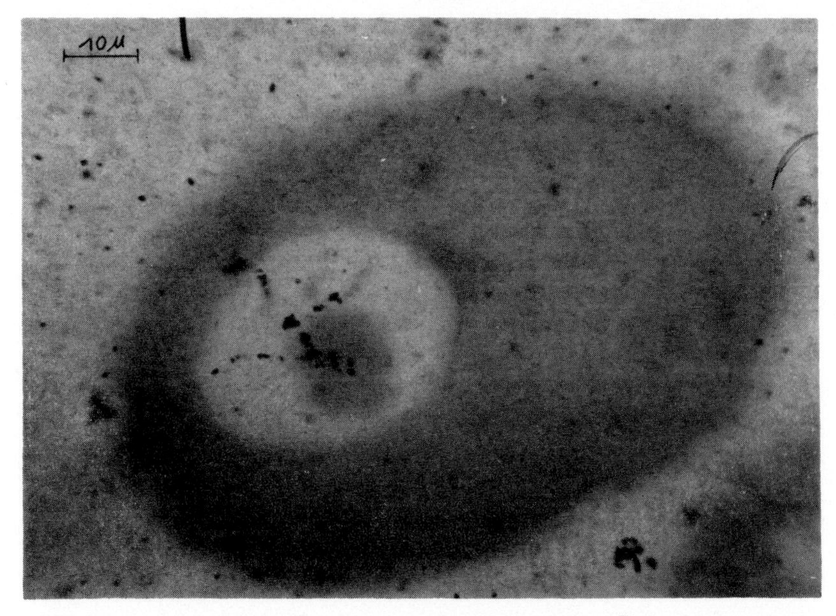

Fig. 16. [8–C^{14}]-Adenine in an oocyte of *Asterias rubens* (courtesy Dr. A. Ficq).

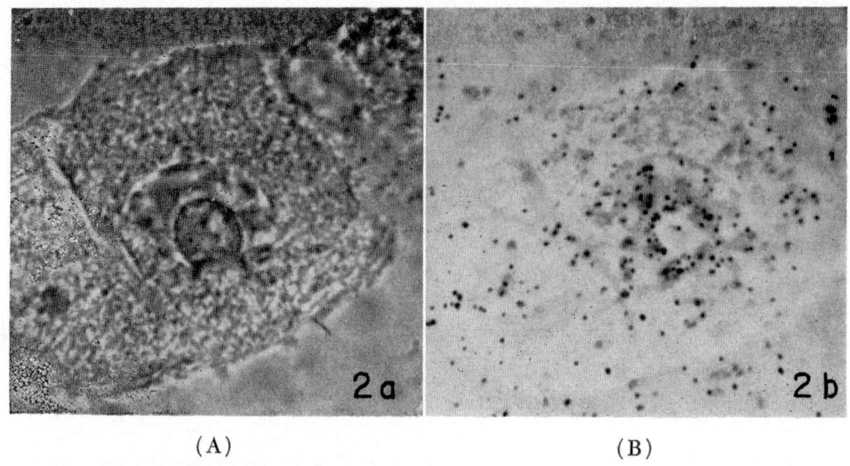

(A) (B)

FIG. 17. Cell from the salivary gland of larvae of *Drosophila repleta* fixed 1½ hours after P^{32} as phosphate. A: Phase contrast; B: P^{32} autoradiograph. (Courtesy of Dr. T. H. Taylor.)

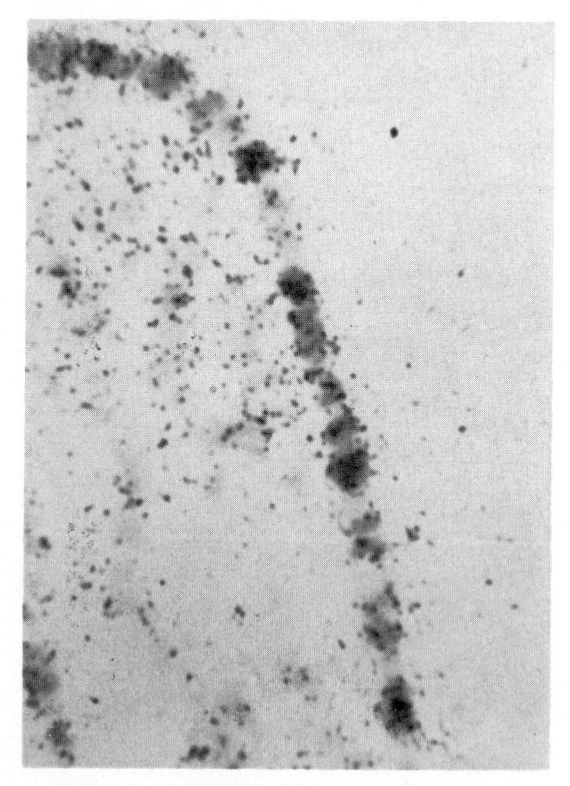

FIG. 18. Incorporation of [8–C^{14}]-adenine in salivary gland chromosomes of *Drosophila melanogaster* (hemotoxylin).

the form of inorganic phosphate (120 μc. per ml. of water containing 30 mg. of gelatin and 30 mg. of brewer's yeast (Taylor, 1953). Larvae were fixed in Carnoy's fluid and, after washing in cold TCA, autoradiographs of serial sections prepared by the stripping film technique (Fig. 17). Accumulations of P^{32} incorporated into RNA were observed over the nucleolus, cytoplasm, and chromosomes of cells of the salivary gland. Autoradiographs of labeled RNA in single sectors of the chromosomes in squashes of salivary glands (Fig. 18) of *D. melanogaster* were observed after feeding with 20 μc. per ml. of [8–C^{14}]-adenine for 24 hours (Pelc and Howard, 1956a).

E. Proteins in Plants

Seedlings of *Vicia faba* were grown in tap water containing 1 μc. per ml. of S^{35} as sodium sulfate; autoradiographs were prepared as described in Section VI, A *(1)*. Resolving power and strength of autoradiograph after 70 days' exposure were sufficient to analyze different concentrations of S^{35} in parts of prophase chromosomes (Fig. 19) (Howard and Pelc, 1951c; Pelc and Howard, 1952).

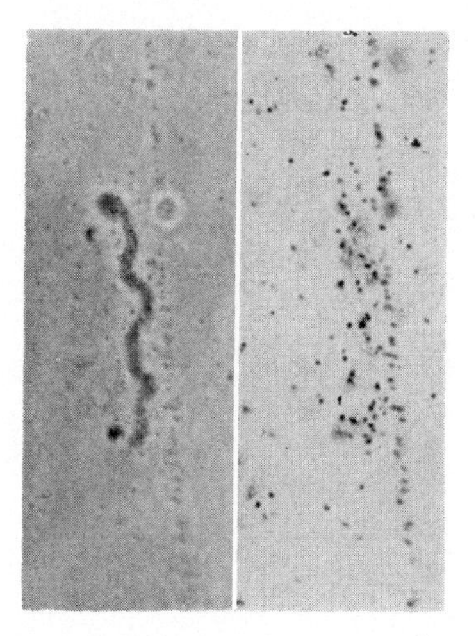

FIG. 19. Incorporation of S^{35} as Na_2SO_4 in prophase chromosomes of meristematic cell of *Vicia faba*.

F. Proteins in Animal Tissues

(1) Mice were killed 2–8 hours after a single intraperitoneal injection of 2 μc. per gram weight of S^{35}-DL-methionine, or 1 μc. per gram for longer times. Stripping film autoradiographs were prepared as described in Section VI, B *(2)*. Exposure times of 20–70 days were used. The resolving power is the same as with C^{14} (see Fig. 10), i.e., intracellular

detail can be observed unless, as is frequently the case when proteins are labeled, the nuclear concentration in small cells is similar to that in the cytoplasm (Glücksmann *et al.*, 1955).

(2) Female frogs were injected with 4 μc. of [1–C^{14}]-glycine or [2–C^{14}]-DL-3-phenylalanine (Fig. 20). Electron-track autoradiographs of ovarian tissues were prepared as described in Section VI, D *(1)* (Ficq, 1955).

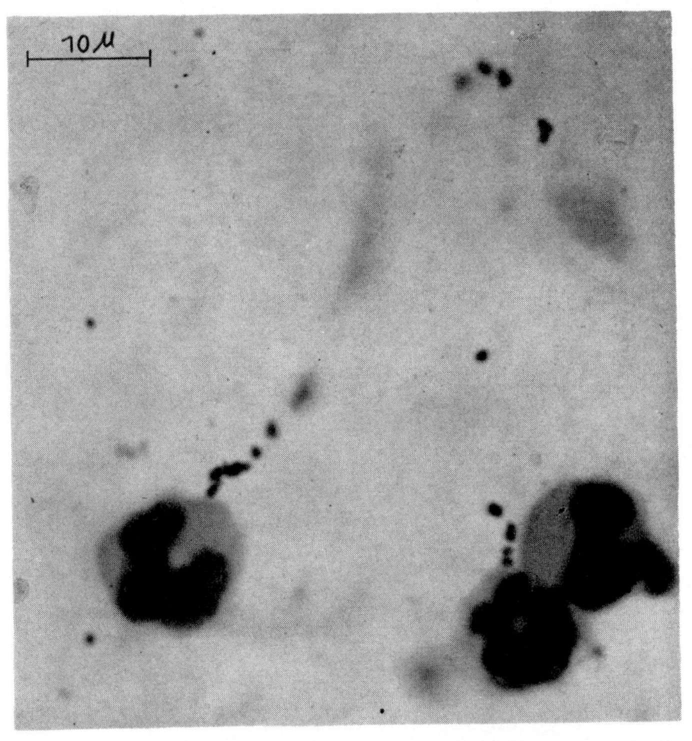

Fig. 20. [1–C^{14}]-glycine in rat bone marrow cells (courtesy Dr. A. Ficq).

(3) Human bone marrow was cultured in a medium containing 2μc. per ml. of S^{35}-DL-methionine; autoradiographs were prepared as in Section VI, C *(1)* (Lajtha, 1954).

(4) Explants of chick ectoderm were grown in tissue culture and labeled by applying Tyrode's solution containing 200 μc. per ml. of S^{35}-DL-methionine as described in Section VI, C *(3)* (Fell *et al.*, 1954). The exposure time was 4 days, therefore the dosage might have been much lower.

(5) Larvae of *Drosophila melanogaster* were grown in food contain-

ing approximately 50 μc. of S^{35}-DL-methionine per ml. for 24 hours. Autoradiographs of parts of salivary gland chromosomes were observed after 60 days' exposure (Pelc and Howard, 1956a).

G. Various Applications

(1) Food organisms in cultures of *Chaos chaos* were labeled by exposure to fluorescent light in Østerlind solution, in which 4% of the available carbonate was present as C^{14}. Stripping film autoradiographs

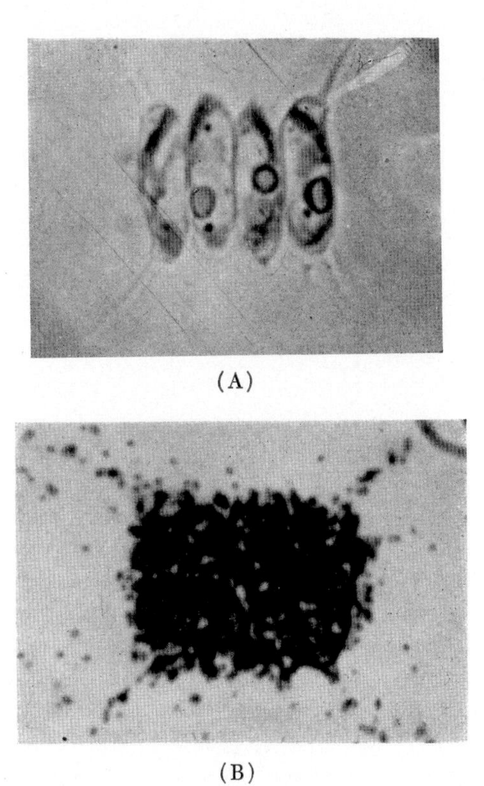

(A)

(B)

Fig. 21. A: Photomicrograph of living *Scenedesmus;* note the processes at the corner of the parcel of 4 cells. B: Autoradiograph after 48 hours in tritiated water, (courtesy Dr. C. Chapman-Andresen).

of sections, sometimes after centrifuging of organisms, were prepared. Intracellular distributions were observed and quantitated by grain counts (Andresen *et al.*, 1952). Similar preparations of ciliates and algae obtained after labeling organisms with tritium (3–33 mc. per ml. of Østerlind's solution) showed autoradiographs of single ciliae in *Scenedesmus* (Fig. 21) (Chapman-Andresen, 1953b).

(2) Various mammals were given intraperitoneal injections of the order of 10 μc. per gram weight of S^{35} as sulfate and autoradiographs prepared by the stripping film technique. This compound is utilized by mammals for incorporation into sulfated mucopolysaccharides, the con-

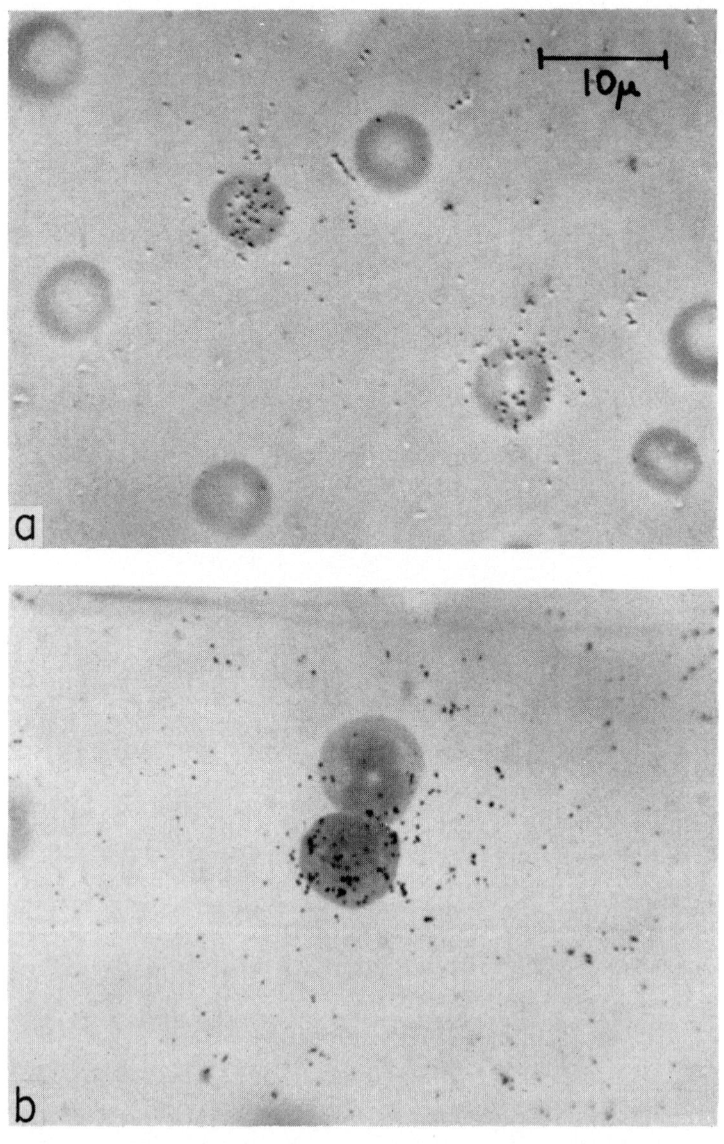

FIG. 22. Autoradiograph of blood smear from rat taken A: 6 days after injection of 11 μc. Fe^{59}. Photograph of red cells taken under phase contrast. B: Early normoblast 4 hours after injection of 20 μc. Fe^{59} (May-Grünwald-Giemsa) (courtesy Drs. Lamerton and Harriss).

centration and metabolic activity of which varies widely for different tissues (see Fig. 8). Exposure times of 10–60 days were used. Distribution of S^{35} at histological and cytological levels was observed (Jennings and Florey, 1956; Pelc and Glücksmann, 1955).

(3) Human bone marrow was grown for 1 day in media containing 10 μc. per ml. of S^{35} as sulfate or Fe^{59}. The labeling of single cells was observed in stripping film autoradiographs after exposure times of 14–20 days (Lajtha, 1952, 1954).

(4) Labeled normoblasts and erythrocytes were observed in stripping film autoradiographs of blood smears (Fig. 22) taken 6 days after rats had been injected with 11 μc. of Fe^{59} (Lamerton and Harriss, 1954).

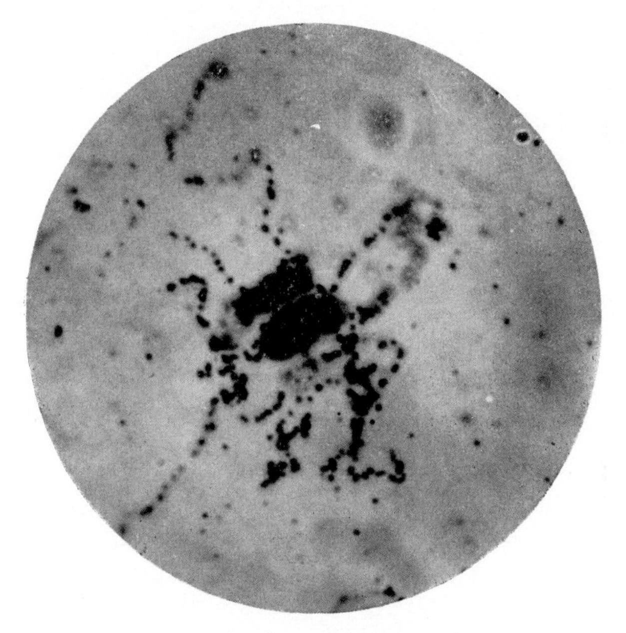

FIG. 23. C^{14}-labeled *Scenedesmus* suspended in Ilford G-5 emulsion (courtesy Dr. H. Levi).

(5) Explants of chick ectoderm and limb buds were grown in normal media and media containing excess vitamin A. S^{35} as sulfate was added and autoradiographs prepared by the methods described in Section VI, C *(3)* (Fell *et al.*, 1954).

(6) Various algae and ciliates were labeled with C^{14}, S^{35} and P^{32}, and electron-track autoradiographs produced by immersing the organisms in molten electron-track emulsion (Ilford G5), pouring the emulsion 100–150 μ thick and developing after exposure (Fig. 23). Single tracks are seen to emerge from the organisms (King *et al.*, 1951; Levi, 1954).

(7) Beta-track autoradiographs were obtained by floating on the emulsion sections of the liver of a rat killed 25 hours after injection of 3 μc. of C^{14}-labeled glycine. After exposure and deparaffinizing, the emulsion was developed and fixed. It is difficult to assess the resolving power from the published photomicrographs, though there is no reason why this should be worse than approximately 1 μ (Boyd and Levi, 1950).

VII. APPENDICES

Appendix 1: Decay of Isotopes

Radioactive decay follows an exponential law, and the number of radioactive atoms remaining after a time will be:

$$N = N_0 e^{-\lambda t} \tag{4}$$

where N is the number of atoms at time t and λ the decay constant. Since $e^{-0.693} = 0.5$, the value of λ can be calculated from Eq. 5 from the half-life H, the time after which a given isotope has decayed to one-half of its original value:

$$H\lambda = 0.693 \qquad \lambda = \frac{0.693}{H} \tag{5}$$

Values of the exponential function can be found in many handbooks. Usually a simple graph on semi-log paper (Fig. 24) will be found sufficiently accurate. The time in days is plotted on the abscissae on a linear scale, the amount of tracer remaining on a logarithmic scale on the ordinate. A straight line is drawn from the full value (1.0) at time 0 to the value of 0.5 at 87.3 days, the half-life of S^{35}. Taking the amount of tracer retained after fixation and extraction in a specimen referred to the time of fixation as 1.0, the amount will be f_A at the time autoradiographs are set up, e.g., 8 days and 0.95, respectively, as marked by arrow A in Fig. 24. After the time of exposure, the amount will be f_B, e.g., 40 days' exposure, and 0.69 in Fig. 24. Decay during the exposure therefore amounts to $f_A - f_B$, $0.95 - 0.69 = 0.26$ in Fig. 24. Bearing in mind that one-half of the particles hits the slide, the number of grains or tracks counted in an autoradiograph represents one-half of the fraction $N_0/\delta(f_A - f_B)$ of the total number of radioactive atoms present in the specimen. If δ is the grain-yield in grains per particle hitting the film for the tracer used, and n the number of grains counted, the number of atoms present in the compounds retained at time 0 will be:

$$N_0 = \frac{2n}{\delta(f_A - f_B)} \tag{6}$$

To convert number of radioactive atoms to curies (c.) or μc. or moles, one uses Eq. 7, which is based on the definition of the unit of radioactivity as "that amount of radioactivity in which the same number of atoms decay as in one gram of Radium;" the figure is 4.3×10^{10} disintegrations per second for 1 c. or 4.3×10^{4} for 1 μc.

$$N_0 = 4.64 \times 10^9 \times \text{H atoms per } \mu\text{c.} \tag{7}$$

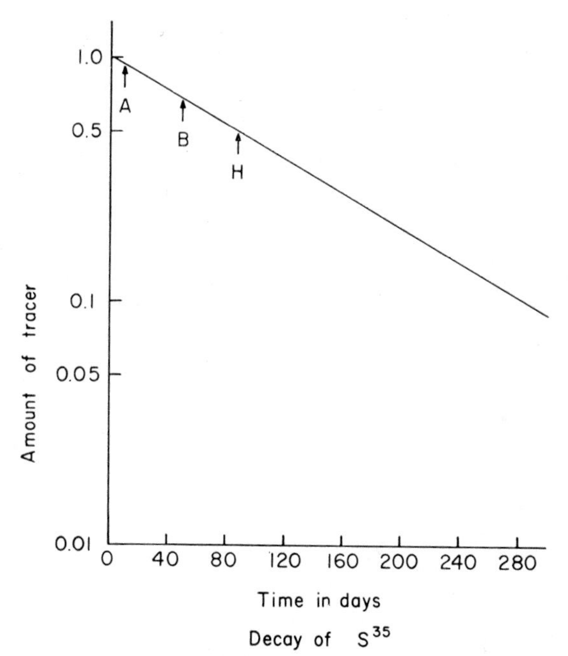

FIG. 24. Decay of radioactive substances.

(H is the half-life in days.) If A is the atomic weight, their weight (w) will be:

$$w = 6.9 \times 10^{-15} \times H \times A \text{ gm. per } \mu\text{c.} \tag{8}$$

The actual values for some isotopes are given in Table V.

TABLE V
QUANTITY OF TRACER PER μC.

Isotope	N_0 atoms per μc.	Weight per μc. in μgm.	μmoles per μc.
I^{131}	3.5×10^{10}	7.3×10^{-6}	5.5×10^{-8}
P^{32}	5.9×10^{10}	3×10^{-6}	9.4×10^{-8}
S^{35}	4.1×10^{11}	2×10^{-5}	5.7×10^{-7}
C^{14}	9.3×10^{15}	0.21	1.48×10^{-2}

The specific activity of labeled compounds is usually stated in terms of mc. per mmole (millicuries per millimole). The figures in the last column of Table V can be applied without conversion. Thus, if an organic compound is labeled with C^{14}, and 1 mc. per mmole is the specific activity, an average of 1.48% of the molecules contains one atom of C^{14} if carbon atoms in all positions are labeled or 1.48% of the molecules if only 1 carbon atom in a specified position is labeled.

Appendix 2: Calculation of Radiation Dose

The following mathematical treatment is based on a previous communication (Pelc, 1951). For the calculation of radiation dose it is assumed throughout that the radioactive material is evenly distributed in a volume which is large compared with the range of particles emitted by the tracer, for C^{14} and S^{35} volumes of 400 μ diameter or more.

If a chemical compound is applied to a living organism, the concentration (a) in an organ will usually increase for some time (τ) and then decrease, as shown schematically in Fig. 25. Some of the applied

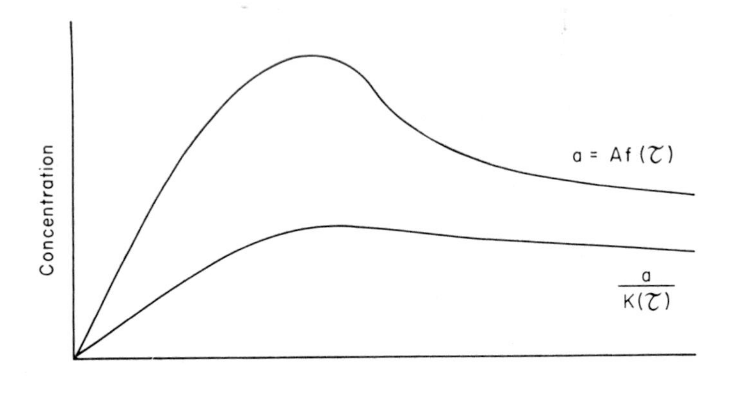

Fig. 25. Change of concentration of tracer.

material will be incorporated into the compounds which are finally to be retained in the specimen in a concentration of $C = a/K(\tau)$. To obtain an autoradiograph the concentration of desirable tracer in the specimen has to be C (Eqs. (1) or (2) at the time (t) the exposure starts. If radioactive decay between the time of fixation and setting up can be neglected and $K(t)$ is the ratio of the concentration of all tracer to that which is retained after fixation and extraction:

$$\frac{a}{K(t)} = C \text{ or } a = CK(t) \tag{9}$$

If an amount of A μc. per ml. is applied at time 0, it will have undergone radioactive decay as well as the various reactions which change its distribution. In general terms this can be written as:

$$a_\tau = af(\tau)e^{-\lambda\tau} \tag{10}$$

Combining Eqs. 9 and 10, we find that the minimum amount to be applied is:

$$CK(t) = A_0 f(t)e^{-\lambda t} \tag{11}$$

$$A_0 = \frac{CK(t)e^{\lambda t}}{f(t)}$$

Bearing in mind that the calculations are meant to be approximate, Eq. 11 can usually be simplified since radioactive decay during the experiment might be neglected. For example, in applications to synthesis of deoxyribonucleic acid a_t/A_0 is the fraction of tracer found in a given organ at time (t) after application, and $K(t)$ is the ratio of activity in DNA to the total activity, data which might be gleaned from the biochemical literature. The factor (f) in Eqs. 1 or 2 would be the fraction of the total volume taken up by the cell nuclei which can be estimated from sections.

The rate of radiation dose to an organ at a time τ is given by Eq. 12 (see Section III, B) if a μc. per ml. are evenly distributed.

$$\text{d.r.} = 61a\bar{E}_\beta e^{-\lambda\tau} \text{ reps per day} \tag{12}$$

The total dose to the organ can be calculated by integrating Eq. 12 over the time of application t:

$$\text{d} = 61\bar{E}_\beta \int_{\tau=0}^{t} a_\tau \, d\tau \text{ reps} \tag{13}$$

If the minimum amount has been applied, the dose will be:

$$\text{d} = 61\bar{E}_\beta \int_{\tau=0}^{t} A_0 f(\tau)e^{-\lambda\tau} \, d\tau \text{ reps} \tag{14}$$

Detailed data to use for the evaluation of Eq. 13 are usually not available and approximations will have to be used. This integral has been evaluated (Pelc, 1951) for a number of simplified functions of $f(\tau)$ such as linear increase or constant concentration of tracer as an approximation for experiments of short duration and for long-term experiments assuming that $f(\tau)$ can be approximated by a function of the form $f(\tau) = b\tau e^{-a\tau}$ where e and α are free parameters.

Relatively simple equations are obtained whenever it can be assumed that the concentration in an organ can be assumed to be constant during the experiment, in which case the minimum dose will be:

$$d = 1000 f K \bar{E}_\beta (e^{\lambda t} - 1) \text{ reps} \tag{15}$$

if the organism is killed at the time (t), or if the organism is kept alive (Figs. 26 and 27):

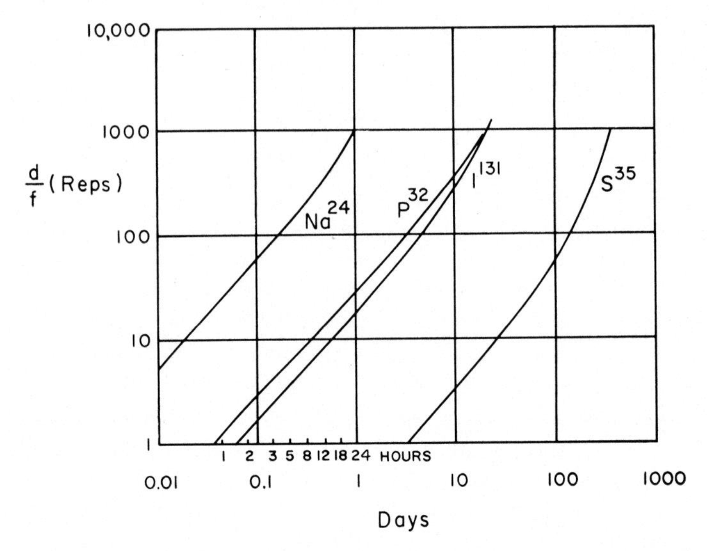

FIG. 26. Minimum dose to tissue for autoradiograph (organism killed).

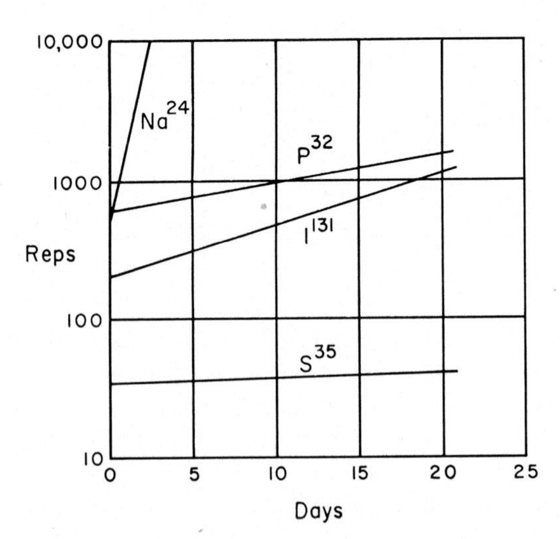

FIG. 27. Minimum dose to tissue for autoradiograph (tracer remains in organism).

$$d = 1000fK\bar{E}_{\beta}e^{\lambda t} \text{ reps} \quad (16)$$

For work with S^{35} or C^{14}, compilation shows that autoradiographs can be obtained with radiation doses of the order of 1 rep per day.

GENERAL REFERENCES

Bourne, G. H. (1952).*Biol. Revs. Cambridge Phil. Soc.* **27,** 108
Boyd, G. A. (1954). "Autoradiography." Academic Press, New York.
Doniach, I., Howard, A., and Pelc, S. R. (1953). *Prog. in Biophys. and Biophys. Chem.* **3,** 1.
Fitzgerald, P. J., and Engström, A. (1952). *Cancer* **5,** 643.
Gross, J., Bogoroch, R., Nadler, N. J., and Leblond, C. P. (1951). *Am. J. Roentgen. Radium Therapy Nuclear Med.* **65,** 420.
Norris, W. P., and Woodruff, L. A. (1955). *Ann. Rev. Nuclear Sci.* **5,** 297.
Pelc, S. R., and Howard, A. (1952). *Brit. Med. Bull.* **8,** 132.

REFERENCES

Andresen, N., Chapman-Andresen, C., and Holter, H. (1952). *Compt. rend. trav. lab. Carlsberg Sér. chim.* **28,** 189.
Bélanger, L. F. (1950). *Anat. Rec.* **107,** 149.
Bélanger, L. F., and Leblond, C. P. (1946). *Endocrinology* **39,** 8.
Blank, H. (1949). *J. Invest. Dermatol.* **12,** 2.
Blank, H., and McCarthy, P. L. (1950). *J. Lab. Clin. Med.* **36,** 776.
Boyd, G. A., and Levi, H. (1950). *Science* **111,** 58.
Canny, M. J. (1955). *Nature* **175,** 857.
Chapman-Andresen, C. (1953a). *Compt. rend trav. lab. Carlsberg Sér. chim.* **28,** 529.
Chapman-Andresen, C. (1953b). *Exptl. Cell Research* **4,** 239.
Clowes, F. A. L. (1956). *New Phytologist* **55,** 29.
Cormack, D. V. (1955). *Brit. J. Radiol.* **28,** 450.
Doniach, I., and Pelc, S. R. (1950). *Brit. J. Radiol.* **23,** 184.
Dudley, R. A. (1954). *Nucleonics* **12,** 24.
Dudley, R. A., and Dobyns, B. M. (1949). *Science* **109,** 327.
Dudley, R. A., and Pelc, S. R. (1953). *Nature* **172,** 992.
Endicott, K. M., and Jagoda, H. (1947). *Proc. Soc. Exptl. Biol. Med.* **64,** 170.
Evans, T. (1947). *Radiology* **49,** 206.
Fell, H. B., Mellanby, E., and Pelc, S. R. (1954). *Brit. Med. J.* **2,** 611.
Ficq, A. (1955). *Exptl. Cell Research* **9,** 286.
Glücksmann, A., Howard A., and Pelc, S. R. (1955). *J. Anat.* **89** (P. I), 13.
Gurney, R. W., and Mott, M. F. (1938). *Proc. Roy. Soc. (London)* **A164,** 151.
Herz, R. H. (1950). *Med. Radiograph. and Phot.* **26,** 46.
Herz, R. H. (1951). *Nucleonics* **9,** 24.
Hornsey, S., and Howard, A. (1956). *Ann. N. Y. Acad. Sci.* **63,** 915.
Hornsey, S., and Pelc, S. R. (1954). Unpublished.
Howard, A., and Pelc, S. R. (1951a). *Exptl. Cell Research* **2,** 178.
Howard, A., and Pelc, S. R. (1951b). *Nature* **167,** 599.
Jennings, M. A., and Florey, H. W. (1956). *Quart. J. Exptl. Physiol.* **41,** 131.
King, D. T., Harris, J. E., and Tkaczyk, S. (1951). *Nature* **167,** 273.
Lajtha, L. G. (1952). *Exptl. Cell Research* **3,** 696.

Lajtha, L. G. (1954). *J. Phot. Sci.* **2,** 130.
Lamerton, L. F., and Harriss, E. B. (1954). *J. Phot. Sci.* **2,** 135.
Lasnitzky, I., and Pelc, S. R. (1951). *Exptl. Cell Research* **13,** 140.
Leblond, C. P., Stevens, C. E., and Bogoroch, R. (1948). *Science* **108,** 531.
Levi, H. (1954). *Exptl. Cell Research* **7,** 44.
Mazia, D., Plaut, W. S., and Ellis, G. W. (1955). *Exptl. Cell Research* **9,** 305.
Mitchell, J. W. (1949). *Phil. Mag.* **40,** 249.
Moses, M. J., and Taylor, J. H. (1955). *Exptl. Cell Research* **9,** 474.
Nadler, N. J. (1951). *Can. J. of Med. Sci.* **29,** 182.
Odeblad, E. (1952). *Acta Radiol. Suppl.* **93.**
Pelc, S. R. (1947). *Nature* **160,** 749.
Pelc, S. R. (1951). *In* "Isotopes in Biochemistry" (G. E. W. Wolstenholme, ed.), p. 122. Churchill, London.
Pelc, S. R. (1956a). *Intern. J. Appl. Radiation and Isotopes* **1,** 172.
Pelc, S. R. (1956b). *Nature* **178,** 359.
Pelc, S. R., and Glücksmann, A. (1955). *Exptl. Cell Research* **8,** 336.
Pelc., S. R., and Howard, A. (1952). *Brit. Med. Bull.* **8,** 132.
Pelc, S. R., and Howard, A. (1956a). *Exptl. Cell Research* **10,** 549.
Pelc, S. R., and Howard, A. (1956b). *Exptl. Cell Research.* In press.
Pelc, S. R., and Scott, O. C. A. (1952). Unpublished.
Pelc, S. R., and Spear, F. G. (1950). *Brit. J. Radiol.* **23,** 287.
Plaut, W. S. (1953). *Hereditas* **39,** 438.
Ray, R. C., and Stevens, G. W. W. (1953). *Brit. J. Radiol.* **26,** 362.
Stevens, G. W. W. (1950). *Brit. J. Radiol.* **23,** 723.
Taylor, J. H. (1953). *Exptl. Cell Research* **4,** 164.
Taylor, J. H., McMaster, R. D., and Caluta, M. S. (1955). *Exptl. Cell Research* **9,** 460.

THE CYTOCHEMICAL DEMONSTRATION AND MEASUREMENT OF SULFHYDRYL GROUPS BY AZO-ARYL MERCAPTIDE COUPLING, WITH SPECIAL REFERENCE TO MERCURY ORANGE

H. STANLEY BENNETT AND RUTH M. WATTS

Department of Anatomy, University of Washington, Seattle, Washington

I. INTRODUCTION

It is easier to apply a cytochemical reaction than to find out what the reaction means. The full significance of a cytochemical reaction cannot be understood solely on the basis of qualitative tests for specificity. Quantitative exploration carried out on a cytochemical and microchemical level can add greatly to one's understanding of a reaction, thus improving one's basis for interpreting results. Yet careful quantitative studies have heretofore been regarded generally as neither necessary nor important steps in the development of a cytochemical reaction. We have nevertheless carried out a quantitative analysis of a cytochemical reaction for sulfhydryl groups in which a colored azo mercurial is coupled with the mercaptan groups of proteins. We present here our findings in some detail, believing that the value of such an analysis will be sufficiently evident to stimulate others to study their reactions quantitatively before applying them extensively. We believe quantitative studies of this general sort should become standard in working out and evaluating cytochemical methods.

II. THEORY OF METHOD AND TEST OF THEORY

A. General Considerations

A method for demonstration and measurement of sulfhydryl groups in tissues can be based on a well-known equilibrium which can be written in the following form:

$$\underset{\text{thiol}}{\text{R–S–H}} + \underset{\text{organic mercurial}}{\text{X–Hg–R}'} \rightleftarrows \underset{\text{mercaptide}}{\text{R–S–Hg–R}'} + \text{H}^+ + \text{X}^- \qquad (1)$$

Here "R" and "R′" represent organic radicals. In the special cases important in cytochemistry, "R" can represent a tissue protein molecule containing cysteine. "R′" represents an alkyl or aryl group. In the method used here, "R′" is the colored *p*-phenylazo-β-naphthyl radical, which is also found in Orange G and in Orange II. "X" in Eq. 1 represents an anion-forming group attached originally to the mercury. The group "X" is liberated into solution as a consequence of the mercaptide-forming reaction. In the method stressed in this discussion "X" is represented by chloride.

The cytochemical method described here is merely a special example of a more general class of mercaptide forming reactions which have been utilized extensively for study of sulfhydryl groups during the last quarter-century. All of the reactions in this group can be represented by the equilibrium in Eq. 1. Within the group the various methods differ in the nature of "R′," and less importantly, in the anion-forming group, "X." Acetate, nitrate, iodide and hydroxide have been used for "X" in Eq. 1, in addition to chloride. Phenyl, *p*-carboxyphenyl, and methyl groups have been used for "R′." "R" has been represented by soluble or insoluble proteins, by glutathione, by cysteine or by other mercaptans.

Hellerman *et al.* (1933) introduced this general group of reactions into biochemical studies of sulfhydryl groups. In their original work phenyl mercuric chloride was the reagent. Later Hellerman and Perkins (1934) used phenyl mercuric hydroxide as a sulfhydryl reagent for protein thiols. In 1943 Hellerman *et al.* introduced *p*-chloromercuribenzoate for reaction with protein mercaptans, and in 1946 Cook *et al.* reported the use of phenyl mercuric nitrate for this purpose. Hughes (1950) used methyl mercuric iodide in a two-phase system and methyl mercuric nitrate in aqueous solution for measuring the sulfhydryl groups of serum albumin. He determined the equilibrium constants for the reaction,

$$CH_3\text{–}Hg\text{–}I + Alb\text{–}S\text{–}H \rightleftarrows CH_3\text{–}Hg\text{–}S\text{–}Alb + H^+ + I^- \quad (2)$$

(where Alb signifies serum albumin). He reported a corrected pK of about 4.45, indicating Eq. 2 is driven strongly to the right.

There is available a considerable body of experience with reactions of this type (see Olcott and Fraenkel-Conrat, 1947; Gurd and Wilcox, 1956). It can be said that no reason has been found to doubt the sulfhydryl specificity of the reaction of the organic mercurical groups mentioned when the mercurials are used under conditions which keep the activity of the ionized forms at low values (see Gurd and Wilcox, 1956; also p. 368). However, Hughes (1950) has warned that the presence of the carboxyl group on *p*-chloromercuribenzoic acid provides a mechanism whereby this mercurial might bind to proteins through groups other than thiols. He stressed the importance of using a monofunctional mercurial when a reagent of reliable specificity for sulfhydryl groups is desired.

A number of monofunctional colored organic mercurials have been considered for possible application as optical tracers for sulfhydryl groups. The following specifications were set down with the view that the compound selected should conform to them as closely as possible if it were to serve as a satisfactory sulfhydryl reagent in cytochemistry.

B. Specifications for a Satisfactory Cytochemical Sulfhydryl Reagent

Specifications for a satisfactory cytochemical sulfhydryl reagent are as follows:

(1) The reagent should be capable of binding specifically to sulfhydryl groups by stable covalent linkage formed in a single coupling reaction.

(2) It should be incapable of binding to any other groups occurring in biological tissue or, if capable of such binding, convenient means should be available for detaching molecules bound nonspecifically without removing any from thiol groups.

(3) It should be monofunctional. That is, each molecule should possess only one group through which the molecule might be bound to a tissue component.

(4) It should have a high molar absorptivity in the visible range.

(5) It should have a spectral absorption band in the visible which does not change when the molecule is bound, nor should it change importantly with pH, nor with different solvents, nor with different binding conditions.

(6) Its reaction with sulfhydryl groups should be stoichiometric.

(7) The equilibrium constants of the binding reaction should favor strongly the bound state.

(8) It should have a relatively small molecular size, so that steric hindrance to binding would be minimized.

(9) It should be available in very pure form.

(10) The reaction should result in a stable preparation which can be used as a permanent record of sulfhydryl content of the tissue sample.

C. A Consideration of Various Organic Mercurials as Possible Sulfhydryl Reagents

The alkyl, aryl, or heterocyclic mercuric chlorides were selected as offering the most promise from the point of view of specificity and stability of binding, as set down in specifications 1 and 2. In view of recent work cited by Gurd and Wilcox (1956), a choice of organic mercuric iodides might have offered some slight advantages over the chlorides. The requirement for color narrowed the choice to aryl or heterocyclic mercuric halides. The requirement of small size focused attention on mono azo compounds and on thiazine and oxazine compounds. Thiazine and oxazine dyes contain polar groups capable of binding to proteins by salt linkage. They were hence discarded as unpromising from the viewpoint of specificity, as they would fail with respect to specification 2.

The simplest and smallest azo dye is azobenzene. A mercurical derivative of azobenzene was considered, but was rejected because of the low molar absorptivity of azobenzene in the visible spectral range. A mercury derivative of phenylazonaphthalene was next considered, as this organic radical has a stronger color than azobenzene. An attempt was made to synthesize *p*-chloromercuriphenylazonaphthalene. This effort failed.

Attention was next turned to the compound, 1-(4-chloromercuriphenylazo)-naphthol-2, also called *p*-chloromercuriphenylazo-β-naphthol. In order to avoid cumbersome use of these attenuated chemical terms, a colloquial name for the compound, "red sulfhydryl reagent," was introduced by Bennett (1951). This colloquial term is no longer applicable uniquely to the compound, as other red sulfhydryl reagents have been synthesized. Thus we believe a more specific brief designation of the mercurical is desirable, and with this in mind we have chosen to refer to it here and henceforth as "Mercury Orange." This new colloquial name is meant to indicate the chemical kinship of the mercurial to the familiar azo dyes, Orange II and Orange G, which possess the same chromophore.

The reagent thus designated had not been reported in the literature prior to its synthesis by Bennett and Yphantis (1948). It was created specifically for the purpose of exploring its possibilities as a sulfhydryl reagent. Of all the cytochemical sulfhydryl reagents so far introduced, it appears to conform best to the specifications set down above. In many ways, it has characteristics of an ideal cytochemical reagent. It is the reagent discussed specifically in this chapter.

D. Properties of Mercury Orange

The properties of Mercury Orange are as follows:

(1) *Molecular weight.* The molecular weight of 1-(4-chloromercuriphenylazo)-naphthol-2 or Mercury Orange is 483.33.

TABLE I
Calculated and Measured Values for Elements in Sample of Mercury Orange Used in This Study [a]

Element	Calculated (%)	Found (%)
Carbon	39.76	39.36
Hydrogen	2.29	2.24
Chlorine	7.34	7.12
Mercury	41.5	42.0
Nitrogen	5.80	6.01

[a] From Bennett and Yphantis (1948).

(2) *Melting point.* 291.5–293° C. (corrected) (melts with blackening).

(3) *Purity.* The sample used in the cytochemical studies gave values in Table I upon analysis for constituent elements. Calculations are on the basis of $C_{16}H_{11}ClHgN_2O$.

The sample gave only one spot on paper chromatography, using more than one solvent.

(4) *Solubility.* Mercury Orange is not detectably soluble in water, ligroin, or petroleum ether.

Approximate values for solubility of Mercury Orange in various organic solvents have been determined, and are cited in Table II. Values are for unspecified room temperature, and are probably correct to within ± 10% for 22° C. Notes on apparent relative solubility of Mercury Orange in a few other solvents have been made, and are listed in Table III.

TABLE II
APPROXIMATE SOLUBILITY OF MERCURY ORANGE IN VARIOUS ORGANIC SOLVENTS [a]

Solvent	Approximate solubility in moles/liter
Methanol	2.0×10^{-6}
Isopropanol	5.8×10^{-6}
sec-Butanol	6.3×10^{-6}
Isobutyric acid	3.1×10^{-5}
n-Propanol	3.9×10^{-5}
n-Butanol	4.4×10^{-5}
n-Butyric acid	6.2×10^{-5}
Toluene	8.6×10^{-5}
Formic acid	1.2×10^{-4}
Propionic acid	1.4×10^{-4}
Acetic acid	1.8×10^{-4}
Aniline	7.3×10^{-3}
Dimethylformamide	9.9×10^{-3}
Pyridine	7.4×10^{-2}

[a] Data by Mrs. Lucy Jache.

TABLE III
QUALITATIVE ESTIMATE OF SOLUBILITY OF MERCURY ORANGE IN VARIOUS ORGANIC SOLVENTS [a]

Solvent	Solubility
Water, petroleum ether	Insoluble
Cyclohexane	Trace soluble
Ethanol (95%)	Slightly soluble
Isobutanol	Slightly soluble
Ether (absolute)	Somewhat soluble
Benzene	Somewhat soluble
Nitromethane	Somewhat soluble
Xylene	Somewhat soluble
Piperidine	Somewhat soluble
Decahydronaphthalene	Somewhat soluble
Cyclohexylamine	Somewhat soluble
Dioxane	Quite soluble
Nitrobenzene	Very soluble
Morpholine	Very soluble
α-Picoline	Very soluble

[a] Data by Mrs. Lucy Jache.

(5) *Molecular size.* The crystal structure of Mercury Orange and of phenylazo-β-naphthol have not been reported. However, one can construct a model of the molecule, assuming that one can use with validity Robertson's (1939) description of azobenzene and the descriptions by

Robertson (1933) and of Abrahams *et al.* (1949a, b) of naphthalene. From these papers a reasonable structure for phenylazonaphthalene and of Mercury Orange can be derived (Fig. 1). The –OH group in the 2 position on the naphthalene ring can be constructed on the basis of data in Pauling (1945). Chelation of the naphthol –OH group with one of the azo nitrogens is deduced from the work of Mason (1932). Van der Waals radii for mercury, carbon, hydrogen, and nitrogen are taken from Pauling (1945). So is the value for the van der Waals thickness of the benzene and naphthalene rings.

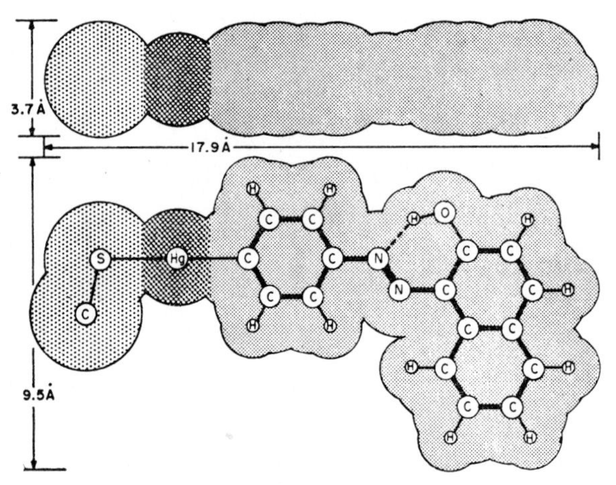

FIG. 1. Diagram of probable molecular shape and dimensions of a Mercury Orange mercaptide. The upper figure shows the molecule as seen edge on. The lower figure represents it in plan view. The limits of the molecule are meant to represent the limits of the van der Waals radii of its constituent atoms. The relative thickness of the lines connecting various atoms in the lower figure are intended to represent roughly the number of electrons implicated in the bond. The dotted line between the naphtholic hydrogen and an azo nitrogen represents a hydrogen bond. The three types of shadings are meant to imply that the molecule has a colored portion, a dense portion, and a portion which was part of the original thiol.

These considerations have led to the construction of a scale model of Mercury Orange as shown in Fig. 1. This represents the molecule as if attached to a thiol group by mercaptide linkage. In the chloromercuri form, Mercury Orange could be represented similarly, but with the alkyl carbon missing and with a chlorine replacing the sulfur. The van der Waals radius of the chlorine atom would be slightly smaller than that of the sulfur, and the mercury to chlorine bond distance slightly less than the mercury to sulfur bond represented. The upper portion of Fig. 1 shows the model of the molecule edge on. The lower portion shows it in plan view. One can assume the molecule to be planar. It can be seen that the mercury atom through which the reagent binds to mercaptans is attached to a benzene ring over 6 Å wide and about 3.5 Å thick, to which a larger naphthalene ring is coupled. Hence the mercury cannot bind to a protein sulfhydryl group unless the steric configuration of the protein molecule close to the thiol will admit the mercury atom with this encumbrance to a position sufficiently close to the sulfur for reaction to occur. This would be within a distance of about 2 Ångstrom units.

(6) *Spectral characteristics.* The absorption spectrum of Mercury Orange resembles that of phenylazo-β-naphthol, which has been determined previously by Brode (1943). An absorption spectrograph of Mercury Orange is shown in Fig. 2. The broad peak in the visible with a

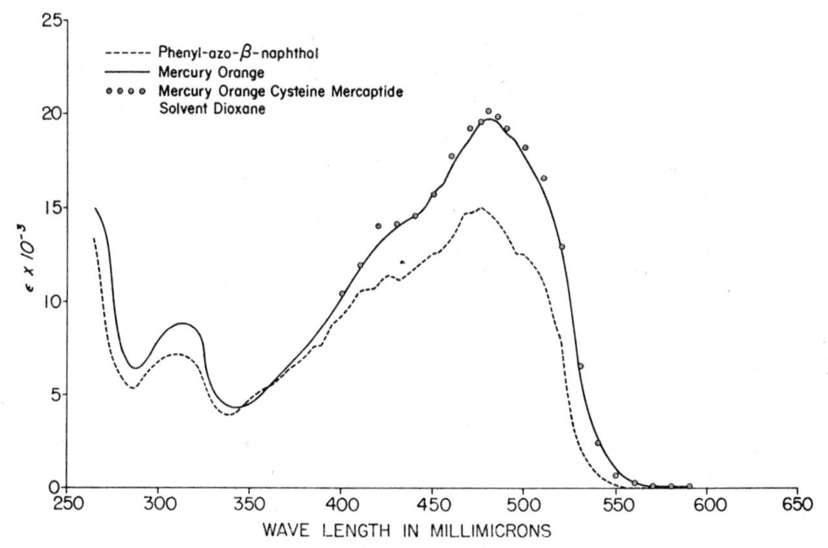

Fig. 2. Absorption spectrophotometric curves for Mercury Orange, for its cysteine mercaptide, and for phenylazo-β-naphthol. Note that the absorption curve of the cysteine mercaptide of Mercury Orange resembles closely that of Mercury Orange itself in the visible range.

maximum at about 484 mμ is of principal interest. This peak is also present with much the same height and wavelength value in many of the derivatives of phenylazo-β-naphthol with substituent groups in the *para* position of the benzene ring. Values for the wavelength and molar absorptivity (ϵ_{max}) of this peak for phenylazo-β-naphthol and several of its derivatives with substituent groups in the *para* position of the benzene ring are shown in Table IV.

It is notable that ϵ_{max} and λ_{max} change rather little with the substituent groups listed. It is particularly important to note that conversion of the

TABLE IV

VALUES FOR ε_{max} AND λ_{max} FOR PHENYLAZO-β-NAPHTHOL AND VARIOUS DERIVATIVES WITH SUBSTITUENT GROUPS IN THE *para* POSITION OF THE BENZENE RING, INCLUDING MERCURY ORANGE AND ITS CYSTEINE MERCAPTIDE [a]

Substituent group in p-position on benzene ring of phenylazo-β-naphthol	ε_{max}	λ_{max} (mμ)	Solvent
–H	1.50×10^4	476	Dioxane
–HgCl	1.97×10^4	480	Dioxane
–HgCl	1.95×10^4	484	Toluene
–Hg-cysteine • HCl (in presence of crystals of cysteine • HCl in excess)	1.92×10^4	480	Dioxane
–Hg-cysteine • HCl (cysteine • HCl not in excess)	2.02×10^4	480	Dioxane
–Hg-cysteine • HCl	1.90×10^4	480	Dimethylformamide
$-HgSCH_2CH_2OH$	1.90×10^4	484	Toluene
$-SO_3H$	2.29×10^4	488	Water
$-SO_3Na$	2.29×10^4	482	Water
–COOH	2.18×10^4	482	Dioxane
$-N(CH_3)_3I$	1.87×10^4	478	Water
$-N(CH_3)_3I$	1.82×10^4	474	95% ethanol

[a] Data by Mrs. Lucy Jache.

chloromercuri form of Mercury Orange to the cysteine or β-mercapto-ethanol mercaptides of Mercury Orange produces a negligible change in the absorption peak in the visible. This means that in quantitative cytochemistry the value for the molar absorptivity (ϵ_{max}), and for the wavelength of the absorption peak (λ_{max}) for the bound mercaptide form of Mercury Orange can be taken to be the same as in the free chloromercuri form.

However, it cannot be assumed that all substituent groups on the *para* position of the benzene ring of phenylazo-β-naphthol have a negligible effect on ϵ_{max} and λ_{max} in the visible. For when the substituted group is an amino group, an alkyl amino group, or a dialkyl amino group, and when the compound is in acid solution and the amino or substituted amino group is coordinated with a proton, a significant change in ϵ_{max} and λ_{max} occurs.

It is also possible to change appreciably the values for ϵ_{max} and λ_{max} by subjecting phenylazo-β-naphthol or one of its derivatives to the action of strong alkali, such as NaOH. In that case the chelate ring between the naphthol OH group and one of the azo nitrogens is broken, a salt of the naphthol derivative is formed, and the color is changed.

These considerable changes which can be introduced into the color of the molecule by strong base or by the action of acids on amino-substituted dyes need not concern one as far as the sulfhydryl reaction with the chloromercuri group is concerned, as strong base can easily be avoided in this case, and converting the chloromercuri group to the mercaptide has no effect or only a very slight effect on ϵ_{max} and λ_{max}. Whatever effect there is can be neglected in quantitative cytochemical work, as other sources of error are of greater importance.

Crystals of phenylazo-β-naphthol and its derivatives, including Mercury Orange, are dichroic in the visible, and dichroism has been detected in mercaptides formed by coupling Mercury Orange to protein sulfhydryl groups of myofibrils of striated muscle (Bennett, 1952). If one streaks out dry Mercury Orange crystals on a glass slide with the finger, the resulting streaks are dichroic in the visible, and absorb most strongly in the 484 mμ band when the electric vector of the illuminating linear polarized light is parallel to the direction of the streak. It would seem reasonable to assume that the shearing forces during streaking would tend to orient the Mercury Orange molecules with their long axes parallel to the direction of streaking. If this assumption be correct, it would follow that the axis of the azo linkage would be oriented roughly parallel to the streaks. From this one would infer that within any given Mercury Orange molecule, the molecule would absorb within the 484 mμ band only light with a component of the electric vector parallel to the axis of the azo linkage. Land and West (1946) reached similar conclusions with respect to the absorption characteristics of a related azo dye, Congo red. However, a more rigorous examination of the geometry of the absorption vectors with respect to the bond axes of the molecule might yield a more reliable correlation between the pathways of excitation available to the electrons responsible for absorption and the steric form of the molecule. This matter is of some importance in interpreting the significance of colored mercaptides bound dichroically to tissue components.

Mercury Orange does show some variation in ϵ_{max} and λ_{max} in the visible band, in accordance with the organic solvent in which the substance is dissolved. ϵ_{max} at 484 mμ in toluene has been found to be $1.95 \pm 0.03 \times 10^4$ (Mean $\pm$ S.D.). In organic acids ϵ_{max} is increased somewhat, and in some organic bases it is decreased, but the changes in value for ϵ_{max} so induced do not exceed ten per cent either way. In most solvents the value for ϵ_{max} is closer to its value in toluene. Table V lists some of the values of ϵ_{max} and λ_{max} as determined for Mercury Orange dissolved in various organic solvents.

Since Michaelis and Granick (1945) have pointed out that many dyes in solution deviate considerably from Beer's law as concentration is varied, one may ask how much ϵ_{max} for Mercury Orange varies with concentration. This has been determined for Mercury Orange dissolved in toluene, in pyridine, and in dimethylformamide, over the concentration ranges which could be accommodated in the absorption spectrometer available for this work (Hilger Uvispec). The values of ϵ_{max} at $\lambda_{max} = 484$ mμ are given in Table VI.

TABLE V

VALUES FOR ϵ_{max} AND λ_{max} FOR MERCURY ORANGE IN VARIOUS ORGANIC SOLVENTS [a]

Solvent	ϵ_{max}	λ_{max} (mμ)
Acetonitrile	1.96×10^4	483
n-Butanol	1.96×10^4	486
Furan	2.17×10^4	488
Nitrobenzene	1.84×10^4	484
n-Propanol	1.92×10^4	484
Toulene	1.95×10^4	484
Aniline	2.04×10^4	496
Cyclohexylamine	1.67×10^4	486
Dibenzylamine	1.88×10^4	490
Dimethylformamide	1.86×10^4	484
Morpholine	1.90×10^4	486
α-Picoline	1.84×10^4	486
Piperidine	1.76×10^4	486
Pyridine	1.82×10^4	486
Propionic acid	2.18×10^4	488

[a] Data by Mrs. Lucy Jache and Dr. James C. M. Li.

At low concentrations the absorbance values read from the spectrophotometer are very low and considerable error can be expected. For this reason significance is not attached to the low values cited for ϵ_{max} in the most dilute solutions. Allowing for this error, it can be seen that no evidence for appreciable deviation from Beer's law is evident over the concentration ranges studied.

However, these data are not altogether reassuring from the cytochemical viewpoint, since concentrations about 50 times greater than the maximum concentrations appearing in Table VI have been encountered in tissues studied cytochemically. Whereas Mercury Orange can be dissolved in pyridine in concentrations up to about 7×10^{-2} molar (see Table II), we have not overcome the difficulties associated with measuring the molar absorptivity of Mercury Orange at these concentrations. The difficulties of measuring accurately absorbing path lengths of the order of 1 to 10 μ, or of dealing with very high absorbance values

TABLE VI
VALUES FOR ε_{max} OF MERCURY ORANGE FOR VARIOUS CONCENTRATIONS IN SOLVENT INDICATED [a]

Molarity	A	b (cm.)	ε_{max}	λ (mμ)
		SOLVENT: TOLUENE		
0.4138×10^{-4}	0.401	0.5	1.94×10^{4}	482
0.2069×10^{-4}	0.403	1.0	1.95×10^{4}	482
0.1035×10^{-4}	0.409	2.0	1.97×10^{4}	482
0.0518×10^{-4}	0.390	4.0	1.88×10^{4}	482
		SOLVENT: PYRIDINE		
4.94×10^{-4}	0.854	0.1	1.74×10^{4}	486
0.989×10^{-4}	0.893	0.5	1.81×10^{4}	486
0.494×10^{-4}	0.884	1.0	1.80×10^{4}	486
0.395×10^{-4}	0.356	0.5	1.80×10^{4}	486
0.247×10^{-4}	0.887	2.0	1.80×10^{4}	486
0.198×10^{-4}	0.353	1.0	1.78×10^{4}	486
0.124×10^{-4}	0.873	4.0	1.76×10^{4}	486
0.0989×10^{-4}	0.361	2.0	1.83×10^{4}	486
0.0494×10^{-4}	0.090	1.0	1.82×10^{4}	486
0.0247×10^{-4}	0.091	2.0	1.84×10^{4}	486
0.0124×10^{-4}	0.057	4.0	1.15×10^{4}	486
		SOLVENT: DIMETHYLFORMAMIDE		
4.94×10^{-4}	0.881	0.1	1.78×10^{4}	484
0.989×10^{-4}	0.899	0.5	1.82×10^{4}	484
0.494×10^{-4}	0.905	1.0	1.83×10^{4}	484
0.395×10^{-4}	0.373	0.5	1.89×10^{4}	484
0.247×10^{-4}	0.894	2.0	1.81×10^{4}	484
0.198×10^{-4}	0.365	1.0	1.84×10^{4}	484
0.124×10^{-4}	0.942	4.0	1.90×10^{4}	484
0.0989×10^{-4}	0.363	2.0	1.84×10^{4}	484
0.0494×10^{-4}	0.094	1.0	1.90×10^{4}	484
0.0247×10^{-4}	0.092	2.0	1.86×10^{4}	484
0.0124×10^{-4}	0.071	4.0	1.43×10^{4}	484

[a] Data by Mrs. Lucy Jache.

have been compelling. Thus we do not know how well Mercury Orange obeys Beer's law in concentrations equivalent to those appearing in cytochemical work.

E. TESTS OF SPECIFICITY AND STOICHIOMETRY OF THE MERCURY ORANGE METHOD

1. *Qualitative Considerations*

Basically, the specificity of the reaction can be tested qualitatively by determining if the binding of Mercury Orange to tissues depends on the presence of sulfhydryl groups on the tissue protein on the one hand, and on the presence of the chloromercuri group on the reagent on the other. These matters have been examined in detail, and have been found to be satisfactory. Bennett (1951) showed that blocking sulfhydryl groups on tissue proteins by oxidation or by alkylation prevented the binding of Mercury Orange. Moreover, he showed that phenylazo-β-naphthol, which is essentially the Mercury Orange molecule without the chloromercuri group, did not bind to sulfhydryl containing tissue proteins under conditions similar to those under which the chloromercuri compound was bound. Experience with Mercury Orange since the work published in 1951 has continued to be satisfactory, and no evidence for lack of specificity has been detected, if procedures and precautions are observed as outlined later in this chapter under "Procedures" (pp. 350–366).

One may ask if the naphthol –OH group or the azo nitrogens on the Mercury Orange molecule could act as functional groups by which the molecule could be bound to tissues, either by salt linkage or by hydrogen bonding. The –OH group ionizes only in highly alkaline solutions, and avoidance of high pH eliminates danger of binding Mercury Orange to protein by salt linkage involving the naphtholic group. Basic organic solvents such as pyridine and picoline do not cause ionization of this group. The naphtholic –OH group is chelated so firmly to the azo nitrogen by a ring-forming hydrogen bond that successful competition for this hydroxyl by an electronegative site on a nearby protein molecule is not reasonably to be expected. There is perhaps more justification for suspecting that a protein –OH or –NH group might form a hydrogen bond to the oxygen or to one of the azo nitrogens of Mercury Orange, or that Mercury Orange might bind to tissues by weak van der Waals forces involving the electrons of the ring systems. Indeed, it can be expected that such bonds might form, and one must predicate the procedure so that any Mercury Orange molecules bound nonspecifically can be removed. This is easily done by utilizing a solvent which will compete successfully for the phenylazo-β-naphthol portion of the molecule against attractive sites which might be present in tissue proteins. Any solvent in which Mercury Orange is readily soluble will serve this purpose. It is not necessary that the solvent molecule have hydrogen bonding sites on it, as toluene is satisfactory from this point of view.

One can draw on a wealth of experience with Sudan III and Sudan IV in this context. These dyes contain the same azo nitrogens, naphtholic –OH, and ring systems present in Mercury Orange, and possess similar solubility properties. Yet it is well known that they do not bind to tissue protein. In dilute alcohol solvents the Sudan dyes dissolve in tissue lipid components in accordance with solubility coefficient principles.

This can be regarded as a binding of the Sudan dye to lipoidal tissue components by van der Waals forces. Subsequent treatment of the tissues with a solvent like toluene removes the lipids and the Sudan dye so bound, and no trace remains attached to tissue components. The dilute alcohol itself appears to be a satisfactory competitor for Sudan dye molecules which might be inclined to bind to tissue by hydrogen bonds. This experience with Sudan dyes leads to the conclusion that Mercury Orange cannot be used safely as a sulfhydryl reagent in dilute alcohol solution if lipid elements are present in the tissue which might bind the reagent by nonspecific van der Waals forces, unless the tissue is treated subsequently with a solvent which will remove all the Mercury Orange which might be nonspecifically bound. Pearse (1954) called attention to this situation, which was likewise fully recognized in the earlier work of Bennett (1951), whose procedures avoided risk of nonspecific binding by this mechanism.

Recently Gurd and Wilcox (1956) have pointed out that organic mercurials can bind to carboxyl or amino groups of proteins if the activities of the organic mercurial cation reach certain values. Conditions wherein such nonspecific binding occurs are not approached when Mercury Orange is used according to the procedures outlined in this chapter. A further discussion of this type of non-specific binding occurs under "Critique of Method" on p. 366.

1. *Quantitative Considerations*

a. General Considerations.

A test of the stoichiometry of the reaction of Mercury Orange with tissue sulfhydryl groups can be based on a measurement of the number of sulfhydryl groups engaged by the reagent in a given tissue sample and a comparison of that number with the number of sulfhydryl groups detected in the sample or in an aliquot by some other method. Such a comparison is faced with a number of pitfalls and uncertainties which have been recognized by biochemists working with sulfhydryl groups during recent decades. These difficulties may be stated briefly to be based on the following circumstances:

(a) Sulfhydryl groups are labile and are readily removed from the analytical field by oxidation, by heavy metal combination or by alkylation.

(b) A known or unknown fraction of the sulfhydryl groups in a sample may be inaccessible to an analytical reagent as a result of steric obstruction due to molecular groupings in the vicinity of the thiol group or of the active group of the reagent.

(c) Analytical methods for determining sulfhydryl groups quantitatively are apt to be capricious and unreliable, and are difficult to standardize.

We have encountered difficulties falling under each of these headings, but have nevertheless concluded that our findings are worth reporting, as they have improved considerably our understanding of the reaction.

The reaction of Mercury Orange with tissue sulfhydryl groups has been evaluated in two types of systems:

First, it has been studied in systems of soluble proteins, such as serum albumin.

Second, it has been studied in systems of insoluble proteins in suspension. Isolated myofibrils from skeletal muscle have been used extensively in this connection.

Advantage can be taken of the color of Mercury Orange. One can base a quantitative method on spectrophotometric measurements of Mercury Orange removed from solution by a sulfhydryl-containing substance in another phase, or on the amount of Mercury Orange bound by a sulfhydryl-containing substance in a solid or liquid phase. The amount of Mercury Orange bound by small thiol-containing tissue components on a slide can be measured under the microscope by microspectrophotometric techniques.

b. Tests with Serum Albumin.

The reaction of Mercury Orange with serum albumin was studied in a two-phase system similar to that utilized by Hughes (1950). A solution of commercial bovine serum albumin adjusted to pH 8.30 with 0.1 *N* NaOH was permitted to equilibrate with an overlying solution of Mercury Orange in toluene. The albumin solution became orange in the course of a few hours. The absorbance of the toluene solution was measured from time to time until no further loss of mercurial was evident. The final measurements were made eleven days after initiating the reaction. The loss of Mercury Orange from the toluene was then calculated from the change in absorbance at $\lambda = 484$ mμ. This loss was assumed to be due to binding of mercurial by sulfhydryl groups in the albumin in the aqueous phase. The number of sulfhydryl groups per gram of serum albumin so determined was compared to those determined on aliquots of the same serum albumin sample by the two-phase methyl mercuric iodide method of Hughes (1950) and by the silver amperometric method of Benesch and Benesch (1948). The results are given in Table VII. It can be seen that the three methods determine the same number of sulfhydryl groups in the sample, within experimental error.

It was concluded that Mercury Orange can be used as a satisfactory

TABLE VII
SULFHYDRYL GROUPS IN A SAMPLE OF BOVINE SERUM ALBUMIN MEASURED BY THREE DIFFERENT METHODS [a, b]

Method	Mercury Orange	CH_3HgI	Ag-Amperometric
Moles –SH × 10^5 per gram of serum albumin	1.02 ± 0.023	1.01 ± 0.091	1.00 ± 0.045

[a] Values are expressed as the Mean ± S.D.
[b] Data by Mrs. Lucy Jache and Dr. James C. M. Li.

sulfhydryl reagent for denatured proteins in solution, and that it can react stoichiometrically and completely with all available sulfhydryl groups in a solution of denatured albumin.

c. Tests with Muscle Fragments.

(i) *Sulfhydryl values as determined with Mercury Orange.* The reaction of Mercury Orange with the sulfhydryl groups of isolated myofibrils and with muscle fragments has been studied extensively. The myofibrils have been considered as comprising a model system of insoluble protein, comparable to sections on a slide, but offering an advantage of lending themselves to ready analysis by microchemical "test tube" methods.

We have used the procedures outlined in later sections of this chapter (pp. 350–353). Values which we have obtained for available –SH content of muscle in moles –SH per gram of dry muscle are cited in Table VIII. It will be noted that the standard procedures were varied in most cases by using solvents other than pure toluene. In obtaining the data cited in Table VIII the procedure was also varied as mentioned on p. 340, by permitting the samples to dry from toluene before placing them in the solutions of Mercury Orange. This drying has been shown to have an effect on the number of –SH groups in muscle accessible to Mercury Orange when the –SH groups are measured in toluene (see p. 340). Table VIII shows a wide variation in the number of –SH groups accessible to Mercury Orange as solvents are varied. This variation has been investigated, and evidence is brought forth that it is due to differences in competitive hydrogen bonding properties of the solvents (see p. 341). For reasons cited later, we deem the higher values for accessible –SH groups appearing under Samples 1 and 2, Table VIII, to approach most closely the true value for total –SH content of muscle. The lower values cited in Table VIII are interpreted as representing titrations of –SH under conditions where only a part of the total –SH content of the muscle is engaged by Mercury Orange. The cytochemical

TABLE VIII

MOLES OF ACCESSIBLE SULFHYDRYL PER GRAM OF AIR-DRIED EXTRACTED FIXED MUSCLE AS MEASURED BY MERCURY ORANGE IN VARIOUS SOLVENTS, CORRECTED FOR THE COMBINED MERCURY ORANGE [a]

Sample No.	Solvent	Moles –SH × 10^5 per gram dried fat-free muscle		
		1st aliquot	2nd aliquot	3rd aliquot
1	Dimethylformamide saturated with urea	7.1	7.0	7.7
2	α-Picoline	6.5	6.4	6.9
3	Dimethylformamide	5.6	5.4	5.7
4	33% phenol in toluene	6.0	5.5	5.4
5	19% phenol in toluene	5.8	5.7	5.7
6	Pyridine	5.0	5.1	5.2
7	Cyclohexylamine	4.5	3.9	3.7
8	Piperidine	2.4	2.2	2.8
9	Acetonitrile	2.4	2.3	2.3
10	Furan	1.5	1.5	1.2
11	Toluene	1.3	1.4	1.3
12	Toluene	0.35	0.55	0.36
13	Nitromethane	No measurable uptake		
14	*sec*-Butanol	No measurable uptake		

[a] Data by Mrs. Lucy Jache and Dr. James C. M. Li.

implications of these variations have been explored extensively and are developed in later passages in this chapter.

(ii) *Comparison of Mercury Orange sulfhydryl values with those in the literature.* The values cited in Table VIII can be compared with the values for the –SH content of purified muscle proteins and for muscle tissue reported by other workers. Some of these values have been calculated from data in the literature and are cited in Table IX. The most accurate determinations of the sulfhydryl content of actin and myosin are probably those of Tsao and Bailey (1953). These values, and the values for thiol content of muscle determined by Mirsky (1936) are close to the values of –SH content of muscle cited in Samples 1 and 2, Table VIII. This concurrence leads us to postulate that the values in Samples 1 and 2, Table VIII, may approach valid figures for total sulfhydryl content of muscle fragments as prepared by our method.

(iii) *Comparison of Mercury Orange sulfhydryl values with those determined by the amperometric method.* The number of moles of –SH groups per gram of dry fat-free fragmented muscle determined by a Mercury Orange method as cited in Table VIII can be compared with the corresponding value of –SH groups for the same material determined by some other method. We have found the amperometric method of Benesch and Benesch (1948) to be helpful for this purpose, though in our hands it has been somewhat capricious, and standard conditions capable of yielding reproducible and consistent results of good accuracy were not achieved. Yet the silver amperometric titrations appeared to be as satisfactory for suspensions of myofibrils as for denatured proteins in aqueous solutions of guanidine or urea. The galvanometer came to rest after adding an increment of silver nitrate solution to the myofibril suspension just about as fast as if a similar increment were added to a solution of serum albumin. Thus it was concluded that the structure of the proteins of the myofibrils in suspension was sufficiently open to admit silver ions readily to positions of access to thiol groups. The results of a series of silver amperometric titrations of sulfhydryl groups in suspensions of trichloroacetic acid and dimethylformamide-fixed rabbit psoas muscle are cited in Table X. The method used is based on that of Benesch and Benesch (1948) in which the titration is carried out in a buffer containing alcohol. The procedure is discussed later in this chapter on pp. 348–350. In the work with myofibril suspensions the alcohol content of the suspending buffer was varied some in a search for optimal conditions for analysis. No such optimum was recognized, as the data in Table X reveal.

Samples 1–15 in Table X yielded determinations based on trichloroacetic acid-fixed muscle, which can be compared legitimately with values in Table VIII. The values of Samples 1–15 in Table X range from 6.0 to 11×10^{-5} with an average of 7.8×10^{-5} moles –SH per gram. This average is close to the values for Samples 1 and 2 in Table VIII, and to Mirsky's and to Tsao and Bailey's values in Table IX.

Special attention is directed towards Sample 16, Table X. This sample is an aliquot of Sample 15. Sample 16 was treated with a Mercury Orange solution in dimethylformamide prior to argentometric titration, whereas Sample 15 was treated with dimethylformamide alone. In Sample 16 a very low value for –SH groups available to the silver ion was obtained. This is interpreted as indicating that most of the –SH groups had previously been occupied by the orange mercurial during immersion in the Mercury Orange solution. These data can be interpreted as evidence supporting a concept of specificity for the reaction of Mercury Orange with sulfhydryl groups under these conditions, and suggest that the Mercury Orange mercaptide reaction has engaged most of the –SH groups in this sample.

Samples 17, 18, and 19 represent sulfhydryl titrations of muscle fixed by immersion in dimethylformamide. Hence we do not feel justified in comparing them with values for trichloroacetic acid-fixed muscle, though we know of no reason why the respective values should differ.

TABLE IX

TOTAL SULFHYDRYL CONTENT OF MUSCLE AND MUSCLE PROTEINS BASED ON DATA RECORDED IN THE LITERATURE

Protein	Moles $-SH \times 10^5$ per gram protein or muscle	Calculated average molecular weight of unit containing 1 mole cysteine	Method	References
(Acto) myosin	5.5	18,200	Iodoacetate-ferricyanide	Mirsky (1936)
Muscle (frog)	6.0	16,700	Iodoacetate-ferricyanide	Mirsky (1936)
Muscle (rabbit)	6.5	15,400	Iodoacetate-ferricyanide	Mirsky (1936)
(Acto) myosin	5.3	19,000	Porphyrindin-urea	Greenstein and Edsall (1940)
(Acto) myosin	9.6	10,400	Porphyrindin-guanidine	Greenstein and Edsall (1940)
Myosin A	5.4	18,500	Ferricyanide	Fredericq (1945)
Actomyosin	9.6	10,400	Ferricyanide	Godeaux (1946)
Myosin T[a]	7.0	14,300	N-ethylmaleimide-guanidine	Tsao and Bailey (1953)
	7.4	13,500		
Actin (F or G)	6.6	15,200	N-ethylmaleimide-guanidine	Tsao and Bailey (1953)
	7.0	14,300		

[a] Myosin T: Bailey's (1954) term for the actin-free myosin of Szent-Gyorgyi's myosin A as prepared by Tsao.

TABLE X
SULFHYDRYL CONTENT OF FIXED RABBIT PSOAS MUSCLE BY THE AMPEROMETRIC METHOD[a]

Sample No.	Fixation	Special treatment of sample prior to titration	Modifications of the ammonia buffer	Moles –SH × 10^5 per gram dried fat-free extracted muscle
1	Trichloroacetic acid	Dried from toluene *in vacuo*	No ethanol	6.2
2	Trichloroacetic acid	Dried from toluene *in vacuo*	No ethanol	6.0
3	Trichloroacetic acid	Dried from toluene *in vacuo*	30% ethanol	7.1
4	Trichloroacetic acid	Dried from toluene *in vacuo*	47% ethanol	6.2
5	Trichloroacetic acid	Dried from toluene *in vacuo*	60% ethanol	8.4
6	Trichloroacetic acid	Dried from toluene *in vacuo*	80% ethanol	6.1
7	Trichloroacetic acid	Dried from toluene *in vacuo* overnight	90% ethanol	8.1
8	Trichloroacetic acid	Air dried from toluene and then to constant weight over P_2O_5 *in vacuo*	0.36 *M* guanidine HCl; 3.4 × 10^{-5} *M* EDTA; no ethanol	9.1
9	Trichloroacetic acid	Same as above	Same as above	9.1
10	Trichloroacetic acid	Same as above	0.37 *M* guanidine HCl; 1.8 × 10^{-5} *M* EDTA; no ethanol	7.9
11	Trichloroacetic acid	Same as above	No ethanol	8.4
12	Trichloroacetic acid	Same as above	No ethanol	7.2
13	Trichloroacetic acid	Same as above; homogenate very fine	No ethanol	11.0
14	Trichloroacetic acid	Same as above; sample teased only	No ethanol	9.5
15	Trichloroacetic acid	Dimethylformamide overnight, then dried from toluene	40% ethanol	6.1
16	Trichloroacetic acid	Dimethylformamide, containing Mercury Orange, 4 days (to bind –SH groups) then dried from toluene	47% ethanol	0.07
17	Dimethylformamide	Dimethylformamide for 3 weeks at 4° C. and dried from toluene	0.2 *M* guanidine HCl saturated with EDTA; no ethanol	5.2
18	Dimethylformamide	Same as above	No ethanol	5.0
19	Dimethylformamide	Same as above	0.075 *M* $NaHSO_3$ saturated with EDTA; no ethanol	5.2

[a] Data by Mrs. Lucy Jache and Dr. James C. M. Li.

(iv) *Summary and evaluation of comparisons of the various methods.* In summarizing our comparison, bearing in mind the appreciable discrepancies in the values for muscle thiol content which we and others have obtained, but nevertheless attempting a considered judgment in evaluating the results, we would surmise that a reasonable value for the total protein sulfhydryl content of muscle fragments or of myofibrils prepared by our method would lie between $6\text{–}8 \times 10^{-5}$ moles –SH per gram dry muscle. This corresponds on the average to approximately one –SH group for every 16,700–12,500 molecular weight unit of the myofibril. If this evaluation be correct, then Samples 1 and 2, Table VIII, indicate that the Mercury Orange method can be used for measuring total protein sulfhydryl content of muscle tissue, whereas the other samples on Table VIII indicate that the total sulfhydryl content of muscle tissue is not recognized by Mercury Orange under all conditions.

The next few sections of this chapter deal with attempts to analyze the conditions which lead to variability in the fraction of muscle –SH groups engaged by Mercury Orange. We lead into this section by inviting once more a study of Table VIII, bearing in mind the following comments.

(1) The values for –SH content within any given set of three aliquots of a sample determined in the same solvent are in reasonable agreement, and betoken a degree of consistency of the method.

(2) Only the first set of three samples cited, which were determined in dimethylformamide saturated with urea, yielded values averaging 7.3×10^{-5} moles –SH per gram of dry muscle, which is close to the average of 7.8×10^{-5} for total sulfhydryl content of myofibrils as determined by the silver amperometric method as cited in Table X.

(3) The values of sulfhydryl groups in myofibrils recognized by Mercury Orange vary considerably in accordance with the solvent used, and are higher in toluene with phenol than in toluene alone, and in dimethylformamide with urea than in pure dimethylformamide.

One may then ask if the variation in myofibrillar sulfhydryl content recognized in the various solvents, and as given in Table VIII, reflect a total capriciousness and unreliability of the method for quantitative analysis, or if the variations represent true differences of accessibility of sulfhydryl groups to the reagent as the solvent is varied. This question has been explored, and the resulting findings have led to the view that the variations do represent differences in accessibility of –SH groups in accordance with variations in the solvent used and with the previous history of the sample. This is altogether reasonable, as different solvents and denaturing agents would be expected to be effective to different degrees in opening up hydrogen bonded or ionically bonded obstructions to the ready access of a reagent to the protein sulfhydryl groups. Our specific evidence on this topic, however, is derived from two approaches: First, on an exploration of the influence of molecular size of the sulfhydryl reagent as related to the number of thiol groups recognized, and second, on the effect of drying the myofibrils on the number of sulfhydryl agents detected.

(v) *Effect of molecular size of the sulfhydryl reagent.* The effect of molecular size of the sulfhydryl reagent with respect to the number of protein thiol groups detected by the reagent was investigated in the following way: Mercury Orange and methyl mercuric iodide were chosen as two specific sulfhydryl reagents which reacted with thiols by identical mechanisms but which differed a good deal with respect to the size of the organic radical attached to the mercury. The dimensions of the Mercury Orange molecule are indicated in Fig. 1. The methyl group of methyl mercuric iodide is about the size of a chloride residue, and can be regarded as approximating in size a sphere less than 3 Å in diameter. Thus it is reasonable to postulate that methyl mercuric iodide might achieve access to sulfhydryl groups from which Mercury Orange would be barred. To test this postulate two rat leg muscles were fixed in trichloroacetic acid and prepared as outlined on pp. 350–351. After drying the samples from toluene, two samples from each muscle were placed in a standard solution of Mercury Orange in toluene and two into a standard solution of methyl mercuric iodide in toluene. In each of the eight samples so prepared the mercurial bound by the muscle samples was determined by the appropriate procedure, which is outlined for Mercury Orange on pp. 351–352, and for methyl mercuric iodide on pp. 353–354. The number of moles of sulfhydryl per gram of dry muscle accessible to each reagent was calculated for each of the eight samples. The results are shown in Table XI. These results are interpreted as indi-

TABLE XI
MOLES OF —SH PER GRAM OF FIXED MUSCLE TITRATED BY MERCURY ORANGE AND BY METHYL MERCURIC IODIDE [a]

Muscle No.	Sample No.	Moles —SH $\times 10^5$ per gram dry muscle	
		Mercury Orange method	Methyl mercuric iodide method
1	1	1.2	4.8
	2	1.3	4.7
2	3	1.3	4.4
	4	1.2	3.8

[a] Data by Mrs. Lucy Jache.

cating that the smaller methyl mercuric iodide molecule has access to more protein thiol groups in the muscle samples than does the larger Mercury Orange molecule. Hence it is concluded that the size of the reagent molecule can be a factor in determining the number of sulfhydryl groups titrated under given conditions. From this it is inferred that the steric configuration in the vicinity of the protein thiol determines whether or not it can react with a sulfhydryl reagent of a certain size and shape.

(vi) *Effect of drying of the muscle sample on the measurement of sulfhydryl.* A further test of the hypothesis that steric considerations in muscle are important in cytochemical sulfhydryl determinations was essayed by attempting to manipulate the steric configuration of the protein molecule in the vicinity of –SH groups, and observing the changes in reactivity to Mercury Orange as related to such manipulation. It was deemed that a satisfactory approach to such manipulation might lie in drying the protein molecule. All the data cited in Table VIII, X, and XI are based on muscle samples air dried from toluene. It was therefore decided to study the influence of drying on the uptake of Mercury Orange by muscle tissue. It was reasoned that during drying new hydrogen bonds might form in the muscle protein and create additional barriers to the access of Mercury Orange to muscle mercaptan groups. In that case, dried muscle would bind less mercurial than muscle protected from drying. It was reasoned that it would be best to test the effect of new hydrogen bonds formed on drying in a solvent which would be unlikely to open hydrogen bonds. Toluene was chosen for this purpose, as it is not capable of forming hydrogen bonds, and hence cannot be regarded as a hydrogen bond competitor.

Rabbit psoas muscle was fixed in 2.7% trichloroacetic acid and prepared according to the standard procedure (p. 351). Some samples of the fragmented muscle were blotted briefly to remove the excess toluene and placed immediately into a standard solution of Mercury Orange in toluene without permitting the samples to dry. Other samples were air dried for designated periods before being placed in the Mercury Orange solution. The –SH determinations were carried out according to the standard procedure (p. 351). The moles of –SH per gram of dry muscle found are shown in Table XII.

Scrutiny of Table XII reveals that the samples of muscle protected from drying bound several times as much mercurial as those permitted to dry. It is concluded that drying from toluene can result in a loss of 70–90% of the sulfhydryl groups in muscle accessible to Mercury Orange in toluene. This is interpreted as evidence that drying creates new steric barriers to the access of Mercury Orange molecules to protein thiols,

TABLE XII
EFFECT OF DRYING OF FIXED RABBIT PSOAS MUSCLE PRIOR TO TREATMENT WITH MERCURY ORANGE IN TOLUENE

Test No.	Rabbit No.	Approximate sample weight (mg.)	Drying period	Moles –SH × 10^5 found per gram dry muscle		
				1	2	3
1	2	3	None	4.73	5.17	4.85
2	3	6	None	5.71	5.58	5.51
	3	6	None	5.92	5.52	5.74
3	1	4	None	1.79	2.52	2.94
	1	3.5	1 hr.	0.44	0.43	0.36
4	2	3	None	3.67	3.53	3.69
	2	3	65 min.	1.50	2.18	
5	1	5.5	None	2.96	2.81	2.97
	1	6	2 hr.	0.65	0.57	0.66
6	1	4	40 min.	0.70	0.65	0.73
7	1	2.5	45 min.	0.74	0.65	0.73
8	1	1.5	2 hr.	0.57	0.75	0.55
	1	5	2 hr.	0.79	0.75	0.73
	1	3	5 hr.	0.74	0.69	0.70

and indicates that the previous history of the specimen has an influence on the number of sulfhydryl groups which can be detected. The variation in the sulfhydryl content detected in the undried specimens, however, shows that drying may not be the only mechanism whereby sulfhydryl groups can be rendered inaccessible to the reagent.

In Table VIII, the available –SH content of Sample 11 is three or four times that of Sample 12, though both are dried and measured in toluene without hydrogen bond competitor. Variations of this type have been encountered from time to time, and it has been ascertained that extreme dehydration produced by prolonged subjection of fixed muscle fibers extracted with toluene to conditions in a vacuum desiccator over P_2O_5 can reduce the available —SH groups to values well below those shown in Table XII or in Sample 12, Table VIII. Indeed, some muscle tissue has been dried over P_2O_5 to the point where it took up no detectable amount of Mercury Orange in toluene, remaining perfectly colorless even after prolonged immersion in the reagent. Since the drying conditions imposed on the samples cited in Table VIII were not carefully controlled, the discrepancy between the apparent sulfhydryl content of Samples 11 and 12 are deemed to be due to differences in degree of drying, with accompanying differences in hydrogen-bonded steric obstructions to the mercaptide coupling reaction.

(vii) *Effect of hydrogen bond competitors.* One may question the hypothesis that the loss of available sulfhydryl groups on drying the muscle samples was due to the formation of hydrogen-bonded barriers to the access of the sulfhydryl reagent rather than to oxidation or some other process destructive of thiol groups. In order to test this hypothesis, muscle fibers fixed, extracted, and dried like those described in the previous section were placed after drying in a solution of Mercury Orange in toluene containing a strong hydrogen-bonding agent. It was reasoned that such a hydrogen-bonding reagent might compete with intra-protein hydrogen bonds and open up some of the barriers to the access of the mercurial to protein thiols. In that case, muscle fragments exposed to Mercury Orange in toluene after drying should bind more Mercury Orange in the presence of a hydrogen-bonding reagent than in its absence. Phenol was chosen as a competitive hydrogen-bonding reagent, and was added to toluene solutions of Mercury Orange in which muscle fragments were immersed. The number of moles of –SH per gram of dry muscle tissue was then determined in the usual way for samples immersed in toluene solutions of Mercury Orange with phenol, with and without previous drying. The results are cited in Table XIII.

TABLE XIII

EFFECT OF 19% PHENOL IN TOLUENE SOLUTION OF MERCURY ORANGE IN RESTORING AVAILABILITY OF –SH GROUPS RENDERED UNAVAILABLE BY DRYING.
(Samples are of trichloroacetic acid-fixed rabbit psoas muscle. Compare with Table XII.)

Test No.	Rabbit No.	Approximate sample weight (mg.)	Drying period	Moles –SH $\times 10^5$ per gram dry muscle		
				1	2	3
1	2	3.5	None	5.39	5.74	5.79
2	1	3	None	2.72	2.86	2.83
	1	4	1 hr.	2.69	3.09	2.83
3	1	3	None	3.50	3.19	3.12
	1	3	1 hr.	3.08	2.98	3.20
	1	3	None	3.08	3.11	3.16
	1	3	1 hr.	3.15	3.31	3.23

The sulfhydryl content of dried muscle tissues immersed in toluene containing phenol, cited in Table XIII, might be compared with the corresponding values for dried muscle titrated for –SH in toluene without phenol, seen in Table XII. This comparison is valid, for the Rabbit Numbers used in Tables XII and XIII refer to the same rabbits, respectively. The drying times in Tables XII and XIII are not quite the same, nor were the determinations cited in the two tables performed on the same day. Yet it is evident that the sulfhydryl content of the muscle as titrated by Mercury Orange is not reduced after drying if phenol is present in the toluene. From this it is concluded that phenol renders accessible to the reagent those –SH groups which were blocked from the reactive field by drying, so that muscle tissue after drying reveals the same number of –SH groups to Mercury Orange in toluene phenol solution as does undried muscle. Since phenol is a strong hydrogen bond competitor, it is inferred that the blockading of –SH groups on drying is indeed largely due to formation of new hydrogen bonds in the protein during the drying process.

In Table VIII are seen other examples of the effect of hydrogen bond competitors on the content of available –SH groups of muscle after drying. Samples number 11 and 12, titrated in toluene without phenol, show one-quarter to one-tenth or less of the sulfhydryl content of Samples 4 and 5, immersed in toluene solution of Mercury Orange in the presence of phenol. Likewise Sample 1, measured in dimethylformamide in the presence of urea, reveals one-fifth to one-quarter more thiol groups than fibers determined in dimethylformamide alone. Thus it seems evident that neither dimethylformamide alone, which is a vigorous hydrogen-bonding competitor, nor toluene with large amounts of phenol, set conditions wherein all the –SH groups of the muscle protein are available to the Mercury Orange reagent. On the contrary, they permit perhaps a quarter or a third of the total –SH content to remain cryptic.

(viii) *Effect of storage on fixed muscle.* One may raise the question as to the effect of storage of tissue samples on the accessible sulfhydryl content. This had been explored in the following way:

Rabbit psoas muscle was fixed in 2.7% trichloroacetic acid at 4° C., washed in distilled water after 24 hours' fixation, and run into 70% ethanol at 4° C. It was stored in 70% ethanol at 4° C. until sampled for test. Samples were taken at intervals of one to 145 days. They were then assayed for available –SH groups in toluene without previous drying. The results are reproduced in Table XIV.

Table XIV shows that there is no detectable loss on storage up to 12 days, and no great loss of available –SH groups as titrated in toluene upon storage up to more than four months, although a disappearance of perhaps 10% of the available groups may take place in this time. The nature of this loss has not been explored. There is no reason to ascribe it to oxidation. Presumably it is a consequence of slow formation of new hydrogen-bonded barriers to the reaction.

(ix) *Effect of fragmentation of the sample.* One may ask if the molecular cagework of a specimen of fixed, extracted muscle provides a

TABLE XIV

EFFECT OF STORAGE ON SULFHYDRYL VALUE OF FIXED PSOAS MUSCLE PRIOR TO TREATMENT WITH MERCURY ORANGE IN TOLUENE

Storage (days)	Number of tests	Moles –SH × 10^5 per gram dry muscle	
		Mean ± S.D.	Range
1	6	5.66 ± 0.16	5.51–5.92
12	6	5.79 ± 0.06	5.70–5.88
26	6	5.27 ± 0.10	5.16–5.40
40	6	4.37 ± 0.13	4.18–4.57
54	6	4.99 ± 0.20	4.61–5.16
82	6	5.35 ± 0.16	5.18–5.54
89	6	5.09 ± 0.30	4.83–5.54
145	6	4.94 ± 0.19	4.71–5.15

barrier to the diffusion of the Mercury Orange molecule such that the central portions of relatively large pieces of muscle are appreciably less accessible to the mercurial than peripheral portions, or than the centers of very small pieces. This question has been approached by comparing the available –SH content of large bundles of muscle fibers, about 1–2 mm. in diameter and 10–25 mm. long, with the corresponding values for the same weight of muscle finely fragmented into isolated myofibrils and groups of fibrils. The muscle tissue was not permitted to dry. The results are given in Table XV.

TABLE XV

EFFECT OF FORM OF SAMPLE OF FIXED RABBIT PSOAS MUSCLE ON SULFHYDRYL VALUE WITH MERCURY ORANGE IN TOLUENE

Test No.	Moles –SH × 10^5 per gram dry muscle							
	Single piece				Finely teased			
	1	2	3	Av.	1	2	3	Av.
1	4.96	5.23	4.81	5.00	5.03	5.01	5.16	5.07
					5.16	4.61	4.96	4.91
2	5.18	5.25	4.88	5.10	5.18	5.37	5.54	5.36
					5.53	5.29	5.21	5.34
3	5.17	4.99	5.26	5.14	5.18	5.05	4.90	5.04
					5.02	4.83	5.54	5.13
Average				5.08				5.14

It is seen that the degree of fragmentation within these limits has no significant effect on the sulfhydryl content recognized by Mercury Orange in toluene.

(x) *Inadvertent loss of tissue sulfhydryl groups.* From time to time we have encountered erratic results or failures in efforts to measure quantitatively the available –SH content of tissue samples. In a number of instances we have explored the accompanying conditions sufficiently to bring us to an awareness of some of the pitfalls which confront one attempting to use the method.

Impurities in solvents and reagents can interfere with the coupling of Mercury Orange to tissue sulfhydryl groups. Four types of impurities have been recognized and avoided in this connection.

First, heavy metal impurities are often found as contaminants in organic solvents obtained through commercial channels, in water, or in reagents such as trichloroacetic acid. Recognition of this important source of loss of –SH groups has led to precautions designed to free all solvents and reagents of even traces of heavy metals before permitting contact with tissue destined for sulfhydryl cytochemistry. Trichloroacetic acid was distilled *in vacuo.* Water and organic solvents were carefully distilled, and, in many instances, further purified by passage through sulfonic acid ion exchange resin columns, or slurried with a sulfonic acid ion exchange resin. Use of metal instruments for handling tissue has been avoided, or has been reduced to a minimum, with precautions to use clean, uncorroded instruments. Volatile heavy metal compounds such as tetraethyl lead and osmium tetroxide have been kept entirely away from the refrigerator and containers used for the work. Diphenylthiocarbazone was used as a reagent for detecting traces of heavy metals in solvents.

Second, peroxides have been detected as impurities in some of the solvents used, particularly some of the alcohols. Since peroxides can oxidize –SH groups, an accurate titration of total sulfhydryl cannot be expected in the presence of peroxides. Solvents used for –SH work should be free of peroxides.

Third, mercaptans or hydrogen sulfide present as impurities in the solvent can interfere with the reaction by competing with tissue thiols for the available Mercury Orange. In the absence of Mercury Orange, thiols or H_2S can convert tissue disulfide groups to sulfhydryl. For these reasons, mercaptans and H_2S impurities in solvents must be avoided if accurate measurement of tissue –SH is desired.

Fourth, acids can interfere with the binding of Mercury Orange by driving the equilibrium indicated in Eq. 1 (p. 319) in the reverse direction. The equilibrium constants are such that substantial amounts of acid are necessary in order to make this reversal appreciable. Trace amounts of acid, therefore, are of no concern. Concentrations above 0.1 to 1 *N* might begin to have an effect.

Although the four types of impurities mentioned here are especially important to consider in quantitative cytochemical work involving thiols, they must also be avoided if reliable results are to be achieved in a qualitative way.

(xi) *Time course of binding of Mercury Orange by muscle.* In order to achieve a better understanding of the mercaptide binding reaction involved in this cytochemical method, we have followed the changes in absorbance at 484 mμ of supernatant toluene solutions of Mercury Orange into which pieces of fixed muscle were placed and permitted to react with the mercurial. In general, the loss of orange mercurial from the solution was rapid over the first few hours, as the muscle became orange. If the muscle had been fragmented into small pieces, a minimum absorbance of the solvent was usually reached within 24 hours (Fig. 3B, D). If the pieces were larger, three or four days might elapse before

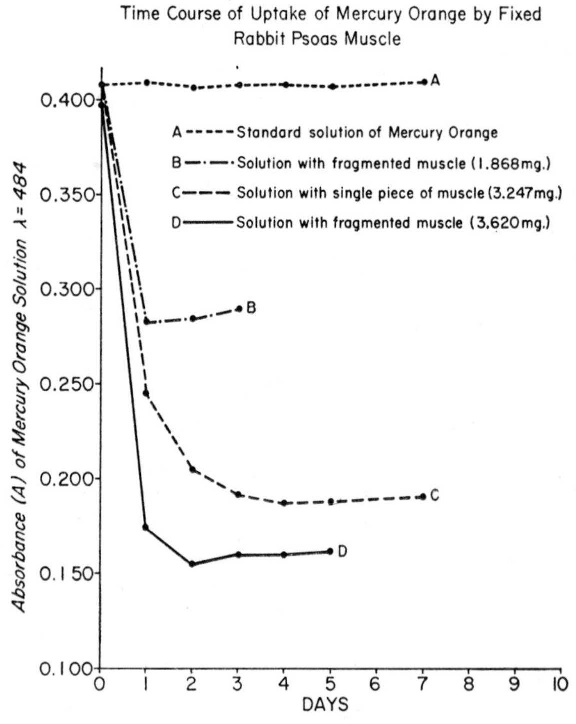

FIG. 3. Curves B, C, and D show absorbance of a toluene solution of Mercury Orange in which muscle fragments were immersed, plotted against days after initial immersion. Curve A shows absorbance of a standard solution of Mercury Orange in which no tissue was immersed, as read at successive intervals for a span of seven days. For further explanation, see text.

a minimum was achieved. A typical uptake curve of this type is reproduced in Fig. 3C.

On some occasions, however, a curious apparent reversal of the reaction took place. In such instances the light absorption of the solvent at 484 mμ reached a minimum within 24 hours and then increased again for several days. Such a curve is shown in Fig. 3B. On the face of it, such curves would suggest that in these instances Mercury Orange is bound by the tissue and then subsequently some of it is released again to the solvent. We have not arrived at a satisfactory explanation of this puzzling behavior. We report the findings nevertheless, hoping that future work will bring about an understanding of it, and hoping that it will serve to warn readers to the complexities of cytochemical reactions, even of relatively simple ones like the example discussed here.

(xii) *Equilibrium considerations.* Since specification 7 (p. 321), relating to the characteristics of a satisfactory cytochemical reaction for —SH groups, requires that a favorable equilibrium be utilized for the reaction, we have attempted to determine the equilibrium constants for the reaction

$$\text{Mu–S–H} + \text{Cl–Hg–R}' \rightleftarrows \text{Mu–S–Hg–R}' + H^+ + Cl^- \quad (3)$$

where Mu represents the insoluble proteins of fixed extracted myofibrils and R′ represents the organic portion of Mercury Orange. Our attempts to obtain a value for the equilibrium constant of the reaction failed, as consistent values for the constant were not obtained as $[H^+]$ was varied experimentally. However, as mentioned previously, Hughes (1950) obtained a satisfactory value of 4.45 for the pK of the related sulfhydryl reaction between serum albumin and methyl mercuric iodide.

Any attempt to determine the pK of Eq. 3 faces a number of complicating factors: First, the Mu—S—H is in a solid phase, and a value for Mu—S—H cannot be obtained in a manner analogous to a value for the concentration of a solute in solution. Second, the reaction must take place in an organic solution, where the activity of $[H^+]$ is not known, and where the interactions between H^+ and Cl^- on the one hand and the solvent on the other are not well understood. Third, the muscle proteins in the insoluble phase can be assumed to possess anion- and cation-binding groups which may cause them to act like ion exchange resins, introducing equilibria which will tend to remove from solution the ions liberated by the binding of the mercurial, thus favoring the progress of Eq. 3 to the right. Fourth, it is possible that engagement of one thiol group by Mercury Orange may have an influence on the accessibility of other —SH groups.

Without countenancing any conclusions relating to the equilibrium constant, we have found that HCl may be added to dimethylformamide solvent up to a concentration of more than 1 N before any noticeable removal of Mercury Orange from muscle mercaptides in suspension in the solvent is achieved. This indicates that under all ordinary conditions the course of Eq. 3 to the right is substantially complete.

F. Theory of the Use of Mercury Orange for the Demonstration and Determination of Disulfide Groups, and of Disulfide and Sulfhydryl Groups

By utilizing the well-known equilibrium,

$$R'\text{-}S\text{-}S\text{-}R'' + 2R'''\text{-}S\text{-}H \rightleftarrows R'\text{-}S\text{-}H + H\text{-}S\text{-}R'' + R'''\text{-}S\text{-}S\text{-}R''' \quad (4)$$

it is possible to convert disulfides to sulfhydryls. Simple applications of the mass action principle can be used to load the equilibrium on either side so it can be pushed to virtual completion in either direction.

Thus if one treats a tissue specimen containing disulfide groups with a suitable mercaptan in sufficient concentration, the tissue disulfide groups can be converted into sulfhydryls, and thus rendered available for reaction with Mercury Orange, along with the –SH groups originally present. Thioglycolic acid, glutathione, and β-mercaptoethanol have been used successfully as mercaptans for converting disulfide to sulfhydryl. After the tissue has been treated in this manner, the subsequent final reaction with Mercury Orange reveals the locations where sulfhydryls and disulfides were present originally, without distinguishing between them.

If one wishes to demonstrate or measure the disulfide groups in a tissue specimen without revealing the sulfhydryl groups, one can proceed by first blocking the sulfhydryl groups with a suitable alkylating agent such as iodoacetamide. This converts the thiols into thioethers, and renders them immune to attack by extraneous mercaptans. If the tissue is then treated with an appropriate mercaptan, the disulfide groups can be converted to sulfhydryl and rendered available for reaction with Mercury Orange. The resulting cytochemical display reveals the original site of tissue disulfide without showing original locations of tissue thiols.

III. INSTRUMENTATION

A. Quantitative Determination of Sulfhydryl in Tissue Suspensions

1. *By the Amperometric Method*

This method is applicable to the measurement of accessible –SH groups in soluble proteins or insoluble suspension of myofibrils, hair, or tissue sections. The instrumentation described by Kolthoff and Harris (1946) and adapted by Benesch and Benesch (1948) has proved to be reasonably satisfactory. Further description is not necessary.

2. *By the Mercury Orange Method*

The procedure is outlined on p. 350. As for instrumentation, any good spectrophotometer can be used. The Hilger, Beckman, or Carey instruments are suitable. No modifications of the standard instrument are necessary. The precautions are those needed for accurate absorption spectrophotometry. Concentrations of Mercury Orange and path length through absorption cuvettes should be chosen so as to yield absorbance readings in the more accurate portions of the scale, which lie roughly between the absorbance (A) values of 0.2 to 0.6.

B. Quantitative Determination of Sulfhydryl in Tissue Components by Microspectroscopy

Sectioned or fragmented tissue components to which Mercury Orange has been bound in mercaptide linkage can be subjected to microspectroscopic analysis. In this way the available –SH content of small tissue components such as single cells or cell components can be measured. The measurements can be made conveniently in the visible, using the absorption peak of Mercury Orange and its mercaptides at 484 mμ, taking the molar absorptivity at that wavelength to be $\epsilon_{max} = 1.95 \times 10^4$. Any good microscope with glass optics can be used. One can use either photographic or photoelectric methods of measurement. Principles and suitable instrumentation have been discussed adequately by others, and need not be repeated here. Readers are referred in this connection to Caspersson (1950), Loofbourow (1950) and Pollister and Ornstein (1955).

In this laboratory very simple photographic methods have been used. An ordinary light microscope with apochromatic lenses has been fitted to a camera in which a hole has been cut to admit a small air-driven step sector which can be placed about 1 mm. under the emulsion of the plate. The camera, step sectors and turbine have been described by Bennett *et al.* (1953b). Plates taken at desired wavelengths and magnifications were scanned on a Kipp and Zonen Moll recording microphotometer modified by Bennett *et al.* (1953a). The voltage from the thermocouple was amplified in a breaker amplifier and fed into a strip-chart pen recording galvanometer.

Tungsten ribbon filament lamps, Pointolites, or zirconium arcs have proved to be satisfactory light sources. We have used a well-corrected 2 inch motion picture projection lens fitted with an iris diaphragm for a lamp condenser. A narrow spectral band centering at 481 mμ was isolated by means of an interference filter and a Wratten 75 gelatin filter (Eastman Kodak Co., Rochester, N. Y.). Wratten M photographic plates are suitable for this wavelength. A band peaking at 691 mμ was isolated by an interference filter and a yellow glass filter to remove a blue and an ultraviolet band which also traversed the interference filter. Kodak IN plates were used to record this red light. A Polaroid disc was inserted in the microscope substage in a mount which permitted it to be rotated so as to place its axis of transmission in a known relationship to any desired geometric axis of the tissue. The Polaroid could easily be removed from the optical system. A Polaroid disc was also placed on top of the ocular when an analyzer was needed in the optical system.

IV. PROCEDURES

A. Procedure for the Determination of Available Sulfhydryl in Muscle Fragments or in Proteins in Solution by the Silver Amperometric Method

The amperometric method used for the titration of sulfhydryl was essentially that described by Kolthoff and Harris (1946) and adapted later by Benesch and Benesch (1948). Both of these methods are argentometric titrations carried out in an NH_4OH–NH_4NO_3 buffer system using a rotating platinum electrode as the indicator electrode and a Hg/HgI_2 half cell as the reference electrode. In this study the procedures of Benesch and Benesch were not modified except as shown in Table X.

B. Procedure for the Determination of Available Sulfhydryl in Muscle Fragments by the Mercury Orange Method

1. *General Considerations*

The determination of sulfhydryl by the Mercury Orange method can be based on the uptake of the colored mercurial by a sample of muscle immersed in a standard solution of the reagent. A value for available –SH groups can be derived from spectrophotometric measurements of the solution before and after uptake. Any good solvent for Mercury Orange can be used. The procedure will be described with reference to toluene as a solvent, since there is more experience with this solvent than with any other. However, determination of available –SH groups using toluene may not give a value for total –SH. Dimethylformamide saturated with urea is a solvent for Mercury Orange which has yielded values for available –SH groups judged to be close to or equivalent to the true value for total –SH (see pp. 338–339). All solvents, reagents, and fixatives must be free from contamination by traces of heavy metals, peroxides or other oxidizing agents, thiols, or significant amounts of acids.

Solutions of Mercury Orange in toluene at a concentration of approximately 2×10^{-5} M give a value for absorbance (A) of approximately 0.40 when measured in a 1 cm. cuvette. This absorbance value lies within the range yielding accurate spectrophotometric measurements. By appropriate choice of the volume of the solution of Mercury Orange and of the sample size, suitable absorbance values can also be obtained at the end of the determination.

2. *Reagents*

Reagents used in the determination of sulfhydryl by the Mercury Orange method are:

Trichloroacetic acid (Reagent Grade) (distilled *in vacuo*)
95% ethanol, redistilled
Absolute ethanol
Toluene (Reagent Grade)
Mercury Orange, standard solution in toluene
Distilled water, high quality

3. *Fixation and Extraction*

Rabbit psoas or other muscle was removed from anesthetized or freshly killed animals and placed immediately in approximately 10 volumes of 2.7% trichloroacetic acid in distilled water at 4° C. Fixation in this fluid was permitted to continue at 4° C. for 15–20 hours. The tissue was then extracted 5 or 6 times with distilled water, followed by extraction with 25% ethanol and 50% ethanol during about 2 hours. The tissue was then placed in 70% ethanol and stored at 4° C. until used. Prior to the measurement the sample was placed in 95% ethanol, fragmented by teasing, extracted successively with absolute ethanol, with 50% toluene in ethanol, and finally with toluene, during about 2 hours. It was stored in toluene until the actual measurements commenced.

4. *Determination of Sulfhydryl Value*

One to five milligrams of teased fixed extracted muscle was taken from the toluene storage container, blotted briefly with filter paper to remove excess toluene, and placed immediately in 15 ml. of a standard solution of Mercury Orange in a 50 ml. glass-stoppered Pyrex bottle. The samples were kept at 4° C. Daily thereafter a test portion of the solution at room temperature was carefully pipetted into a 1 cm. cuvette and the absorbance (A) at 484 mμ was determined with a Hilger "Uvispek" spectrophotometer. The solution was then returned to the reaction bottle. Great care was used to avoid losses during transfer and measurement. The absorbance (A_1) of the standard solution was also determined daily. The value of the absorbance at the end-point (A_2) was taken when a minimum value of A had been reached, which was usually within 1–7 days. This represented the absorbance of the solution at the time when the sample had taken up a maximum of the reagent. The samples were then removed from the solution, rinsed with toluene, blotted with filter paper, placed in an aluminum foil envelope and dried to constant weight *in vacuo* over P_2O_5. A Roller-Smith Precision Balance of 5 mg. capacity was used. The sample weight was obtained by difference after weighing the empty envelope. The sulfhydryl value for the sample was calculated according to the following formula:

$$\text{Moles —SH per gram dry muscle (uncorrected)} = \frac{(A_1 - A_2) \cdot V \cdot M}{A_1 \cdot 1000 \cdot W} \quad (5)$$

where, M = molarity of the standard Mercury Orange solution; A_1 = absorbance of the standard Mercury Orange solution; A_2 = absorbance of the Mercury Orange solution at end-point; V = volume (ml.) of Mercury Orange solution used; W = weight (in grams) of dry muscle sample.

Since the weight of the sample (W) was obtained after the muscle had reacted with the Mercury Orange the value for W used in Eq. 5 is actually that of the muscle mercaptide. Strictly speaking, a correction should be made for the bound mercurial. This correction can be made in accordance with the following formula:

$$\text{Moles —SH per gram dry muscle (corrected)} = \frac{\text{Moles —SH per gram dry muscle (uncorrected)}}{1 - [(m)(\text{Moles —SH per gram dry muscle (uncorrected)})]} \quad (6)$$

where m refers to the molecular weight of the bound mercurial.

In our calculations we have assumed that the H^+ and Cl^- liberated by the mercaptide-forming reaction are bound to the muscle by ionic bonding. This assumption leads to a value of 483.3 for m in Eq. 6. Using this value for m, the error due to bound mercaptide when 7×10^{-5} moles –SH per gram of dry muscle are recognized by Mercury Orange is 2.86%. Thus in this case the corrected value is greater than the uncorrected value by a factor of 1.0286. When smaller values for –SH content result, the correction factor falls off rapidly. Thus within the ranges of –SH content with which we have dealt, the correction factor for bound mercaptide appears to be well within experimental error. Consequently we have not deemed it necessary to correct for bound mercaptide in our routine work. In this chapter the values cited for moles –SH per gram dry muscle are uncorrected for bound mercurial unless otherwise stated.

5. *Precision of Method*

Thirteen standard solutions of Mercury Orange in toluene were prepared and the ϵ_{max} at 484 mμ determined on different days. As previously stated (see p. 327) the ϵ_{max} at 484 mμ, expressed as Mean $\pm$ S.D., was $1.95 \pm 0.03 \times 10^4$. The precision obtained for the determination of sulfhydryl in muscle is shown in Table XIV. In this evaluation 6 samples were tested on each of 8 days. The average deviation from the mean ranged from 0.04 to 0.18 and the standard deviation, from 0.06 to 0.30.

6. *Variations in Method*

This procedure was varied from time to time by varying the solvent. This has a considerable effect on the value for available —SH content of muscle, as shown in Table VIII. In some cases the procedure was varied by permitting the solvent toluene to evaporate from the sample and by permitting the sample to dry before immersing the sample for measurement in the solvent containing Mercury Orange. This led to reduced values for available —SH content of muscle when measurements were made in certain solvents (see Table XII).

C. Procedure for the Determination of Available Sulfhydryl in Muscle Fragments by the Methyl Mercuric Iodide Method[1]

This method was suggested to us by Dr. W. L. Hughes, Jr. (1950) to whom we are most grateful. The principle of the method is similar to that of the Mercury Orange method, save that methyl mercuric iodide is used instead of Mercury Orange and the content of unreacted methyl mercuric iodide in the solution is determined by titration with diphenylthiocarbazone (dithizone) rather than by direct spectrophotometric measurement.

Muscle was fixed, extracted, and fragmented as described on pp. 350–351. The subsequent measurement procedure adopted in this laboratory was as follows:

A sample of fragmented fixed muscle (4–6 mg.) was placed in 2 ml. of standard methyl mercuric iodide solution (approximately 4×10^{-4} M). The following day, or later, samples of solution were withdrawn and the unreacted CH_3HgI was titrated with a CCl_4 solution of dithizone which had been currently standardized against standard CH_3HgI. Several drops of a dilute solution of *n*-amylamine were added to the titration sample, giving a final concentration of amine of about 0.75×10^{-7} M. The titration was carried out at 0° C. The end-point was the color change of the solution from orange-yellow to green. The titration was carried out in small test tubes (10×75 mm.). For the titration a microburette

[1] Data by Mrs. Lucy Jache.

was improvised from a 0.2 ml. measuring pipette graduated to 0.001 ml. This pipette was calibrated using CCl_4. Upon completion of the reaction the muscle sample was removed from the reaction bottle and dried to constant weight *in vacuo* over P_2O_5. The —SH content available to the methyl mercuric iodide was then calculated by the following formula:

$$\text{Moles —SH per gram dry muscle (uncorrected)} = \frac{(M_1 - M_2) \cdot V}{W \cdot 1000} \quad (7)$$

where, M_1 is the molar concentration of methyl mercuric iodide in the original standard solution; M_2 is the molar concentration of methyl mercuric iodide at the end-point minimum; $V =$ the volume of the solution in ml.; $W =$ the dry weight of the muscle sample in grams. A correction for the bound methyl mercuric iodide can be made by using Eq. 6 on p. 352, if one uses 342.6, the molecular weight of methyl mercuric iodide, for m in Eq. 6. This correction is small and has not been applied in the values cited in this chapter.

Ten determinations on 1 muscle sample ranged from $4.36\text{–}5.46 \times 10^{-5}$ moles –SH per gram of dry muscle. These values gave a Mean $\pm$ S.D. of $4.65 \pm 0.36 \times 10^{-5}$ moles –SH per gram of dry muscle.

D. Procedure for the Cytochemical Determination of Available Sulfhydryl Groups in Tissue Sections, Smears, or Spreads by Mercury Orange

1. *Tissue Preparation*

One can use many of the standard histological procedures in preparing tissue for microspectrophotometric quantitative study of available –SH groups in minute tissue components. In interpreting the results, however, the effect of each of the reagents and fixatives used must be considered. Oxidizing agents such as osmic acid, dichromate ion, and chromate ion may destroy –SH groups. Heavy metal ions such as the mercuric ion can bind to –SH groups, though they can be removed subsequently by treatment with an appropriate mercaptan such as BAL or β-mercaptoethanol.

If one wishes to preserve all protein –SH groups, it is wise to avoid fixatives and reagents containing heavy metals, even in traces, or containing oxidizing agents such as peroxides, or containing alkylating agents such as alkyl halides. If one wishes to avoid risk of creating new –SH groups out of disulfide linkages one must protect the tissue from mercaptans or H_2S.

Fixatives which can reasonably be expected to preserve all protein –SH groups and to create no new –SH groups are trichloroacetic acid,

Carnoy, alcohol, and dimethylformamide. Many –SH groups survive formaldehyde fixation, though the quantitative reliability of formaldehyde fixation has not been studied with respect to –SH groups. One would expect each of the above fixatives to remove glutathione and free cysteine from the tissues.

Freezing-drying preserves all tissue –SH groups, including those in proteins and in glutathione. Subsequent immersion in a solution of Mercury Orange in which glutathione is insoluble would permit a reaction with most of the tissue thiol groups. However, we have not explored this possibility.

Freezing-substitution (see Bennett, 1951) provides a convenient way to prepare tissue for sulfhydryl cytochemistry without treating the tissue with aqueous media.

Dehydration after aqueous fixation can be carried out in graded alcohols, acetone, dioxane, dimethylformamide, or other suitable non-reacting solvent.

One can embed in paraffin, celloidin, carbowax, or esterwax. Methacrylate embedding has not been tried in connection with sulfhydryl cytochemistry, but it would appear to offer little promise, as the free radicals formed in polymerization might react with protein mercaptans, even if peroxide catalysts were avoided and ultraviolet light used to promote polymerization. One can avoid embedding by cutting sections from frozen tissue, teasing, homogenizing, blending, smearing, imprinting, or by using whole mounts.

2. *Mounting*

The preparations can be mounted on slides in the usual manner, though albumin used for sticking specimens to slides may contain –SH groups which might confuse interpretation of the reaction. These albumin –SH groups might be blocked by alkylation prior to mounting the sections on albumin-covered slides, though we have not used this procedure, and have avoided the use of albumin.

One should take precautions at all times to prevent the evaporation of solvent from or drying of the tissue preparation, if one is to avoid risk of rendering some of –SH groups in the sample inaccessible to the Mercury Orange reagent from formation of new hydrogen bonded barriers during drying (see pp. 340–343).

The properties of the mounting medium are important. A medium which differs appreciably from the tissue with respect to refractive index can introduce difficulties and errors into microabsorption measurements from diffraction and scattering effects. Moreover, a mounting medium which contains thiols, acidic groups, or other reactive radicals is unsuitable as it may remove the Mercury Orange from the tissue.

Canada balsam and gum damar are not satisfactory for quantitative work as their refractive indices are lower than that of protein tissue components. However a perfect refractive index match of all tissue components is not to be expected in any single preparation, as different tissue components may differ slightly from each other in refractive index. Moreover birefringent objects present two refractive indices (see Bennett, 1950), both of which cannot be matched exactly by any one medium. For muscle fibers and for many tissues a medium with a refractive index of 1.54–1.56 is satisfactory. Such a medium can be made by mixing two resins, one of which has a refractive index above and another below the desired value, each dissolved in a solvent of the desired refractive index. We have used a resin mixture prepared by Dr. John H. Luft, who has kindly given us permission to describe his medium. He took solutions of Clarite X (Neville Co., Pittsburgh, Pa.) and of H.S.R. Microscopic Mounting Medium (Hartman-Ledden Company, Philadelphia, Pa.), each dissolved in monobromotoluene. The refractive index of the solvent alone is 1.553. The two resin solutions in this solvent were mixed in proportions adjusted to give a refractive index of the mixed solutions of 1.553. This does not change appreciably upon evaporation of the solvent. This mounting medium is satisfactory for microspectrophotometry of isolated myofibrils.

Glass slides and coverslips can be used.

3. *Photographic Photometry*

The following procedure has been used in this laboratory:

Isolated myofibrils or sections of muscle to which Mercury Orange has been bound in mercaptide linkage have been taken from containers in which the uptake of the mercurial has been measured. Thus the available –SH content per gram of dry tissue was known. The tissue was then mounted on glass slides in a medium of matching refractive index. The slide was placed in a microscope fitted with a rotating stage, an apochromatic 2 mm. objective, NA 1.40, and an achromatic condenser of the same aperture. A Polaroid was inserted in the substage and one placed on the ocular in the crossed position. Suitable myofibrils were selected and oriented by visual inspection, contrast being provided by virtue of the intrinsic birefringence of the A band. After focusing, adjusting the camera, inserting the step sector and setting it to spinning, a series of photographs of the specimen were taken with spectral bands and Polaroid settings as given in Table XVI. The specimen was not moved, nor were lenses adjusted or focused, except for refocusing for Photograph No. 5. After taking Photograph No. 5, the slide was removed and for it was substituted a comparable slide of the same tissue, pre-

TABLE XVI
SETTINGS OF POLAROIDS AND OF LIGHT FILTERS FOR SERIES OF PHOTOGRAPHS FOR MICROSPECTROMETRY OF —SH GROUPS IN MYOFIBRILS WITH MERCURY ORANGE

Photograph No.	Filter setting (λ in mμ)	Azimuth of Polaroid setting with respect to axis of myofibril		Mercaptide group
		Analyzer	Polarizer	
1	481	−45°	+45°	Mercury Orange
2	481	Removed	0°	Same
3	481	Removed	90°	Same
4	481	Removed	Removed	Same
5	691	Removed	Removed	Same
6	481	−45°	+45°	Phenyl mercuric chloride
7	481	Removed	Removed	Same

pared and mounted in an identical way, save that the myofibrils had been immersed in a solution of phenyl mercuric chloride rather than in Mercury Orange. The myofibrillar —SH groups in this second preparation were hence bound in colorless phenyl mercaptide linkage, which does not absorb at 481 mμ. After locating suitable myofibrils in this colorless preparation between polarizer and analyzer, the camera and step sector were adjusted and Photographs Nos. 6 and 7 were taken.

The individual photographs in this series of seven were used in the following way:

No. 1, in Table XVI, served as a means for identification of cross bands and for determining the excellence of focus for Nos. 2, 3, and 4. In Photographs Nos. 2, 3, and 4, contrast should be entirely on the basis of absorption by the Mercury Orange mercaptides. Nos. 2 and 3 provided checks for linear dichroism, which is absorption which varies with polarization vectors of incident linearly polarized illumination. When dichroism was present, Photographs Nos. 2 and 3 provided data from which the dichroism could be measured. No. 4 provided a basis for determining the absorbance of isotropic objects. Photograph No. 5, taken in red light at a wavelength which Mercury Orange does not absorb, provided a check for nonspecific scattering of light by the specimen used for the measurements. Thus it can be said to be a detector for certain types of optical noise. Photograph No. 6 was designed to check the focus for No. 7. No. 7 served to detect "optical noise" at 481 mμ. It revealed nonspecific scattering and diffraction, and any absorption which the tissue displayed at this wavelength which was not related to Mercury Orange. It was taken at a wavelength where Mercury Orange absorbs strongly but where the phenyl mercaptide and most tissue proteins do not absorb light appreciably. Under conditions of ideal refractive index match and in the absence of absorption in the visible not due to Mercury Orange, the specimen should be invisible in Photographs Nos. 5 and 7. If the specimen was visible in these photographs, densitometric tracings of the photographic images provided data which could be used for approximate corrrections for nonspecific light losses.

Negatives of the above photographs or positive prints on lantern slide plates were then scanned in the microphotometer. Reference tracings of absorbance standards were obtained by scanning across the shadows cast by the rotating step sector. From the pen recorder tracings obtained at each of the steps of the sector shadow, a curve was constructed where microphotometer pen recorder galvanometer deflections were plotted against the absorbance standards provided by each step of the sector disc. Tracings across any portion of the image of the sample provided pen recorder galvanometer deflections which could be referred to the curve, from which a value for the absorbance of the tissue component could be obtained. All tracings from a single plate were taken at the same microphotometer slit width, the same amplifier gain, and at the same brightness of the microphotometer lamp, which was monitored throughout the tracings. This procedure was followed for each photograph from which quantitative values were desired. The principles of this method are discussed by Thorell (1947).

Since dichroism was encountered in the muscle mercaptides of Mercury Orange (see Bennett, 1952), a significant value for absorbance (A) of muscle mercaptides could not be obtained from Photograph No. 4, taken in unpolarized light. The muscle mercaptides were found to form uniaxial dichroic absorbers, explainable on the assumption that the linear axis of the absorbing chromophore of Mercury Orange in the A bands tended to be arranged radially about the long axes of molecular filaments running longitudinally in the muscle. Land and West (1946) have pointed out that one can calculate a value for $\bar{A}$, which is the value of the absorbance which a uniaxial dichroic absorber would show if its absorbing chromophores were randomly arranged geometrically in unchanged concentration, from values for $A_{\parallel}$ and $A_{\perp}$, which are the absorbance of the body to linear polarized light with its electric vector parallel and perpendicular to the axis of the absorber, respectively, by using the following equation:

$$\bar{A} = \frac{(A_{\parallel} + 2A_{\perp})}{3} \tag{8}$$

$\bar{A}$ so calculated can be used for A in the usual expression of the

Beer-Lambert law relating molar concentration (M) of an absorbing substance to molar absorbtivity (ϵ) and to path length (b), which can be written

$$A = \epsilon Mb \tag{9}$$

A value for b can be obtained by measurement of object thickness in the microscope, in accordance with the method of Caspersson (1950, p. 43). The value for ϵ for Mercury Orange and its mercaptides can be taken to be 1.95×10^4, as indicated on p. 327. However, it must be borne in mind that it is not known how well Mercury Orange obeys Beer's Law at concentrations achieved in some tissue mercaptides.

Photographs 5 and 7 (Table XVI) were also scanned with the microphotometer in order to ascertain if nonspecific scattering, absorbing, or diffraction effects were introducing appreciable error. In our experience the specimens were very nearly invisible in these pictures, and the tracings traversing the images on Photographs 5 and 7 showed no significant fluctuations above the background noise level. These findings led to the conclusion that the refractive index match of the mounting medium was satisfactory, and that no correction for mass or for nonspecific light losses in the specimen was necessary.

By means of these procedures, concentrations of available sulfhydryl groups can, in principle, at least, be determined for small tissue components which are isotropic, or which are uniaxial dichroic absorbing bodies.

But one should not leave this topic without mentioning some of the difficulties and limitations of which we have become aware.

First, if biaxial absorbing bodies were to be encountered, further difficulty would be at hand. We have not explored the theory of biaxial absorbers sufficiently to know if determinations of A in three directions in the body would permit a calculation of a value for $\bar{A}$. We would anticipate that biaxial absorbers would be very difficult to measure cytochemically, if not actually intractable.

Second, circular dichroism must be considered. Bennett (1952) searched for circular dichroism in Mercury Orange muscle mercaptides, and failed to find evidence for it. We have neither searched for nor attempted to derive a theory for the handling of circular dichroism cytochemically. One can search for circular dichroism by comparing the absorbance of a body with respect to dextrorotatory circular polarized light with its absorbance with respect to levorotatory circular polarized light. Bennett (1950) discusses means of producing circular polarized light with either sense of rotation. This can be accomplished conveniently in a microscope by mounting a polarizer in the substage, and placing between it and the object a quarter-wave plate with its principal axes at $\pm 45°$ to the axis of transmission of the polarizer. Rotating either polarizer or quarter-wave plate through 90° reverses the sense of rotation of the circular polarized light. In order to be sure that no circular dichroism is present, one should search for it along more than one geometrical axis of the specimen. These axes should diverge widely from the parallel. In muscle it is sufficient to search for circular dichroism by passing the circular polarized light through the muscle in a direction parallel to the muscle axis (by using cross sections) and again by illuminating the muscle with circular polarized light traversing the muscle at an angle normal to the muscle axis (as by using longitudinal sections).

Third, special consideration must be given to low values for A, which can be encountered in quantitative sulfhydryl cytochemistry. In attempts to measure the available —SH concentration in the individual cross bands of striped muscle, isolated myofibrils have been permitted to react with Mercury Orange under conditions when $5\text{–}6 \times 10^{-5}$ moles of —SH per gram of muscle were engaged in mercaptide linkage. These isolated myofibrils were mounted in a medium of matching refractive index and were subjected to the procedures outlined above on pp. 356–359. Isolated myofibrils were chosen, because the sections used by Bennett (1952) were unsatisfactory for purposes of measuring —SH content of individual cross bands. In relatively thick (3–6 μ) sections there was a possibility of superposition of different cross bands in the light path so that in the sections reliance could not be placed on the assumption that light traversing a given cross band in focus in the microscope was unaffected by previous or subsequent passage through another underlying or overlying out-of-focus cross band. The use of isolated myofibrils forces one to use path lengths (b) through the absorbing specimen which are of the order of 1 μ, which is a fair approximate value for the diameter of a mammalian myofibril. If one assumes a partial volume of 0.25 for the protein of muscle, if one takes 1.33 for the specific gravity (density) of protein, if one neglects shrinkage, and if one takes the value of 6×10^{-5} moles of available —SH per gram of dry muscle, it follows that the average concentration of Mercury Orange mercaptide formed in that muscle is about 0.02 molar. Applying the Beer-Lambert law (Eq. 9), it follows that the absorbance, A, to be derived from a 1 μ path length through a 0.02 M solution of Mercury Orange is about 0.04. Values above and below this would be expected in microspectrophotometric determination of dichroic Mercury Orange muscle mercaptides in the several cross bands of isolated myofibrils.

Absorbance values in the neighborhood of 0.04 fall in a range where accurate spectrophotometric readings are very difficult to attain, even under the best circumstances. These difficulties are multiplied in the microspectroscopic case. Although we did not succeed in overcoming the difficulties sufficiently to permit citation of reliable values for the sulfhydryl concentrations in the separate cross bands of muscle, we anticipate that experiences gained in this effort will be of value to others, and are hence deserving of some mention here.

Once we came to realize the absorbance range in which the desired values lay, we proceeded as follows:

Logarithmic step sectors with low values of increments of absorbance values were calculated and made. The sectors were made with a two-inch over-all diameter, and with steps 0.060 inches deep. The sectors were made to be spun by an air turbine with an over-all diameter of less than 0.7 inch, and are pictured in a publication by Bennett *et al.* (1953b). The step sector with absorbance standard increment steps of 0.005 each was chosen for the study of myofibrillar Mercury Orange mercaptides. This sector had ten steps, providing reference standards for A from 0 to 0.050 in increments of 0.005.

High-contrast photographic plates were chosen, and plates, exposures, developers and developing schedules were worked out which gave contrasting density increments on the plate for the shadows of each of the steps of the sector. This required a photographic system which would amplify the very small absorbance increments into detectable signals in terms of increments of plate blackening. Acting like any high gain amplifier, this photographic system also amplified drift and noise as well as the signals. Difficulties were encountered which were photographic analogs of each at these sources of difficulty in high gain electronic direct current amplifiers.

Photographic drift became apparent in the form of unevenness of illumination of the plate characterized in the absence of any specimen by a bright area in the center fading off very gradually as one followed any radius away from the center. If one placed such a plate in the microphotometer and scanned it from one edge to the other across the center, one did not obtain a straight line tracing, but instead one saw a shallow arc representing a center brighter than the edges by an amount equivalent to absorbance differences of perhaps 0.02 to 0.05, when compared to standard shadows cast by the rotating step sector. Thus the sector and the specimen might receive different amounts of radiant energy per unit area during exposure. In order to reduce error from photographic drift of this nature we attempted always to place the step sector close to the image of the specimen, and to place the specimen and the anticipated matching sector step at symmetrical distances, one on each side of the center of brightness of the arc mentioned above. But there are limits to what one can accomplish by this approach. For even if one places the sector so that its edge is close to the image of the object, the matching sector step may still cast its shadow at an appreciable distance from the image. Moreover, if there are several portions of an object to measure photometrically, each of which differs from another with respect to location and absorbance, one cannot readily place all of them symmetrically with respect to the particular sector step which matches each portion. In the face of this situation, after placing the sector disc and specimen image as favorably as possible with respect to the center of symmetry of the arc of unequal illumination, we resorted to a practice of scanning all the way across the plate several times, taking parallel tracings through the sector and specimen and close to them, obtaining a family of curves from which a reasonable baseline could be interpolated over areas obstructed by specimen or sector, thus providing a baseline corrected at least partially for drift due to uneven illumination of plate. A correction based on these tracings could be applied to the calibration curve mentioned on p. 358. By use of such corrections it was often possible to reduce error from drift to a value equivalent to absorbances of 0.01 or less. However, this remained a very significant error in the face of the low absorbance values with which we were dealing. It must be emphasized that the degree of unevenness of field illumination which was encountered, and which was found to be troublesome, would occasion no difficulty in ordinary microscopy or photomicrography, and would introduce but little error in microspectrometry of objects falling within the desirable ranges of absorbance values, which are above those which we are discussing here by factors of ten to a hundred.

Photographic noise manifested itself in the form of significant variations in photographic blackening of the plate induced for the most part by Fresnel diffraction fringes engendered by imperfections in the optical system such as scratches, dust particles, or streaks on lenses, or on the slide or cover glass, or from small and otherwise insignificant transparent particles in the mounting medium close to the specimen. Additional inhomogenieties in background resulted from stray light reflected from the sides of the tube or camera, or from the edges of the sector. These effects caused fluctuations in the background blackening of the plate which were coarser than the photographic grain, but which were recognized by the microphotometer, and which appeared in the tracings as noise-like fluctuations in the recording pen excursions. This photographic noise often had an RMS amplitude equivalent to absorbance increments of 0.02 or more, when referred to the standard shadows cast by the step sector. We attempted to eliminate it by scrupulous attention to cleanliness of all lens and glass surfaces and of the mounting medium, by meticulous optical alignment, and by careful baffling and aperturing of the optical system. But we did not succeed in reducing the noise much below the value cited above. We traced some of this to small imperfections in lens cement in the light condenser lens, to minute scratches on lens surfaces, and to small fragments of tissue introduced into the mounting medium along with the myofibrils.

Thus in spite of considerable success in reducing the effects of the background drift and noise, they remained as significantly large in relation to our signal, the strength of which, it will be recalled, lay in the neighborhood of the absorbance value of 0.04. Since satisfactory determination of $\bar{A}$ in a dichroic specimen depended upon determining $A_{\parallel}$ and $A_{\perp}$, and since the muscle mercaptides were unequally concentrated within the cross bands, values for A substantially below 0.04 were expected and were encountered. However, the lower values were often pretty well smothered by the background noise fluctuations.

A further factor contributed to the difficulty of our endeavor and to the error of the results. This factor can be regarded as being analogous to signal distortion, if one is to draw another parallel to an electronic amplifier. This optical distortion derives from the resolving limitations of the light microscope, the wavelength of the light used in the absorbance measurement and its ratio to the sizes of the objects, the —SH content of which is to be measured. It will be recalled that the limits of resolution of the light microscope with light of wavelength 480 mμ and with an optical system of numerical aperture 1.40 lies in the neighborhood of 0.2 μ, whereas the width of the Z band of muscle is about 0.1μ, and of each half I band about 0.3 μ. It is evident that in such an optical system significant fractions of the light traversing one band of a sarcomere are imaged with a neighboring band, so that the differences in absorbance characteristic of the two bands appears to be less than is truly the case. Thus if one photographs an image of a striated muscle myofibril wherein the contrast between the various cross bands in the image depends on differential light absorbance in the cross bands of the specimen, a photometric tracing longitudinally along the photographic image of the fiber reveals a sequence of wave-like fluctuations corresponding to the different absorbances of the several cross bands. But such a wave sequence contains distorted wave forms, with peaks and troughs reduced in amplitude from the true ratios of amplitude prevailing in the specimen. In the cases of objects as small as the cross bands of mammalian striated muscle the distortions assume significant proportions and introduce considerable error.

Although photographic drift, noise, and distortion summed to defeat efforts to advance our knowledge of the sulfhydryl content of the several cross bands of muscle beyond that reported by Bennett (1952), we would not wish this experience to discourage one from attempting quantitative sulfhydryl cytochemistry with Mercury Orange on more suitable objects. If one chooses specimens, the size of which do not approach the limit of resolution of the microscope, and which are characterized by absorbance values in more favorable ranges, either as a consequence of greater thickness of specimen or of greater concentration of available –SH groups or both, cytochemical measurement of –SH groups with Mercury Orange may well achieve considerable accuracy. In efforts to achieve such accuracy, the experience reported above with respect to drift, noise, and distortion might be invoked by a worker with some advantage.

E. Procedure for the Qualitative Cytochemical Study of Sulfhydryl Distribution in Tissue Components with Mercury Orange

Whereas a reader could prepare suitable preparations for qualitative study of sulfhydryl distribution in tissues on the basis of what has been said earlier in this paper, a few comments may aid in the interpretation and use of such material.

In interpreting observations on such material, it should be borne in mind that differences in intensity of absorption due to bound Mercury Orange mercaptides may not necessarily be due to differences in total concentration of protein –SH groups, but may be due to differences in accessibility of the –SH groups due to molecular steric differences in the two areas in question. If differences in concentration of total –SH groups are to be detected reliably, one must have a reliable indication that all –SH groups are engaged in Mercury Orange mercaptide linkage.

A second comment relates to facilitation of detection of bound Mercury Orange in tissue components where the color is faint, either due to low concentration of –SH groups in the tissue component, or to incomplete reaction with the –SH groups present. Dr. Shinya Inoué and Professor Katsuma Dan were kind enough to call our attention to an observation of theirs which facilitates study in such cases. If, according to their suggestion, one places between the light source and the microscope a light filter consisting of a plane parallel-sided vessel containing a solution of toluidine blue, the bound Mercury Orange is seen with enhanced contrast against the blue background provided by the light filtered through the toluidine blue. This simple filter can be used to improve the sensitivity of qualitative observations on sulfhydryl distribution in tissue components as studied cytochemically with Mercury Orange.

F. Procedure for Blocking Tissue Sulfhydryl Groups

We have not found it necessary to modify the blocking procedures of Bennett (1951) in which the tissues were immersed in 0.001 *M* solution of iodine in propanol, or in a 1.0 *M* solution of H_2O_2 and 0.001 *M* solution of $FeCl_3$ in water or propanol for purposes of oxidizing tissue —SH groups, and in 0.1 *M* solutions of iodoacetamide or of iodoacetic acid in *n*-propanol for alkylation. Immersion of tissues for a few hours or overnight in these solutions prevented completely binding of Mercury Orange.

Bennett (1951) found phenyl mercuric chloride and other organic mercurials to be unreliable blocking agents for tissue —SH groups when using Mercury Orange, as the blocking mercurials participated in an equilibrium reaction wherein they competed for tissue sulfhydryl groups with Mercury Orange.

Barrnett and Seligman (1952) and Barrnett (1953) have found N-ethylmaleimide to be a satisfactory sulfhydryl-blocking agent for cytochemical work with their method. Whereas we have not used this reagent, we would anticipate that it would be a satisfactory —SH blocking agent with respect to Mercury Orange. Barrnett (1953) reported successful blocking when he treated tissues with 0.1 *M* N-ethylmaleimide in Sorensen's phosphate buffer at pH 7.4 for 4 hours at 37° C.

G. Procedure for Demonstrating Disulfide and Sulfhydryl Groups

We have not exploited the possibility of using Mercury Orange for demonstration of disulfide and sulfhydryl. However, we anticipate that it could be used successfully for this purpose, though a study of the quantitative significance of the reaction would be necessary before the reaction could be understood or evaluated very well. From the qualitative standpoint, and as a starting point for quantitative work, we would anticipate that Barrnett and Seligman's (1954) procedure for converting tissue disulfides to sulfhydryl would be satisfactory for subsequent sulfhydryl demonstration by Mercury Orange. Barrnett and Seligman (1954) covered deparaffinized tissues with a thin layer of ½% celloidin, permitted it to dry, and then incubated their tissue sections for 1 to 2 hours at 50° C. in 0.2 *M* to 0.5 *M* fresh thioglycolic acid titrated to pH 8 with 0.1 *N* sodium hydroxide just before use. The sections were then washed in water, in 1% acetic acid, and again in water. This procedure was deemed to have converted disulfide groups to sulfhydryl. If successful, it would render the sulfurs of original disulfide and sulfhydryl groups available for reaction with Mercury Orange. The sulfhydryl groups then available in the tissue could be demonstrated or measured according to the procedures outlined on pp. 354–364.

Barrnett and Seligman used tap water for the washing procedures cited in the preceding paragraph. We deem it preferable to use distilled water of high quality, as heavy metal or other impurities in tap water might reduce the number of sulfhydryl groups available for reaction with the cytochemical sulfhydryl reagent.

H. Procedure for Demonstrating Disulfide Groups Only

We have not attempted to use Mercury Orange for this purpose. But here again we believe it could be used successfully, though the quantitative significance of the reaction might bear considerable study. If one first blocked all the sulfhydryl groups by alkylation with iodoacetamide or iodoacetate, or by N-ethylmaleimide, as outlined on p. 365, then converted disulfide groups to sulfhydryl as outlined in the preceding passages, and finally coupled the tissue thiols so created to Mercury Orange as directed on pp. 351–356, the original sites of localization of disulfides would be revealed specifically by Mercury Orange, and would be available for qualitative or quantitative study.

V. CRITIQUE OF METHOD

The Mercury Orange method for cytochemical demonstration of –SH groups has attracted some criticism in the literature. The criticisms have been based on alleged "weak color" or lack of contrast, and on hints as to lack of specificity. It is well to examine these criticisms in turn.

Taking up first the question of color contrast, one can start out by quoting Barrnett and Seligman (1952), who state in reference to Bennett, 1951 that "A method based on a new mercaptide-forming agent, 1-(4-chloromercuriphenylazo)-naphthol-2 gives weak color reactions in tissue sections.

"In order to improve the sensitivity of sulfhydryl histochemistry by increasing the color value of the final compound and by increasing the specificity of the reaction for sulfhydryls, a reagent was developed which contained a disulfide linkage, the specific oxidative group, and a naphthol moiety for coupling to form an azo dye." Barrnett and Seligman then go on to describe their reagent, which is 2,2′-dihydroxy-6-6′-dinaphthyl disulfide. In Barrnett and Seligman's method this compound is permitted to bind to tissue protein by disulfide linkage. The location of the bound naphthol disulfides is then demonstrated by azo-coupling with tetrazotized diorthoanisidine. The authors state that when a mono-azo compound is formed a red azo dye appears, whereas if bis-azo coupling occurs, the resulting compound is blue. The authors present no absorption curves of either the mono- or bis-azo compound formed by their reagents, and present no evidence that the molar absorbtivities of either their mono- or bis-azo compounds are any higher than that of Mercury Orange. Thus there is no evidence that Barrnett and Seligman have in fact succeeded in developing a sulfhydryl reagent which, on a mole-for-mole basis, has an appreciably higher absorbance than Mercury Orange.

We have shared Barrnett and Seligman's awareness of the desirability of a specific sulfhydryl reagent with a higher molar absorbtivity than Mercury Orange, and have explored the absorption characteristics of several bis-azo compounds with this in mind. Some of the absorption curves are reproduced in Fig. 4. In general, we have found bis-azo dyes

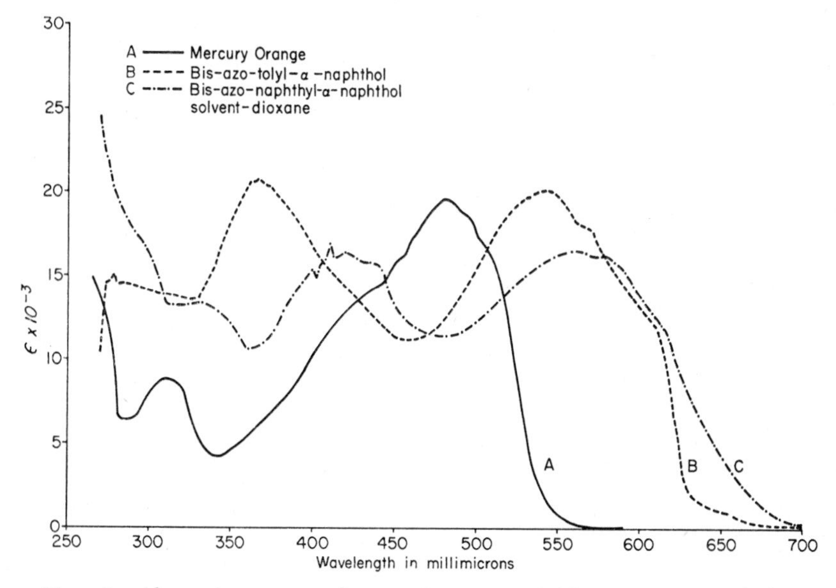

Fig. 4. Absorption spectrophotometric curve of Mercury Orange (A) compared with that of two bis-azo compounds. Note that the bis-azo compounds have broader peaks and absorb over a greater fraction of the visible range than does Mercury Orange, but the heights of the peaks in the visible of the bis-azo compounds are not significantly higher than that of Mercury Orange.

to have broader peaks than Mercury Orange, or to have more peaks in the visible, and to absorb over a wider fraction of the visible spectrum. But we have not found bis-azo compounds with peaks appreciably higher than the 484 mμ peak of Mercury Orange. Thus we concluded that bis-azo compounds offered little promise of advantage over Mercury Orange with respect to quantitative sulfhydryl cytochemistry, and that any small advantages which bis-azo compounds with their broader absorption ranges provided for qualitative work with the eye were offset by greater difficulties of synthesis and purification. We have not, however.

explored the absorption characteristics of the mono- or bis-azo compounds advocated by Barrnett and Seligman (1952).

Since Gomori (1956) has reported that Mercury Orange gives a stronger color with sulfhydryl groups than does Barrnett and Seligman's reagent, we are tempted to think that factors other than the color intensity of the reagent itself might explain the "weak color reaction in tissue sections" attributed to Mercury Orange by Barrnett and Seligman (1952). Barrnett and Seligman do not state the conditions under which they carried out coupling of Mercury Orange with protein sulfhydryl groups in tissue sections. If they permitted the slides to dry while transferring them during the staining procedure, or if they failed to use hydrogen-bonding competitors in the solvent, or if they permitted solvents contaminated with traces of heavy metals to come into contact with the tissues, they might have unwittingly attempted to carry out the reaction with only a small fraction of the total –SH groups available for reaction with Mercury Orange, and thus have been misled as to the significance of the weak reaction they obtained.

From the point of view of molecular size, Barrnett and Seligman's reagent would appear to offer no advantage over Mercury Orange with respect to its ability to seek out –SH groups sequestered by steric encumberances of tissue protein molecules. It is possible that their coupling reaction, which is carried out in dilute alcoholic buffer solution in water, would afford opportunity to use a variety of water-soluble, hydrogen-bonding competitors such as guanidine or urea as aids in rendering tissue protein –SH groups accessible to the reagent. However, the authors do not report the use of such denaturing agents, and are silent on the question of the proportion of –SH groups which their reagent can detect. Since Cafruny *et al.* (1955) were able to detect a binding of some Mercury Orange by tissue proteins after they had been treated with Barrnett and Seligman's reagent, one must consider the possibility that Mercury Orange may be able to engage some protein sulfhydryl groups which are inaccessible to the other reagent.

Turning now to the question of specificity, the quotation cited above from Barrnett and Seligman suggests that they sought a reagent which would be an improvement from the point of view of specificity over those developed previously. This implies that they may have felt that Mercury Orange has shortcomings from this standpoint. However, in a later paper, Barrnett (1953) states, referring to Mercury Orange as a sulfhydryl reagent, "Although organic mercury compounds may combine with groups in proteins other than thiols, under the controls established by Bennett (1951) for each tissue examined, the method appears to be specific." Barron (1951) has indeed called attention to the possibility that mercuric chloride, phenyl mercuric nitrate, and perhaps certain other organic mercurials might under certain circumstances combine with protein carboxyl or amino groups. More recently Gurd and Wilcox (1956) have reviewed the circumstances under which such combination can occur. They point out that, with concentrations of R′-Hg-X below 10^{-3} molar or thereabouts in water, organic mercurials appear to combine specifically with –SH groups, whereas at higher activities of the organic mercurial ion, nonspecific combination with carboxyl or amino groups may take place. The nitrates, which ionize readily, have higher activities than the halides. Among the halides, the chloride tends to ionize the most readily. This tendency to ionize can be suppressed by the presence of iodide, which hence would promote specificity when the activity of the organic mercurial might otherwise be high enough to result in nonspecific binding.

Examining now the use of Mercury Orange from this standpoint, it is pointed out on p. 351 that we used concentrations of our orange organic mercurial of about 2×10^{-5} molar. These solutions were in organic solvents, in which the ionization of the mercurial would be much less than in water. Thus the activities of the solutions of Mercury Orange were well below those in which nonspecific binding to protein has been demonstrated. Moreover, the measurement of bound mercurial reported in this chapter indicates specific reaction with thiols, and the amounts bound are lower than would be expected if reactions with amino and carboxyl groups were taking place. Furthermore the complete inhibition of uptake of Mercury Orange by prior alkylation or oxidation of the –SH groups of proteins provides good evidence that amino and carboxyl groups do not bind the mercurial under the conditions we achieved.

Thus we see no evidence which casts doubt on the specificity of Mercury Orange as a sulfhydryl reagent when used according to the methods developed by us, and we feel that considerable confidence can be placed in Barrnett's statement that the Mercury Orange reaction "appears to be specific" under these circumstances.

It is difficult to see, therefore, any reason why an improvement in specificity over that of Mercury Orange is reasonably to be expected, since it appears to be a satisfactory specific agent as already developed and used. The margin of safety, however, might be increased by using the reagent in the form of the iodide, or by carrying out the coupling in the presence of iodide. It is evident further that Barrnett and Seligman's (1952) intention of creating a sulfhydryl cytochemical method more specific than Mercury Orange has not succeeded, for in their paper the authors point out that their reagent can produce spurious positive reactions in elastic fibers in which no protein –SH groups occur. Barrnett and Seligman (1952) reasonably attribute this nonspecific reaction to an affinity of elastic tissue for naphthols. Apparently their naphthol disulfide reagent binds nonspecifically to elastic tissue and is subsequently revealed by azo-coupling, giving rise to false positives. Mercury Orange does not suffer from this defect.

Whereas there is a possibility that Mercury Orange might be capable of combining with lipid tissue components nonspecifically by van der Waals forces, as pointed out by Bennett (1951), and by Pearse (1954), and as discussed earlier in this review, the conditions which we have set up and which were used by Bennett (1951) remove lipid components in which Mercury Orange might dissolve, and prevent reagent which may have been bound nonspecifically from creating false positive reactions. Thus Pearse's (1954) fears of nonspecific coloring of lipid are not soundly based if the reaction is carried out as we have described. The Mercury Orange sulfhydryl reaction remains without successful challenge as to its specificity.

VI. BRIEF ASSESSMENT OF RESULTS TO DATE

The Mercury Orange cytochemical method has not enjoyed the popularity which has fallen to the Barrnett-Seligman (1952, 1954) method, in spite of the lack of specificity which the latter method can display, and which was recognized by Barrnett and Seligman in their first paper on the method. The failure to exploit fully the possibilities of Mercury Orange may have stemmed from the fact that it is not yet available commercially, and from difficulties some have experienced in synthesizing the reagent (see Gomori, 1956). A further factor may derive from the failure of Bennett (1951) to realize at the time he did the first qualitative work with the reagent that tissue —SH groups could be rendered inaccessible by drying, nor was it realized that only a fraction of the total —SH content of the tissues was revealed by the method used at that time. Rather, it was assumed that once proteins were denatured by trichloroacetic acid or by alcohol, that —SH groups would thereafter remain accessible without danger of loss except through oxidation or other covalent bonding. The quantitative studies reported in this paper betrayed the fallacy of these earlier assumptions and pointed up conditions which must be met if all or most of the —SH groups of a tissue are to be demonstrated cytochemically. Now that some of these conditions have come to be appreciated, Mercury Orange can be used with considerable understanding of its field of usefulness.

In applying the reaction originally, Bennett (1951) discovered sulfhydryl groups to be present in capillary endothelial cells, in retinal rods, and in fibers of the crystalline lens. Bennett (1952) later studied the sulfhydryl distribution in the cross bands of striated muscle by the Mercury Orange method and found the resulting mercaptides to be dichroic. He interpreted this dichroism in terms of the bond angles of the –C–S– bonds in the fixed myofibrils.

No reports dealing with quantitative applications of Mercury Orange to study of organized tissue sections have come to our attention. Cafruny *et al.* (1955) attempted to use the reagent for semiquantitative estimation of –SH groups in kidney tubules as affected by mercurial diuretics, but ultimately turned to a modification of the Barrnett-Seligman method.

There are a few reports of attempts to use Mercury Orange for titration of –SH groups in tissue homogenates. These do not, however, betray a complete understanding of the properties and behavior of the reagent, nor of the properties of protein –SH groups. Thus Flesch and Kun (1950) attempted to use the reagent in amyl acetate solution for determining –SH groups in aqueous tissue homogenates. Cannefax and Freedman (1955) and Rausch and Ritter (1955) have called attention to unsatisfactory features of the procedure of Flesch and Kun, and have recognized some of the pitfalls which Flesch and Kun overlooked. Nevertheless, we are of the view that Mercury Orange can be exploited as a satisfactory reagent for titration of mercaptan groups in soluble or insoluble proteins or in tissue homogenates if the properties of the reagent and of –SH groups are understood fully and exploited.

ACKNOWLEDGMENTS

We wish to acknowledge the devoted and skillful assistance of Mrs. Lucy Jache and of Dr. James C. M. Li, who made many of the measurements reported in this study. We are also grateful to Mr. George Reis for expert photomicrography in connection with our efforts to work out a quantitative sulfhydryl method based on Mercury Orange.

We wish to thank the American Cancer Society (Grant No. C-12) and the United States Public Health Service, Department of Health, Education, and Welfare, (Grant No. B-401) for financial assistance which made the study possible.

VII. APPENDICES

APPENDIX 1: NOTE ON SPECTROSCOPIC TERMINOLOGY

We have chosen to use spectroscopic terminology recommended in Report No. 6 of the Joint Committee on Nomenclature in Applied Spectroscopy (see *Anal. Chem.* **24,** 1349, 1952) and recommended by the Committee on Spectrophotometry of the American Association of Agricultural Chemists (see *J. Assoc. Offic. Agr. Chemists,* **87**, 54, 1954). This terminology is now standard in publications of the American Chemical Society.

APPENDIX 2: ABBREVIATIONS

A	= Absorbance, $\log_{10} T^{-1}$ (formerly referred to as extinction, density, or as optical density, $\log_{10} I_0/I$, abbreviated as d).
$A_{\parallel}$	= Absorbance of a body with respect to linear polarized light with its electric vector parallel to the optic axis of the body.
$A_{\perp}$	= Absorbance of a body with respect to linear polarized light with its electric vector perpendicular to the optic axis of the body.
$\bar{A}$	= Calculated absorbance of a body as it would be if all absorbing molecules were randomly arranged. $\bar{A}$ is equivalent to A in an isotropic absorber.
A_1	= Absorbance of standard solution.
A_2	= Absorbance of solution at end-point.
b	= Path length through an absorbing sample.
EDTA	= Ethylenediaminetetraacetic acid (a metal chelating agent, often sold under the trade name of "Versene" or of "Sequestrene").
M	= Molarity of solution, or concentration of solution in moles per liter.
M_1	= Molar concentration of standard solution.
M_2	= Molar concentration at end-point.
m	= Molecular weight of a mercurial.
Mu	= Muscle.
R	= An organic radical (here used specifically for an organic radical attached to a sulfur atom).
R′, R″, R‴	= Other organic radicals.
T	= Transmittance (the ratio of the radiant power transmitted by the sample to the radiant power incident on the sample when both are measured at the same spectral position and the same slit width).
V	= Volume.
W	= Weight.
X, X′	= Anion-forming groups.
ϵ	= Molar absorptivity (formerly often called molecular extinction coefficient).
ϵ_{max}	= Molar absorptivity at a peak.
λ	= Wavelength.
λ_{max}	= Wavelength at a peak.

REFERENCES

Abrahams, S. C., Robertson, J. M., and White, J. G. (1949a). *Acta Cryst.* **2,** 233.
Abrahams, S. C., Robertson, J. M., and White, J. G. (1949b). *Acta Cryst.* **2,** 238.
Bailey, K. (1954). II. Muscle. *In* "The Proteins" (H. Neurath and K. Bailey, eds.), Vol. 2, Part B, p. 951. Academic Press, New York.
Barrnett, R. J. (1953). *J. Natl. Cancer Inst.* **13,** 905.
Barrnett, R. J., and Seligman, A. M. (1952). *Science* **116,** 323.
Barrnett, R. J., and Seligman, A. M. (1954). *J. Natl. Cancer Inst.* **14,** 769.
Barron, E. S. G. (1951). *Advances in Enzymol.* **11,** 201.
Benesch, R., and Benesch, R. E. (1948). *Arch. Biochem.* **19,** 35.
Bennett, H. S. (1950). *In* "McClung's Handbook of Microscopical Technique," 3rd ed., p. 591. Paul B. Hoeber, New York.
Bennett, H. S. (1951). *Anat. Rec.* **110,** 231.
Bennett, H. S. (1952). *Lab. Invest.* **1,** 96.
Bennett, H. S., and Yphantis, D. A. (1948). *J. Am. Chem. Soc.* **70,** 3522.
Bennett, H. S., Quinton, W. E., and Mueller, V. C. (1953a). *Appl. Spectroscopy* **7,** 127.
Bennett, H. S., Quinton, W. E., Othberg, R., and Reis, G. W. (1953b). *Appl. Spectroscopy* **7,** 129.
Brode, W. R. (1943). *In* "Chemical Spectroscopy," 2nd ed., p. 199. Wiley, New York.
Cafruny, E. J., Di Stefano, H. S., and Farah, A. (1955). *J. Histochem and Cytochem.* **3,** 354.
Cannefax, G. R., and Freedman, L. D. (1955). *Proc. Soc. Exptl. Biol. Med.* **89,** 337.
Caspersson, T. O. (1950). *In* "Cell Growth and Cell Function." Norton, New York.
Cook, E. S., Kreke, C. W., Sister Mary of Lourdes McDevitt, and Sister Mary Domitilla Bartlett (1946). *J. Biol. Chem.* **162,** 43.
Flesch, P., and Kun, E. (1950). *Proc. Soc. Exptl. Biol. Med.* **74,** 249.
Fredericq, E. (1945). *Bull soc. chim. Belges.* **54,** 265.
Godeaux, J. (1946). *Compt. rend. soc. biol.* **140,** 675.
Gomori, G. (1956). *Quart. J. Microscop. Sci.* **97,** 1.
Greenstein, J. P., and Edsall, J. T. (1940). *J. Biol. Chem.* **133,** 397.
Gurd, F. R. N., and Wilcox, P. E. (1956). *Advances in Protein Chem.* **11,** 311.
Hellerman, L., and Perkins, M. E. (1934). *J. Biol. Chem.* **107,** 241.
Hellerman, L., Perkins, M. E., and Clark, W. M. (1933). *Proc. Natl. Acad. Sci. U. S.* **19,** 855.
Hellerman, L., Chinard, F. P., and Deitz, V. R. (1943). *J. Biol. Chem.* **147,** 443.
Hughes, W. L., Jr. (1950). *Cold Spring Harbor Symposia Quant. Biol.* **14,** 79.
Kolthoff, I. M., and Harris, W. E. (1946). *Ind. Eng. Chem., Anal. Ed.* **18,** 161.
Land, E. H., and West, C. D. (1946). *Colloid Chem.* **6,** 160.
Loofbourow, J. R. (1950). *J. Opt. Soc. Am.* **40,** 317.
Mason, F. A. (1932). *J. Soc. Dyers Colourists* **48,** 293.
Michaelis, L., and Granick, S. (1945). *J. Am. Chem. Soc.* **67,** 1212.
Mirsky, A. E. (1936). *J. Gen. Physiol.* **19,** 559.
Olcott, H. S., and Fraenkel-Conrat, H. (1947). *Chem. Revs.* **41,** 151.
Pauling, L. (1945). *In* "The Nature of the Chemical Bond," 2nd ed. Cornell Univ. Press, Ithaca, New York.
Pearse, A. G. E. (1954). *In* "Histochemistry, Theoretical and Applied," p. 71. Churchill, London.
Pollister, A. W., and Ornstein, L. (1955). *In* "Analytical Cytology" (R. C. Mellors, ed.), pp. 1–3. McGraw-Hill, New York.

Rausch, L., and Ritter, S. (1955). *Klin. Wochschr.* **33,** 1009.

Robertson, J. M. (1933). *Proc. Roy. Soc. London* **A142,** 674.

Robertson, J. M. (1939). *J. Chem. Soc.* p. 232.

Thorell, B. (1947). Studies on the Formation of Cellular Substances During Blood Cell Production. Thesis from the Department for Cell Research, Karolinska Institutet, Stockholm, Sweden. Kungl. Boktryckeriet, P. A. Norstedt and Söner, Stockholm.

Tsao, T.-C., and Bailey, K. (1953). *Biochem. et Biophys. Acta* **11,** 102.

INDIGOGENIC STAINING METHODS FOR ESTERASES

S. J. Holt

Courtauld Institute of Biochemistry, Middlesex Hospital Medical School, London, England

> "Although *Nil* or *Indico* be not in form like Woade, yet for the rich blew colour sake I thinke good to make mention of it here with it, not only to shew you what it is, and how made, but to incite some of our nation to be as industrious therein as they have been with the former Woade, seeing no doubt it would bee profitable"
>
> J. Parkinson, *Theatrum Botanicum* (1640)

I. INTRODUCTION

The indigogenic substance indoxyl occurs in nature as the glucoside, in Woad (*Isatis tinctoria*) and in various *Indigoferas.* It also occurs in the form of its sulfate as urinary indican. These derivatives have potential uses as substrates in cytochemical staining procedures for glucosidases and sulfatases, respectively (Holt, 1952), but greatest progress in applying indigogenic reactions to the study of cellular enzymes has been made through the use of substituted indoxyl acetates (I). Systematic studies have shown that certain indoxyl acetates may be utilized as substrates in the formulation of precise staining methods for esterases of fixed tissues (Holt, 1956; Holt and Withers, 1958). A precision of the order of 0.5 μ is achieved by use of 5-bromo-4-chloroindoxyl acetate, the synthesis and use of which is described. Somewhat inferior results are given by 5-bromo-6-chloroindoxyl acetate and 5-bromoindoxyl acetate, syntheses of which are also described. The former is prepared from a byproduct of the synthesis of the 5-bromo-4-chloro- compound, while 5-bromoindoxyl acetate is included because its synthesis is relatively simple.

The cytochemical substrates are not sufficiently soluble for study of enzyme–substrate specificity relationships, or other quantitative biochemical aspects, by use of standard manometric procedures. Such studies, however, are essential for full interpretation of results of cytochemical staining procedures. For this reason, preparation of the more soluble simple esters, indoxyl acetate, propionate, and butyrate, is also described, for these related substrates may be used satisfactorily in quantitative esterase studies (Underhay *et al.*, 1956; Underhay, 1957; Hobbiger, 1957).

II. THEORY

The chromogenic reaction sequence underlying indigogenic staining methods for esterases is represented by the following equation. A substi-

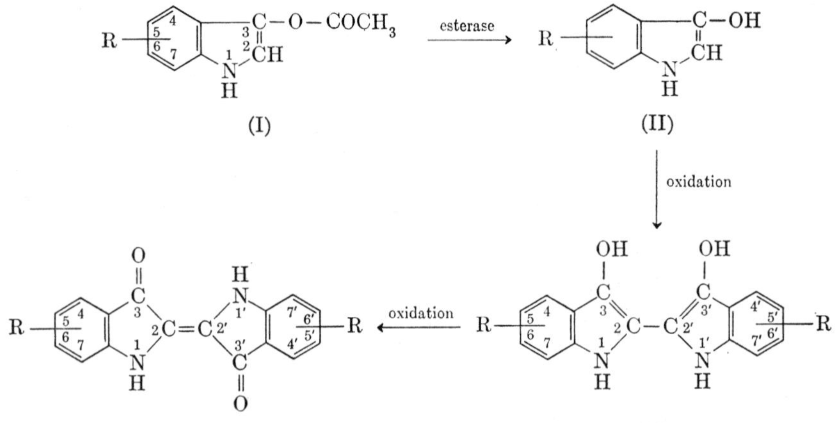

(I) —esterase→ (II) —oxidation→ (III) —oxidation→ (IV)

tuted indoxyl acetate (I) is hydrolyzed by esterases to the free indoxyl (II). This soluble product is oxidized to the corresponding indigoid dye (IV) via its leuco-derivative (III), a reaction discussed in more detail by Cotson and Holt (1958).

The precision of localization given by any chromogenic system, such as the above, in which a soluble diffusing intermediate is converted into a stain, depends upon the diffusion coefficient of the intermediate, the size of the enzyme site, and the velocity constant of a first-order reaction converting the intermediate into stain or dye (Holt and O'Sullivan, 1958). Extremely high rates in the latter reaction ("capture reaction") are required to give precise localization in enzyme sites of radius 1 μ and below, unless the intermediates are adsorbed by cellular substances (Holt, 1956; Holt and O'Sullivan, 1958). In indigogenic reactions, the intermediate *leuco*-compound (III) is strongly adsorbed at alkaline pH values by protein and polysaccharides (Vickerstaff, 1954; Venkataraman, 1952), unless hindered by the presence of bulky substituents in the 4,4′-positions (Holt and Withers, 1958). The chemically related indoxyls (II) also have similar characteristics (Holt and Withers, 1958). With suitably rapid reactions, therefore, indigoid dyes should be deposited in enzyme sites with high precision, particularly if the dyes are insoluble, or better, are strongly bound ("substantive") to cellular components. Relationships between solubility, substantivity, and molecular structure in indigoid dyes have been elucidated by Holt and Sadler (1958b). These considerations have led to the development of staining methods for esterases based upon the use of 5-bromo-4-chloroindoxyl acetate. The derived indoxyl and dye are both strongly bound by protein, and the system fulfills the basic requirements for precise localization to the highest degree that has been achieved at the present time with indigogenic methods (Holt, 1956; Holt and Withers, 1958). Oxidation of the enzymically formed indoxyl is carried out at a high rate with an equimolecular mixture of potassium ferri- and ferro-cyanides. This reagent is more effective at alkaline pH values than under acidic conditions, and is therefore very appropriate in the study of esterases. In addition, the oxidizing reagent has a negligible effect upon the activity of the major types of esterase (Holt and Hobbiger, unpublished), all of which hydrolyze the substrates.

Theoretical studies (Holt and O'Sullivan, 1958) indicate that staining methods may give a direct quantitative representation of levels of enzymic activity, in terms of stain densities, only when the capture reaction is kinetically first order. Evidence has been obtained that this condition is fulfilled in the indigogenic methods (Holt, 1956; Holt and Withers, 1958).

III. PROCEDURE

Indigogenic methods of high precision are, as yet, applicable to *fixed* tissues only (see Section IV). The composition of the preferred formol-calcium fixative, reagent solutions, and incubation media are given in Appendix 1.

A. Tissue Sampling

Tissues must be obtained as fresh as possible to avoid autolytic changes. Biopsy specimens should be fixed immediately, tissues within 1–2 minutes after removal from anesthetized or killed animals. Tissue samples should be as small as possible, consistent with the nature of the investigation, but specimens larger than $10 \times 5 \times 2$ mm. should generally be avoided. Smears are best made on coverslips, so that small volumes of staining solution may be used, and substrate conserved.

B. Fixation and Pretreatment

Tissue samples, obtained as described in (A), should immediately be placed in freshly prepared ice-cold formol-calcium fixative, and allowed to fix for 24–36 hours at 0–2° C. Different fixation times within this period have little effect on the staining results. Use about 100 ml. of fixative for 1 gm. of tissue, which should be placed on a stainless steel gauze platform or cotton wool to allow complete circulation of fluid. After fixation, remove tissue from fixative, but *do not wash,* for morphological changes at the cytological level and loss of enzyme have been observed after washing. Instead, blot the tissues gently and place in ice-cold gum-sucrose solution. Maintain at 0–2° C. for 24 hours. Tissues should become impregnated and sink to the bottom of the vessel. Occasional stirring helps the process. Tissues treated in this way and stored in the ice-cold medium have retained a considerable proportion of their activity for more than two *years,* and the staining pattern is substantially the same as that given by newly prepared tissues. For example, in the case of rat liver, 30% of the initial activity was retained after two years. The other function of the gum-sucrose solution is to facilitate the cutting of frozen sections, as described in (C).

Esterases resist brief fixation in ice-cold buffered osmium tetroxide, and minute pieces of tissue (about 1 mm.^{3}) may be treated with the fixative (Palade, 1952) for about 10 minutes, followed by impregnation in gum-sucrose, and the cutting of frozen sections. Alternatively, the fixed pieces may be incubated whole in the staining solution, dehydrated, cleared, embedded in wax, and then sectioned, as described by Novikoff and Holt (1957).

Acetone-fixed material, although poor morphologically, should be embedded in wax, cleared in toluene, and sectioned. When the enzyme survives this treatment, results are similar to those given by formol-calcium–fixed frozen sections, allowing for differences in structural preservation. Formol-calcium–fixed tissue may also be dehydrated, without washing, in graded alcohols or in absolute acetone, then cleared in toluene, and embedded in wax and sectioned. Smears and dewaxed sections of paraffin-embedded frozen-dried tissue may be fixed by exposure to formalin vapor for 1–5 minutes.

C. Section Cutting

For adequate microscopical resolution of cytological detail, and study of the detailed staining patterns given at the cytological level by the methods described in (E), thin sections are required. These also favor rapid penetration of reagents, and correct staining conditions (Holt and O'Sullivan, 1958). Frozen sections are preferred, because any further prestaining treatment which may result in loss of enzyme is thereby avoided. Thin frozen sections of tissues, impregnated in the gum-sucrose solution as described in (B), are easily cut without recourse to a freezing cabinet. With a robust and mechanically sound freezing microtome, such as the Leitz Model 1310, routine cutting of sections at 2.5 μ may be achieved. On the other hand, for study of certain intact irregular structures, such as motor end-plates in striated muscle, thicker (20 μ) sections may be required.

Sections should be allowed to collect on the microtome knife, and then transferred directly to the staining solution, using a small brush.

D. Staining

Use about fifteen 10 μ sections to 10 ml. of staining solution, and those of other thicknesses *pro rata.* They may either be added to the cold staining solution, or it may be preheated beforehand. The usual incubation temperature is 37° C., but temperatures up to 50° have been used successfully. Highly active tissues may be adequately stained in as little as 5 minutes, but staining times above 1 hour are seldom needed. Sections should be removed when adequate visual contrast has been achieved. Staining times in excess of this requirement cause mechanical blocking of enzyme sites by dye, interfere with the penetration of reagents and maintenance of correct kinetic conditions, and lead to loss of resolution. On the other hand, it may be necessary to sacrifice resolution in highly active sites, by prolonging the incubation time, for the purposes of studying sites of low activity in the same specimen. When the above staining procedure is used with 5-bromo-4-chloroindoxyl acetate, a completely nongranular bluish green stain is produced. This is due to the formation of a protein–dye complex of 5,5′-dibromo-4,4′-dichloroindigo. If staining is overprolonged, the binding capacity for dye of protein at the enzyme sites may be exceeded, when the excess dye will deposit in a microcrystalline form. 5-Bromoindoxyl acetate gives rise to deep blue 5,5′-dibromoindigo, while 5-bromo-

6-chloroindoxyl acetate produces purplish red 5,5′-dibromo-6,6′-dichloroindigo. Neither of these dyes is substantive, and the deposits are microcrystalline in each case. The greatest resolving power is therefore afforded by 5-bromo-4-chloroindoxyl acetate.

No variation in the composition of the staining media given in Appendix 1, or interpretation of the results of such variations, should be made without full regard to the effects upon the parameters discussed briefly in Section II, and more fully elsewhere (Holt, 1956; Holt and O'Sullivan, 1958; Holt and Withers, 1958).

E. Mounting and Counterstaining

Sections should be removed from the staining solutions with glass or stainless steel implements. Ordinary steel needles may cause contamination of sections by deposits of Prussian Blue, due to reaction with ferrocyanide in the staining medium. The sections should be placed in 30% alcohol containing 0.1% acetic acid, to neutralize occluded buffer. Transfer individual sections to the surface of water, where they should rapidly become flat, due to surface tension effects between water and alcohol. They should then be picked up on a slide, drained, and allowed to become almost dry before mounting in glycerogel. Anderson's Carmine is a useful nuclear stain for the bluish dyes, while methyl green gives an effective nuclear contrast to the reddish dye. In the latter case, the nuclear stain may be made more permanent if some methyl green is added to the mountant.

For balsam mounts, the sections should be picked up on gelatinized slides, blotted, and placed for 1 minute in a Coplin dish containing a 1-cm. layer of 40% formaldehyde solution. This period should not be prolonged, since, in time, formaldehyde reacts with the dyes to give soluble products. Rinse the slides in water and counterstain if required. The recommended counterstain for balsam mounts of sections stained with the blue dyes is a combination of *acid* hematoxylin and picric acid, which gives brilliant red nuclei and yellow cytoplasm. Stain the sections in Lillie-Mayer hemalum (Lillie, 1954) for 5 minutes, rinse rapidly in water faintly acidified with acetic acid, and then stain in saturated aqueous picric acid for 2–3 minutes. Pass rapidly through graded alcohols and xylol to balsam. Sections overstained in 5-bromo-4-chloroindoxyl acetate may contain the corresponding indigo not bound to protein (see above). In such cases, considerable redistribution of dye may take place in balsam. Even when correctly stained, a slight migration of dye may occur in the smallest sites, in which there is a tendency for initially uniform stain to migrate to the periphery in a few weeks. This must be borne in mind, however, in assessing the staining results. The other dyes are perfectly stable in balsam.

IV. CRITIQUE OF METHOD

A. General Criteria for Precision of Staining

Cytochemical staining methods, like others used for studying cellular composition and function, are subject to certain errors, the effects of which may be assessed by adequate investigation and control. Unlike the related, but simpler, colorimetric methods for enzyme assay in solutions and homogenates, staining methods involve chromogenic reactions occurring within complex heterogeneous structures. The prime requirement of a staining method is that the stain be formed at the site of the enzyme concerned.

Analyses of model staining systems resembling those occurring in tissues have indicated that certain physico-chemical characteristics must be fulfilled if staining results are to be precise and quantitative (Holt and O'Sullivan, 1958). The first requirement is that the enzymic reaction should be kinetically zero order. This may be achieved by the use of high concentrations of suitably soluble substrates, but the analysis also shows that, even with low substrate concentrations, staining systems rapidly approach a steady state in which zero-order kinetics are automatically maintained. The second requirement is that the product of enzymic action ("stain precursor") should be converted into stain by a first-order reaction ("capture reaction"). If the structure of the stain precursor is such that binding to protein readily occurs, high precision of localization should be possible with capture reaction rates within the range of practical achievement. It is a reasonable assumption that these findings are valid for the basic pattern of events occurring during staining operations. Superimposed upon them, however, are the effects of physical and chemical discontinuities, such as those occurring in membranes. These effects are probably less marked in fixed tissue, and may be considerably reduced by the use of substrates having adequate lipid solubility for them to pass through lipid layers. Substrates of this type are usually nonpolar and have the added advantage of not being strongly adsorbed by protein.

Apart from correct conditions in the staining reaction, maintenance of the enzymic integrity of the specimen is clearly of paramount importance. At present, this is probably the least satisfactory aspect of staining methods for enzyme localization, but staining reactions occurring under conditions which give them *intrinsically* high precision will probably help greatly in the study of tissue treatment prior to staining (Holt, 1956).

B. The Indigogenic System

The staining reactions described in this article have been developed to conform as fully as possible to the above requirements. This has been discussed in outline in Section II, but full details have been presented elsewhere (Holt and O'Sullivan, 1958; Holt and Sadler, 1958a, b; Cotson and Holt, 1958; Holt and Withers, 1958). The methods are applicable to fixed tissues only, due to the nature of the chromogenic stage of the processes. During ferricyanide oxidation of indoxyls, the reaction proceeds in two directions, firstly to the dye, and secondly, to colorless byproducts. The proportion of dye produced decreases as the reaction medium becomes more alkaline. When excess ferricyanide is added to solutions of free indoxyls at alkaline pH values, no dye is formed. Similarly, no dye is deposited in *unfixed* cells or tissue sections when incubated in solutions of indoxyl acetates under the same conditions. With fixed specimens, however, the dye is produced normally, although, once again, the relative amount of dye decreases with increasing pH. Denaturation caused by fixatives is probably responsible for this difference. The additional chemical groups exposed in the protein appear to lead to enhanced binding of indoxyls, and modify the relative proportions of the two alternative oxidation reactions.

At first sight, these observations appear to be a serious indictment of the staining system, but experimental evidence has been obtained which indicates that formation of dye occurs under first-order conditions in the fixed tissues (Holt, 1956; Holt and Withers, 1958). Thus, the dye must everywhere be formed in proportion to the amount of indoxyl liberated by enzymic action. Also, an entirely unrelated chromogenic system, based upon the formation of an azo dye, and conforming to the theoretical requirements discussed above, gives substantially the same results as the indigogenic methods (Holt, 1956). In all staining methods, it is necessary to be clearly aware of the course and stoichiometry of the chromogenic system employed. For example, the *amount* of indigoid dye deposited in unit time in the specimen *decreases* as the pH of the medium is raised, due to increased formation of byproducts during ferricyanide oxidation of indoxyl. This effect is much greater than the corresponding effects upon the activity of esterases. Failure to appreciate this may result in serious misinterpretation of the staining results. Already, the increased formation of dye at low pH values, when tissue sections are incubated in 5-bromoindoxyl acetate solution, has led Pearson and Defendi (1957) to conclude that the *enzymic* reaction has a pH optimum of 4.8.

C. Specificity

Indoxyl acetates are not specific substrates for any particular esterase. Studies with inhibitors, using manometric procedures, have shown that indoxyl acetate is hydrolyzed by most of the common esterases, e.g., acetocholinesterase ("specific cholinesterase"), nonspecific cholinesterase, and aliesterase (Underhay, 1957; Hobbiger, 1957). Preliminary studies indicate that substituted indoxyl acetates, used in the staining procedures, are similarly hydrolyzed by the enzymes. Potassium ferri- and ferrocyanides have a negligible effect upon these enzymes (Holt and Hobbiger, unpublished).

The staining method therefore demonstrates the distribution of a wide range of esterases. It is hoped to extend the scope of the methods by use of selective inhibitors. When studies of this type are to be undertaken with reversible inhibitors (e.g., eserine), which must be included in the incubation medium to maintain their activity, experiments must be carried out to determine whether the inhibitors are destroyed by ferricyanide.

Indoxyl acetates are also hydrolyzed by proteolytic enzymes (peptidases) such as chymotrypsin and trypsin. Chymotrypsin has the higher activity, but the contribution of enzymes of this type to the staining results may probably be ignored. For example, whole rat liver is 100 times more active towards 5-bromoindoxyl acetate than an equal weight of crystalline chymotrypsin (Holt and Hobbiger, unpublished). Suggestions such as that of Pearse and Pepler (1957) that peptidases may be responsible for the staining of carcinoid tumors by 5-bromoindoxyl acetate must therefore be accepted with caution unless suitable quantitative comparisons of esterase and peptidase activity are presented.

D. Tissue Aspects

Interpretation of results of staining procedures for enzyme localization should always be made with due regard to effects of tissue pretreatment. As mentioned earlier, detailed knowledge is seriously lacking concerning these effects upon the enzymic integrity of cells and tissues, although development of staining reactions of intrinsically high precision may clarify this position. The indigogenic methods described above are applicable to fixed tissues only. These, however, retain their morphological relationships to a high degree, but fixation usually results in loss of enzymic activity by extraction or direct inactivation. Mere reduction of enzymic activity may easily be determined by application of standard biochemical techniques before and after fixation, using the staining substrates, or closely related analogs. Such studies should always be carried out in conjunction with similar measurements with definitive substrates for the enzymes. Guidance for carrying out comparative determinations of this type with indigogenic and other substrates for esterases have been given elsewhere (Underhay, 1957; Hobbiger, 1957).

Effects of fixation upon the *distribution* of enzymes in tissues are more difficult to determine, but the intrinsically high precision (Holt, 1956; Holt and Withers, 1958) of the indigogenic reactions may be utilized in such investigations. For example, effects of different fixatives and duration of fixation upon the staining pattern may be observed, bearing in mind any concomitant changes in activity, and results of control experiments for detection of losses of essential tissue substances, such as lipids. If the staining pattern does not change substantially with changes in fixation technique, greater confidence may be placed in the results obtained. Studies of this type, carried out with the esterase techniques described here, and many types of tissue, show that substantially the same staining pattern is obtained with formol-calcium, formol-saline, acetone, and brief osmium tetroxide fixation, although the results of generally brief fixation are more diffuse. The effects of alcohol fixation are variable, but this process seriously inhibits the enzymes.

Quantitative studies (Holt and Hobbiger, unpublished) with rat liver show that the tissue preparation technique given in Section III retains up to 60% of the activity of the fresh tissue. It therefore appears that the over-all method gives a reasonably reliable guide to esterase distribution in cells and tissues. The discrete staining of motor end-plates is confirmatory evidence for this, since independent data indicate intimate association of acetocholinesterase with these structures. It seems unlikely that enzyme is effectively redistributed during staining, for the staining pattern remains substantially the same, whatever the duration of incubation, except for increasing diffuseness due to deposited dye blocking enzyme sites.

There is no doubt that some enzyme is lost from the tissues during impregnation with gum-sucrose and during staining, for it may be detected in these media. Even if all *soluble* enzyme were lost in the process, the results obtained still appear to be of considerable significance as representing sites of bound enzyme, perhaps sites of enzyme production or secretion.

With many staining procedures, loss of enzyme into the medium results in production of stain which may deposit at random in the specimens. With the indigogenic methods, extracted enzyme produces an indoxyl *in solution,* and this, therefore, does not lead to formation of dye, for reasons described in Section III, B.

The combination of rapid capture reaction of bound indoxyl at the enzyme sites, apparent first-order capture reaction kinetics in the specimens, and "clean" working of the staining solution appear to eliminate the possibility of spurious adsorption of chemical components of the staining system. Control experiments directed to investigate this are difficult to carry out routinely. They require, for example, treatment of tissues with appropriate solutions of indoxyls and leuco-indigos, necessitating work in anaerobic systems, with subsequent introduction of oxidizing agents. Experiments of this type have been carried out in this laboratory (Holt, unpublished) with rat liver, kidney, and muscle sections. There was no evidence of selective adsorption of the reaction intermediates, since no selective staining occurred. Similarly, esterase preparations added to the staining medium containing inactivated sections, resulted in pale homogeneous staining, indicating that selective adsorption of enzyme or chromogenic intermediates did not occur.

V. RESULTS

Experience of indigogenic methods for esterases during their development and subsequent applications in the author's laboratory have shown them to be easy to apply, rapid in action, and consistent in results. Esterase distributions revealed by 5-bromoindoxyl acetate have been presented elsewhere (Holt, 1954), but the superior, nongranular stain produced by 5-bromo-4-chloroindoxyl acetate in discrete cytological structures in rat tissues are illustrated in Figs. 1–4. Figure 1 shows esterase activity associated with peribiliary bodies of a size range 0.5–1 μ in rat liver. The possible relationship between these bodies and liver lysosomes found in the same region (Novikoff *et al.*, 1956) has been discussed recently (Underhay *et al.*, 1956). Figure 2 shows strongly esterase-positive spherical bodies in the cells of the proximal convoluted tubules in rat kidney. These bodies have a size range of 1–4 μ and have recently been shown by Strauss (1956) to resemble liver lysosomes in their spectrum of enzymic activities. The stained motor end-plate in Fig. 3 shows greater activity in the 1-μ-wide subneural apparatus. Figure 4 shows esterase distribution in a dividing liver cell (telophase) in regenerating rat liver. The usual peribiliary esterase activity is absent, but the telophase nuclei (*N*) are closely associated with strongly esterase-positive regions between them, and also at opposite poles.

VI. APPENDICES

Appendix 1: Materials and Solutions

Quality

Materials of the following grades should be used:

Anhydrous calcium chloride	Analytical reagent or C.P. grade
Calcium carbonate (heavy powder)	Analytical reagent or C.P. grade
Formaldehyde solution 40% (formalin)	Analytical reagent or C.P. grade
Potassium ferricyanide	Analytical reagent or C.P. grade
Potassium ferricyanide	Analytical reagent or C.P. grade

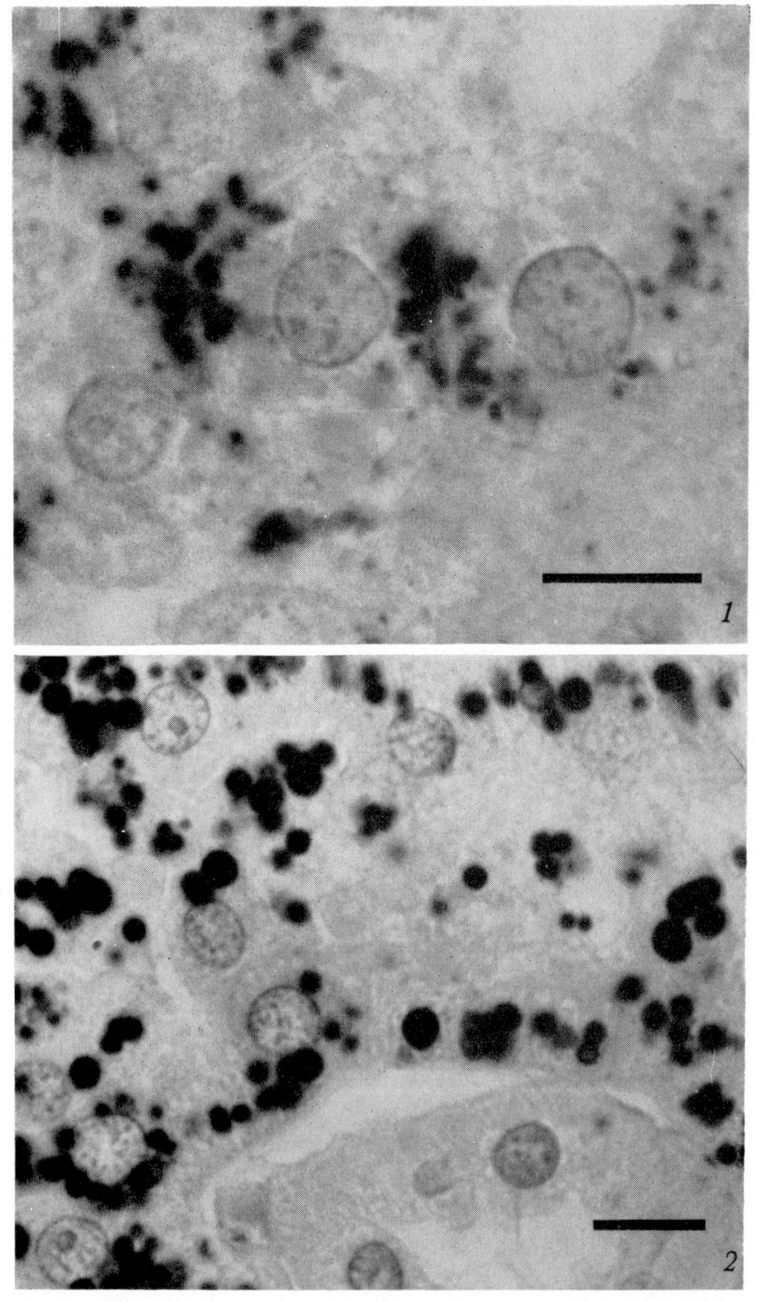

Note for Figs. 1–4: In each case, 5-bromo-4-chloroindoxyl acetate was used as substrate, and the preparations were counterstained with acid hematoxylin-picric acid as described in the text. A 10 μ scale is shown on each figure.

Sodium chloride	Analytical reagent or C.P. grade
Sucrose	Analytical reagent or C.P. grade
Tris(hydroxymethyl)-aminomethane	Pure, or alcohol recrystallized commercial quality
Gum acacia	Best quality
Absolute ethyl alcohol	Not methylated
Thymol	Pure

SOLUTIONS

Buffer

0.1 *M*-Tris(hydroxymethyl)aminomethane-hydrochloric acid buffer, pH 8.5 (Gomori, 1946).

Fixative

Solution containing 4% formaldehyde and 1% calcium chloride. Make up freshly, add about 1 gm. calcium carbonate per 100 ml., shake well and cool to 0–2° C. Decant from calcium carbonate before use. Use same day as preparation. Discard surplus.

Gum-Sucrose

Solution containing 1% gum acacia, and sucrose to a concentration of 0.88 *M*. Dissolve gum first. Add 0.1 gm. thymol per liter and store at 0–2° C.

Oxidant

Solution containing potassium ferricyanide and potassium ferrocyanide, each at a concentration of 5×10^{-2} *M*. Store at 0–2° C., but discard after 1 week.

Staining Solution

5-bromo-4-chloroindoxyl acetate	1.5 mg.
Absolute ethyl alcohol	0.1 ml.

Add the following mixture rapidly with shaking, and make the total volume up to 10 ml. with distilled water.

Buffer	2.0 ml.
Oxidant	1.0 ml.
M-Calcium chloride	0.1 ml.
2 *M*-Sodium chloride	5.0 ml.

FIG. 1. Rat liver. Esterase activity associated with peribiliary bodies just within cell membrane of parenchymatous cells. Magnification: ×2150.

FIG. 2. Rat kidney. Esterase-positive spherical bodies in cells of proximal convoluted tubule, some closely associated with cell nuclei. Note absence of similar bodies in distal tubule. Magnification: ×1515.

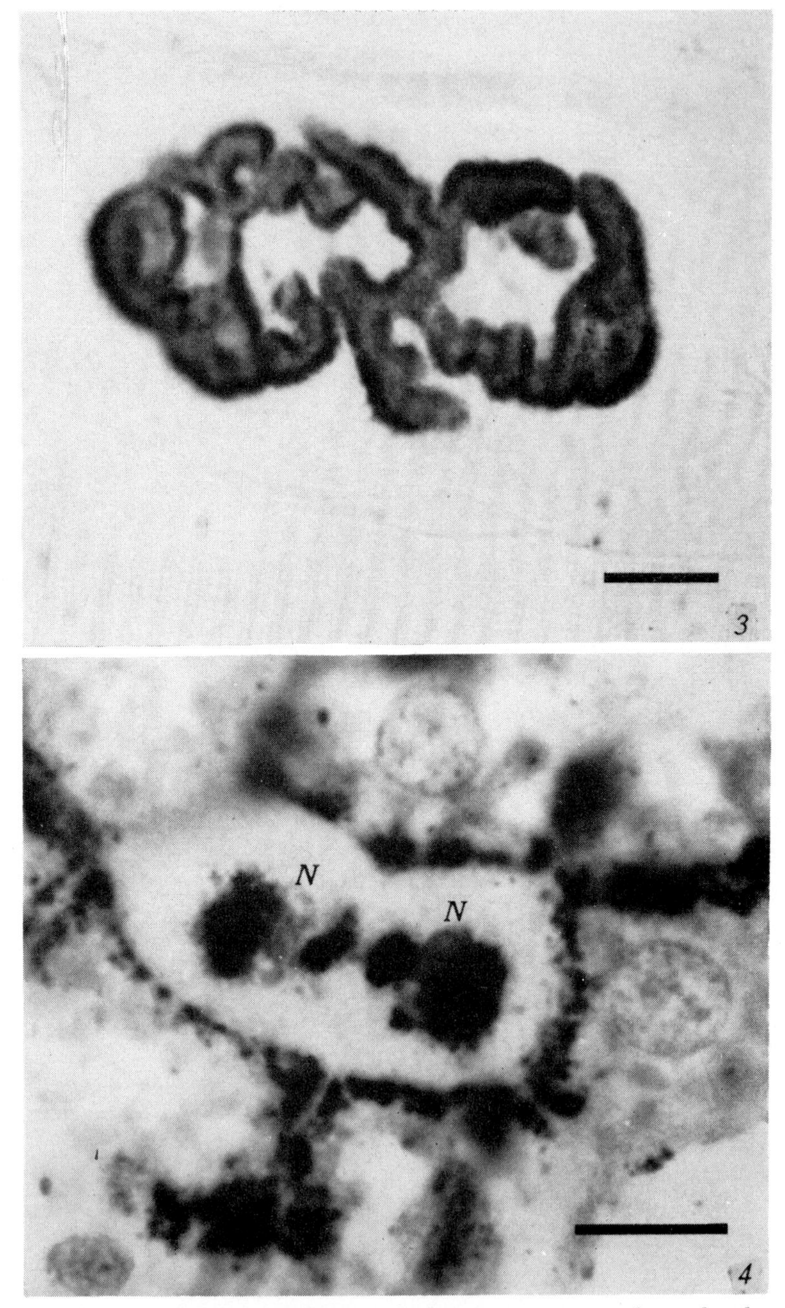

FIG. 3. Discretely stained motor end-plate in rat intercostal muscle, showing darker-stained peripheral subneural apparatus. Magnification: ×2050.

FIG. 4. Dividing parenchymatous cell in rat liver. Strongly esterase-positive regions between nuclei and also in larger regions at opposite poles (*N*, telophase nuclei). Magnification: ×1575.

The following quantities of the alternative substrates should be used:

5-bromoindoxyl acetate	1.3 mg.
5-bromo-6-chloroindoxyl acetate	3.0 mg.

The staining solutions cannot be stored, and should be made up freshly when required.

APPENDIX 2: SYNTHESIS OF SUBSTRATES

Syntheses of six substrates are described. The first three (indoxyl acetate, propionate, and butyrate) are unsuitable for staining purposes, since deposits of unsubstituted indigo are always grossly crystalline. These substrates are included because of their greater solubility, which makes them of use in biochemical studies ancillary to staining results. Syntheses of 5-bromoindoxyl acetate, 5-bromo-4-chloroindoxyl acetate and 5-bromo-6-chloroindoxyl acetate are based upon the methods of Holt and Sadler (1958a), but are given in detail, and with certain improvements.

Indoxyl Acetate (VII, R = CH_3)

Synthesis of this compound is formulated as follows:

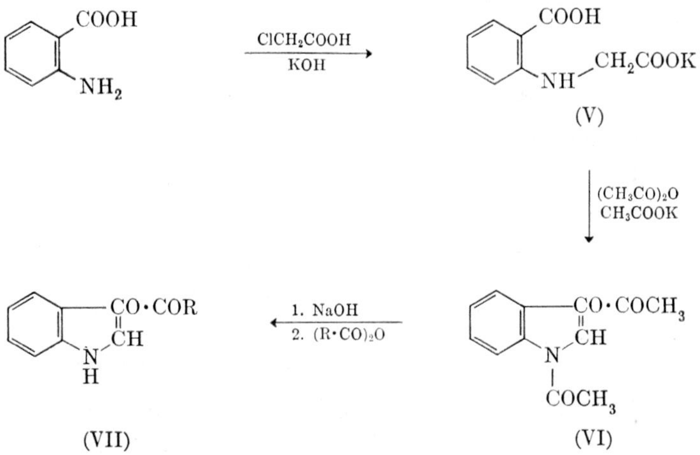

COOH
NH_2
$ClCH_2COOH$
KOH
COOH
CH_2COOK
NH
(V)
$(CH_3CO)_2O$
CH_3COOK
CO·COR
CH
N
H
1. NaOH
2. $(R·CO)_2O$
CO·$COCH_3$
CH
N
$COCH_3$
(VII)
(VI)

(*a*) *Phenylglycine-*o*-carboxylic Acid* (monopotassium salt) (V)

A suspension of anthranilic acid (13.7 gm.) in water (20 ml.), and a solution of monochloroacetic acid (9.5 gm.) in water (20 ml.) are separately treated with 30% potassium hydroxide solution (about 19 ml. in each case) until clear solutions, just alkaline to phenolphthalein are obtained. The cooled solutions are mixed, treated with activated charcoal (1 gm.), and filtered. The filtrate is then incubated in a covered vessel at 50° C. for 48 hours. The resulting suspension is cooled and the precipitate of the required salt collected, washed with water (3×10 ml.) and dried. Yield 19 gm. (80%).

(b) Diacetylindoxyl (VI)

A mixture of potassium acetate (15 gm.) and acetic anhydride (90 ml.) is gently boiled in a 500 ml. beaker (hood!), and finely powdered, dry, monopotassium salt of phenylglycine-*o*-carboxylic acid (15 gm.) added as rapidly as possible with stirring, allowing the reaction to subside between each addition. Afterwards, boil the orange solution for 15 minutes more. Cool, add water (90 ml.), and stir with cooling until excess acetic anhydride has been hydrolyzed and a suspension of the solid diacetylindoxyl is obtained. Collect the solid, wash with 30% acetic acid (20 ml.), then with 50% ethanol (2×15 ml.) and dry. The product is a cream-colored crystalline solid, m.p., 80–82° C. Yield 10 gm. (70%).

Notes

(1) It is essential to use the *potassium salt* of phenylglycine-*o*-carboxylic acid. If the free acid is used, a large amount of tar is formed, and purification of the product is rendered extremely difficult.

(2) If the product is darker, or has a lower melting point, than that given above, it should be recrystallized from ethanol or aqueous ethanol. When pure, diacetylindoxyl is a white crystalline solid, m.p., 83–84° C.

(c) Indoxyl Acetate

2 *N* sodium hydroxide solution (55 ml.) is added to a multinecked flask fitted with a stirrer (magnetic stirring is advantageous), reflux condenser, tap funnel, and delivery tube reaching to the bottom of the flask. A slow stream of oxygen-free nitrogen is passed through the solution and apparatus for 5 minutes to displace air, and diacetylindoxyl (2.17 gm.) added through a further side-neck which is normally kept stoppered. The stopper is replaced and nitrogen passed for a further 5 minutes. The mixture is heated to boiling, and then under reflux, until the diacetylindoxyl has completely dissolved. The amount of indigo formed during this procedure should be negligible. The indoxyl solution is cooled to 0° C., and acetic anhydride (5.3 ml.) slowly added from the tap funnel with rapid stirring (it may be necessary to apply a slight positive pressure of nitrogen to the tap funnel to ensure that the anhydride flows into the flask against the internal pressure). After allowing the mixture to stir for about 20 minutes more, the precipitated indoxyl acetate is collected, washed with water, and crystallized from aqueous alcohol by the following procedure: Dissolve the moist product in absolute alcohol (20 ml.) by applying gentle heat. Add activated charcoal (0.5 gm.), filter, and reheat the filtrate to about 60° C. Add

an equal volume of water at the same temperature. Allow to cool, when pure indoxyl acetate separates as white plates. Yield 1.5 gm. (85%); m.p., 129° C.

Notes

(1) Indoxyl acetate must be decolorized in alcoholic solution. In the presence of water, hydrolysis occurs and the product becomes contaminated with indigo.

(2) Indoxyl acetate is photosensitive, becoming colored when exposed to light and air. It should be stored in the dark, or in black-painted bottles. The impure compound decomposes more rapidly than the pure.

Indoxyl Propionate (VII, R = CH_3CH_2)

This substrate is prepared and purified as described for indoxyl acetate, but propionic anhydride (7.2 ml.) is used instead of acetic anhydride. Yield 1.6 gm. (84%); m.p., 90° C.

Indoxyl Butyrate (VII, R = $CH_3CH_2CH_2$)

Prepared as described for indoxyl acetate, but using *n*-butyric anhydride (9.0 ml.). When addition of the anhydride is complete, the reaction mixture should be stirred for about 1 hour longer to hydrolyze excess butyric anhydride. The product should not be collected until it is converted into a powdery precipitate. It should then be crystallized (preferably twice) from aqueous alcohol as described for indoxyl acetate. Yield 1.4 gm. (68%); m.p., 93° C.

Indoxyl propionate and butyrate should both be stored as recommended for indoxyl acetate.

5-Bromoindoxyl Acetate

This substrate is also prepared from the monopotassium salt of phenylglycine-*o*-carboxylic acid. The synthesis is formulated as follows:

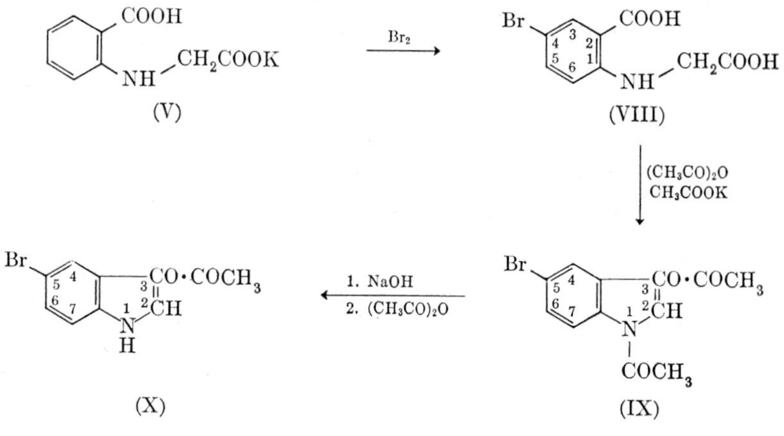

COOH, NH, CH_2COOK (V) —Br_2→ Br, COOH, NH, CH_2COOH (VIII)

(VIII) —$(CH_3CO)_2O$, CH_3COOK→ Br, $CO \cdot COCH_3$, CH, N, $COCH_3$ (IX)

(IX) —1. NaOH, 2. $(CH_3CO)_2O$→ Br, $CO \cdot COCH_3$, CH, N, H (X) (*a*) *4-Bromophenylglycine-*o*-carboxylic Acid* (VIII)

Rapidly add a solution of bromine (4.25 ml.) in glacial acetic acid (15 ml.), with thorough stirring, to a smooth mixture of finely powdered monopotassium salt of phenylglycine-*o*-carboxylic acid (18 gm.) in glacial acetic acid (75 ml.). The salt dissolves, and the product separates almost immediately. Allow to stand for 30 minutes and dilute with water (300 ml.). Collect the solid, wash with water, and dry. The product (18 gm., 87%), forms a cream-colored solid, m.p., 220° C. (decomp.), which is sufficiently pure for the next stage. A sample crystallized from 50% ethanol (charcoal) should form pale cream needles, m.p., 238° C. (decomp.).

(*b*) *5-Bromodiacetylindoxyl* (IX)

4-Bromophenylglycine-*o*-carboxylic acid (15 gm.) is added as rapidly as possible with stirring to a gently boiling mixture of acetic anhydride (150 ml.) and potassium acetate (15 gm.), allowing the reaction to subside between each addition. The resulting dark solution is boiled for a further 15 minutes and then cooled. Water (125 ml.) is then added, and the mixture stirred with cooling until excess acetic anhydride has been hydrolyzed. The collected solid is washed with 50% acetic acid (25 ml.) and with ethanol (15 ml.) before crystallization from ethanol (125–175 ml.) (charcoal). The product forms pale yellow needles, m.p., 120–122° C., yield 12 gm. (75%). A further small quantity may be recovered from the ethanol liquors by evaporation to a small volume. When purified by further crystallization from ethanol or aqueous ethanol, 5-bromodiacetylindoxyl forms white needles, m.p., 123–124° C.

(*c*) *5-Bromoindoxyl Acetate* (X)

Proceed as described for indoxyl acetate, but use 5-bromodiacetylindoxyl (3.0 gm.) instead of diacetylindoxyl. The crude product is purified by dissolving in hot ethanol (30 ml.), filtering (charcoal), and adding water (30 ml.) at 60° C. to the reheated filtrate. On cooling, the required 5-bromoindoxyl acetate separates as glistening white plates (2.0 gm., 78%); m.p., 134° C.

5-Bromo-4-chloroindoxyl Acetate and 5-Bromo-6-chloroindoxyl Acetate

These substrates are both prepared from *m*-chloroaniline by reaction sequences summarized in the chart (Fig. 5) on p. 393. The starting material is converted into a mixture of 4-chloroisatin (XII) and 6-chloroisatin (XVIII) via 3-chloro*iso*nitrosoacetanilide (XI). The isatins are then separated and converted independently into the substrates as described below. These syntheses are more complex and difficult than those described above, but the extra effort is particularly worthwhile

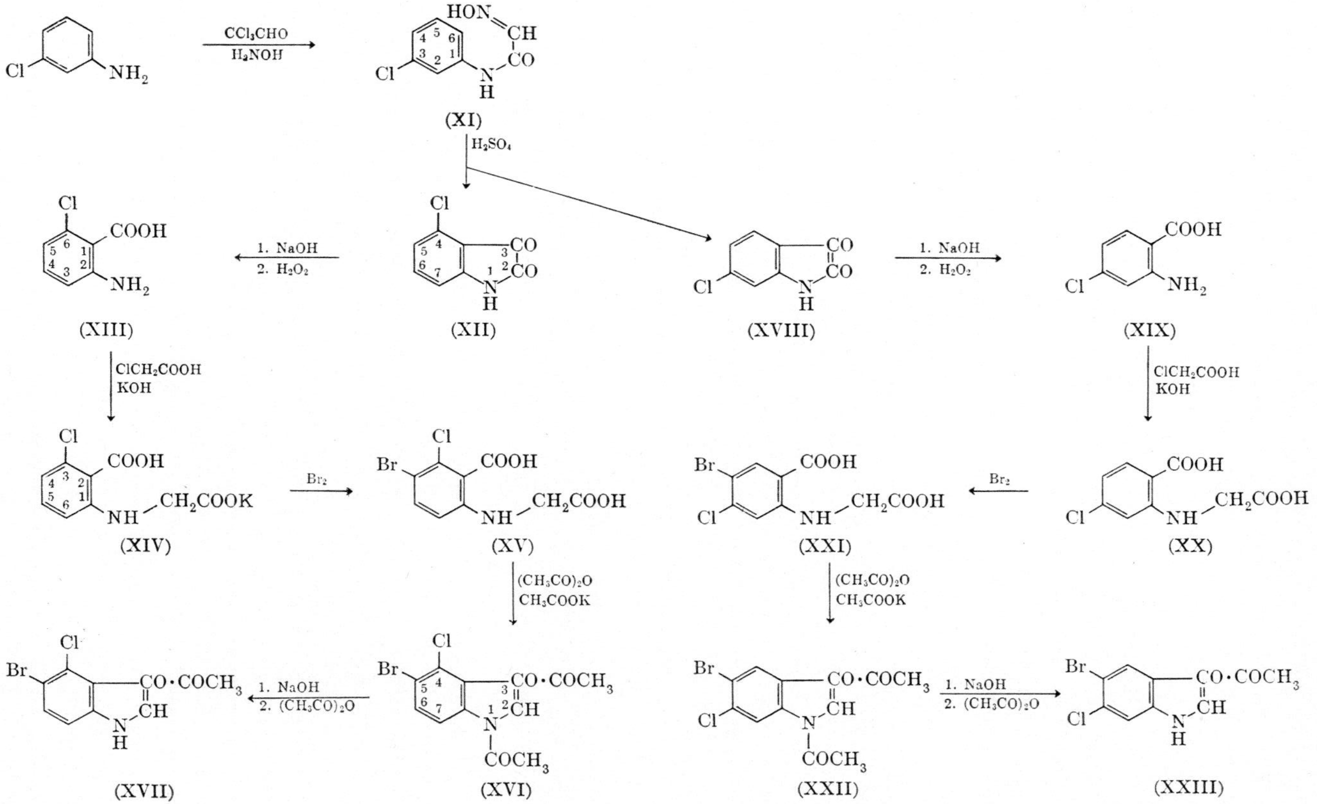

Fig. 5. Reaction sequences in the preparation of 5-bromo-4-chloroindoxyl acetate and 5-bromo-6-chloroindoxyl acetate from *m*-chloroaniline.

in the case of 5-bromo-4-chloroindoxyl acetate, in view of the superior cytochemical resolution given by this substrate. As mentioned in the Introduction, the synthesis of 5-bromo-6-chloroindoxyl acetate is included to make use of the 6-chloroisatin obtained in the early stages of the process, and also because the purplish red color of the derived 5,5′-dibromo-6,6′-dichloroindigo may be convenient in certain applications. The two substrates should be stored as described for indoxyl acetate.

(*a*) *3-Chloroisonitrosoacetanilide* (XI)

Add the following to a 5 l. beaker in the order given, stirring well between each addition: (1) a solution of chloral hydrate (90 gm.) in water (1200 ml.), (2) crystalline sodium sulfate (1300 gm.), (3) a solution of *m*-chloroaniline (64 gm.) in water (250 ml.) containing concentrated hydrochloric acid 43 ml., d. 1.19), (4) a solution of hydroxylamine hydrochloride (110 gm.) in water (500 ml.). Heat the mixture, with stirring, so that it reaches the boiling point in about 45 minutes. During this period, the bulk of the product separates and the mixture becomes very thick and tends to froth. Cool the mixture and collect the cream-colored crystalline 3-chloro*iso*nitroacetanilide. Wash it several times with water (150 ml.) and dry by spreading it in a thin layer exposed to a current of air. Yield 99 gm.; m.p., 147° C. A sample recrystallized from aqueous alcohol (charcoal) melts at 155° C.

(*b*) *4-Chloroisatin* (XII) *and 6-Chloroisatin* (XVIII)

Heat concentrated sulfuric acid (350 ml.) to 50° C. and add 3-chloro*iso*nitrosoacetanilide (98 gm.), with thorough stirring, at such a rate that the reaction temperature is maintained at 60–70° C. After addition is complete, the temperature is raised to 80° C. for 15 minutes to complete the reaction. The dark purple solution is cooled and poured on to crushed ice (2 kgm.). After standing for 30 minutes, the orange solid (mixed isatins) is collected and washed substantially free from acid with cold water.

Separation of the mixture into its two components is effected in the following manner: Suspend the solid in water (350 ml.) and add 2 *N* sodium hydroxide solution, with warming to about 60° C. and stirring, until a clear yellow solution of the corresponding sodium isatinates is formed. Carefully adjust the pH of the warm solution to 6.5 by addition of 2 *N* acetic acid, immediately stir in decolorizing charcoal (20 gm.), and filter at once over a previously prepared bed of "Hyflo-Supercel." Crystals of 4-chloroisatin may already separate from the filtrate at this stage. Add glacial acetic acid (50 ml.) and allow to stand overnight to complete the precipitation of the orange 4-chloroisatin. This should be collected and the filtrate placed on one side. Wash the solid (about 45 gm., m.p., 256–258° C.) with water and dry.

The filtrate reserved from the 4-chloroisatin precipitation is treated with concentrated hydrochloric acid (50 ml.), giving a yellow precipitate of crude 6-chloroisatin (about 40 gm., m.p., 260–262° C.). Collect, wash with water and dry.

For success in the subsequent stages of the syntheses, the crude isatins should be purified as follows: Crystallize the 4-chloroisatin from glacial acetic acid (1250 ml.) (charcoal) to obtain the pure compound as orange-red needles. Yield 40 gm. (44%); m.p., 258–260° C. Similarly, crystallize the crude 6-chloroisatin from glacial acetic acid (1500 ml.) (charcoal), to obtain the purified compound (35 gm., 38%) as flat yellow elongated plates or yellowish-orange needles, m.p., 262–263° C.

(*c*) *6-Chloroanthranilic Acid* (XIII)

4-Chloroisatin (27.2 gm.) is added to 2 *N* sodium hydroxide solution (225 ml.) and the mixture warmed until a pale yellow solution is obtained. Hydrogen peroxide (5 ml., 100 vols.) is then added, when heat is evolved. When the solution begins to cool, add a further 5 ml. of hydrogen peroxide, and continue in this way until a total of 20 ml. has been added. Allow to stand for 1 hour to complete the reaction (a test sample treated with excess 2 *N* hydrochloric acid should not regenerate the orange isatin color) and add glacial acetic acid to adjust the solution to pH 5.0. Filter (charcoal) and bring the filtrate to pH 2.0 with concentrated hydrochloric acid. Cool the suspension in ice, collect the white precipitate of 6-chloroanthranilic acid, and wash with small quantities of chilled water before drying. Yield 19.5 gm. (76%); m.p., 148° C.

(*d*) *3-Chlorophenylglycine-*o*-carboxylic Acid* (monopotassium salt) (XIV)

A suspension of 6-chloroanthranilic acid (17.2 gm.) in water (25 ml.) is treated with 30% potassium hydroxide solution until the acid dissolves to give a solution just alkaline to phenolphthalein. To this is added a similarly neutralized solution of chloroacetic acid (10 gm.) in water (25 ml.). The combined solutions are filtered (charcoal), and the filtrate incubated in a covered vessel at 50° C. until precipitation of the required salt has ceased (48–72 hours). The suspension is then cooled, the product collected, washed with water (2×15 ml.), and dried. Yield 18.8 gm. (70%).

(*e*) *4-Bromo-3-chlorophenylglycine-*o*-carboxylic Acid* (XV)

Finely powdered monopotassium salt of 3-chlorophenylglycine-*o*-

carboxylic acid (16 gm.) is suspended in glacial acetic acid (40 ml.) and a solution of bromine (4 ml.) in glacial acetic acid (10 ml.) added with stirring over 15 minutes. The product is allowed to stand for 30 minutes, diluted with water to 300 ml., and the precipitate of the required compound collected, washed with water and dried, giving a cream-colored solid (12.8 gm., 70%), m.p., 170° C. (decomp.). A sample crystallized from aqueous alcohol melts at 176° C. (decomp.).

(*f*) *5-Bromo-4-chlorodiacetylindoxyl* (XVI)

4-Bromo-3-chlorophenylglycine-*o*-carboxylic acid (10 gm.) is added in portions over 5 minutes to a gently boiling mixture of acetic anhydride (75 ml.) and potassium acetate (10 gm.). The mixture is then boiled for a further 15 minutes, cooled, and excess acetic anhydride hydrolyzed by stirring with water (65 ml.), with cooling. The precipitate is collected, washed with 50% acetic acid, then with water, and dried. Crystallize (charcoal) from ethanol (250 ml.), to obtain the pure diacetyl compound as pale yellow needles (5.5 gm., 50%), m.p., 168° C.

(*g*) *5-Bromo-4-chloroindoxyl Acetate* (XVII)

Proceed as described for indoxyl acetate, but use 5-bromo-4-chlorodiacetylindoxyl (3.3 gm.) instead of diacetylindoxyl. Purify the crude product by dissolving in hot ethanol (30 ml.), filtering (charcoal), and adding water (30 ml.) at 60° C. to the reheated filtrate. On cooling, 5-bromo-4-chloroindoxyl acetate separates as white needles (2.1 gm., 73%), m.p., 107° C.

(*h*) *4-Chloroanthranilic Acid* (XIX)

6-Chloroisatin (27.3 gm.) is added to *N* sodium hydroxide solution (550 ml.) and the mixture warmed until a pale yellow solution is obtained. Add hydrogen peroxide (100 vols.) in 5 ml. portions over 40 minutes until 20 ml. has been added. Allow to stand for 1 hour to complete the reaction (test sample treated with excess 2 *N* hydrochloric acid should not regenerate orange isatin color) and adjust the solution to pH 7.5 with glacial acetic acid. Filter (charcoal) and acidify the filtrate to pH 3–4 with concentrated hydrochloric acid. Collect the precipitated product and wash with water, to obtain 4-chloroanthranilic acid as a white solid (24 gm. 95%), m.p., 234° C.

(*i*) *5-Chlorophenylglycine-*o*-carboxylic Acid* (XX)

2 *N* Sodium hydroxide solution (100 ml.) is added to 4-chloroanthranilic acid (17.2 gm.) and chloroacetic acid (10 gm.) in water (250 ml.). Phenolphthalein indicator is added and the whole heated under reflux while 2 *N* sodium hydroxide solution (about 50 ml.) is run in at such a rate that the solution is maintained slightly alkaline to the

indicator. When consumption of alkali has ceased, the solution is cooled, adjusted to pH 7.5 with 2*N* acetic acid, and filtered (charcoal). The filtrate is acidified to pH 1–2 with concentrated hydrochloric acid and the white product collected, washed with water, and dried. Yield 20 gm. (87%); m.p., 215° C. (decomp.).

(*j*) *4-Bromo-5-chlorophenylglycine-*o*-carboxylic Acid* (XXI)

A solution of bromine (3 ml.) in glacial acetic acid (20 ml.) is added with stirring over a period of 15 minutes to a dispersion of finely powdered 5-chlorophenylglycine-*o*-carboxylic acid (11.5 gm.) in glacial acetic acid (50 ml.). After standing for 30 minutes, the product is diluted with water (200 ml.) and the gel-like suspension heated to 60° C. to convert to a form more easily filtered. The cooled suspension is filtered, the solid washed with water, and dried, giving the required product as a pale cream solid (13 gm., (85%), m.p., 229° C. decomp.).

(*k*) *5-Bromo-6-chlorodiacetylindoxyl* (XXII)

4-Bromo-5-chlorophenylglycine-*o*-carboxylic acid (10 gm.) is added in portions over 5 minutes to a gently boiling mixture of acetic anhydride (75 ml.) and potassium acetate (10 gm.). The mixture is then boiled for a further 15 minutes, cooled, and excess acetic anhydride hydrolyzed by stirring with water (65 ml.), with cooling. Collect the precipitate, wash with 50% acetic acid and then with water, and dry. Crystallize from ethanol (250 ml.) to obtain the pure diacetyl compound as pale yellow needles (5.4 gm., 49%), m.p., 176° C.

(*l*) *5-Bromo-6-chloroindoxyl Acetate* (XXIII)

Proceed as described for indoxyl acetate, but use 5-bromo-6-chlorodiacetylindoxyl (3.3 gm.) instead of diacetylindoxyl. Purify the crude product by dissolving in hot ethanol (35 ml.), filtering (charcoal), and adding water (30 ml.) at 60° C. to the reheated filtrate. 5-Bromo-6-chloroindoxyl acetate (2.2 gm., 76%) separates on cooling as white needles, m.p., 131° C.

REFERENCES

Cotson, S., and Holt, S. J. (1958). *Proc. Roy. Soc. (London)* **B148,** 507.

Gomori, G. (1946). *Proc. Soc. Exptl. Biol. and Med.* **62,** 32.

Hobbiger, E. (1957). *Biochem. J.* **67,** 600.

Holt, S. J. (1952). *Nature* **169,** 271.

Holt, S. J. (1954). *Proc. Roy. Soc. (London)* **B142,** 160.

Holt, S. J. (1956). *J. Histochem. and Cytochem.* **4,** 541.

Holt, S. J., and O'Sullivan, D. G. (1958). *Proc. Roy. Soc. (London)* **B148,** 465.

Holt, S. J., and Sadler, P. W. (1958a). *Proc. Roy. Soc. (London)* **B148,** 481.

Holt, S. J., and Sadler, P. W. (1958b). *Proc. Roy. Soc. (London)* **B148,** 495.

Holt, S. J., and Withers, R. F. J. (1958). *Proc. Roy. Soc. (London)* **B148,** 520.
Lillie, R. D. (1954). "Histopathologic Technic and Practical Histochemistry," pp. 76, 77. McGraw-Hill, New York.
Novikoff, A. B., Beaufay, H., and de Duve, C. (1956). *J. Biophys. Biochem. Cytol.* **2** (suppl.), 179.
Novikoff, A. B., and Holt, S. J. (1957). *J. Biophys. Biochem. Cytol.* **3,** 127.
Palade, G. E. (1952). *J. Exptl. Med.* **95,** 285.
Pearse, A. G. E., and Pepler, W. J. (1957). *Nature* **179,** 589.
Pearson, B., and Defendi, V. (1957). *J. Histochem. and Cytochem.* **5,** 72.
Strauss, W. (1956). *J. Biophys. Biochem. Cytol.* **2,** 513.
Underhay, E. (1957). *Biochem. J.* **66,** 383.
Underhay, E., Holt, S. J., Beaufay, H., and de Duve, C. (1956). *J. Biophys. Biochem. Cytol.* **2,** 635.
Venkataraman, K. (1952). "The Chemistry of Synthetic Dyes," Vol. II. Academic Press, New York.
Vickerstaff, T. (1954). "The Physical Chemistry of Dyeing," 2nd ed. Interscience, New York.

FLUORESCENT ANTIBODY METHODS

ALBERT H. COONS

Department of Bacteriology and Immunology,
Harvard Medical School, Boston, Massachusetts

The uses to which antibody labeled with fluorescein has been put up to the present are described in a recent review (Coons, 1956), which summarizes the findings and discusses some of the theoretical considerations underlying various parts of the methodology. The purpose of this paper is to provide instructions and recipes for carrying out the various steps in the procedures for the cytological demonstration of antigens, and of antibody in tissue sections.

I. THEORY OF THE METHOD

The basic strategy of the method is simple and obvious. Antigenic material present in tissue cells and more particularly in sections of them, will react specifically with antibody complementary to it. This immunological reaction results in the deposit of minute amounts of specific antibody over those areas of a tissue section where the antigen is present. When the antibody molecules have previously been chemically marked with fluorescein, the micro-deposit of fluorescent antibody is visible under the fluorescence microscope. The specific step in the reaction is the selection by the antigen of its specific antibody from a solution of fluorescent proteins. Other serum protein molecules present in the fluorescent solution are washed away leaving the specific deposit *in situ*.

The preparation of tissue sections must be carried through without chemical fixatives which would injure the immunological activity of the antigenic material. This limits the general methods available to the use of smears, or of freeze-dried sections, or of frozen sections of unfixed tissue. In certain cases, where the antigen is chemically hardy standard histological fixation can be employed, but these instances are exceptional. Diffusion during fixation is of course a hazard here.

When a protein solution containing a known antibody has been chemically coupled with fluorescein, when appropriate tissue sections have been prepared, and when a fluorescence microscope with high intensity illumination is available, the actual manipulations necessary to demonstrate antigenic material microscopically are very simple.

The interpretation of the results obtained is made more difficult by the occurrence of interactions between the fluorescent proteins in the antibody solution and substances in the tissue sections unrelated to the specific antigen-antibody system under study. The nature of these nonspecific reactions is at present unknown.

II. DETAILS OF INSTRUMENTATION

A. Fluorescence Microscope

1. *Light Sources*

In the author's laboratory only carbon arcs have been used as light sources. The best one of these is a direct current 20 ampere carbon arc with an automatic electromagnetic feed. Arcs operating on direct current are considerably brighter at the same power level than those operating on alternating current.

Other light sources which have given satisfactory performance in the hands of others or which have been tested and have been found intense enough by the author are high pressure mercury vapor arcs as follows: A-H 6 (General Electric Co.), HBO 200 (Osram), ME/D 26 (Mazda) (Hill and Cruickshank, 1953).

2. *Condensing Lenses*

A single bispheric condensing lens with a diameter of about 6 cm. and a focal length of about 6 cm. (see Appendix) is placed in front of the light source, mounted firmly on an optical bench, and adjusted so that the image of the crater of the horizontal carbon falls on the back lens of a dark field condenser.

3. *Filters*

Into this beam is placed a Pyrex water cell containing a 3.2 cm. thickness of copper sulfate solution (25% $CuSO_4 \cdot 5H_2O$ w/v). In the author's laboratory this cuvette is fitted with a glass cooling coil, immersed in the copper sulfate solution around the light beam, through which running tap water circulates. Without this cooling device the copper sulfate will boil within a few minutes after the arc is turned on. Attached to the cuvette holder on the side away from the arc is a glass filter to remove the middle range of the spectrum (Corning Glass Co., No. 5840, half-standard thickness). The light beam then passes to the mirror of the microscope. This is an ordinary glass mirror silvered on the back. In this laboratory a first-surface aluminized mirror has not shown appreciable superiority.

4. *Microscope*

The microscopes in use in the author's laboratory are standard microscopes equipped with achromatic objectives and glass dark field condensers. (Nonfluorescent lenses, which are important when a transmitting condenser is used, are not required with a dark field condenser.) They are equipped with a monocular 10× eyepiece containing a protective filter. Satisfactory filters to remove wavelengths below 4200 Å are Wratten No. 2A or No. 2B (Eastman Kodak Co. 2A is to be preferred), or filter UV 420 (Polaroid Corp.). These ocular filters are nearly colorless, an important point because yellow filters make it difficult or impossible to detect small amounts of yellow-green fluorescent light. In order not to diminish the available light, monocular microscopes are to be preferred.

5. *Photography*

With the equipment described above and a 35 mm. film holder, the exposure time for a fast film (TriX, Eastman Kodak) at 400× magnification is about 2 minutes. The exposure time must be varied somewhat depending on the amount of fluorescence in the object under examination. It is rarely possible to focus the image on a ground glass plate, as the amount of light is too small. Rather, focusing must be carried out with a lens focused on the ground glass plate, through a clear port in the plate. In preparations with large amounts of blue-gray autofluorescence, it is advantageous to increase the contrast between this blue-white light and the yellow-green light from the fluorescein-labeled antibody with yellow or green filters, with appropriate prolongation of the exposure.

These photographic conditions produce rather "thin" negatives, and require high contrast printing.

B. Preparing Sections

1. *Frozen Sections*

The method used in the author's laboratory is a modification of a method of Linderstrøm-Lang and Mogensen (1938). It consists of cutting sections of unfixed frozen tissue on a microtome placed in a refrigerated cabinet at −20° C. To accomplish this, Linderstrøm-Lang and Mogensen built an insulated wooden box cooled by circulating the air in the interior over crushed dry ice. The box was equipped with gloved armholes, lighting, and an insulated glass window through which to observe the work. In routine use it is somewhat difficult to control the temperature adequately with this arrangement, because the dry ice is so much colder than the temperature desired. It is more convenient, and of course also more expensive, to place the microtome in a cabinet cooled by mechanical refrigeration. The essentials of the construction of this cabinet, together with illustrations, and a drawing of the glass guide which fits on the knife and which is essential to prevent curling of the sections, were described by Coons *et al.* (1951).

The sections as cut are placed on clean, cold slides, quickly thawed on the slide by pressing a finger against the undersurface beneath the section, and withdrawn from the cabinet. They are then dried in a stream of air from a fan.

An adhesive is not necessary provided the slides are clean.

2. *Freeze-drying*

Freeze-drying methods have been carefully reviewed by Malmstrom (1951) and by Bell (1952), and applications of the principles developed are described by Glick and Malmstrom (1952) and by Danielli (1953).

The main principle is to regard freeze-drying as a molecular distillation, and hence to put the watertrap (the cold spot) within such a short distance of the frozen tissue (the warm spot) that, at the pressure employed, the mean free path of the water vapor molecules will exceed this distance. The application of this principle necessitates maintaining the temperature of the tissue at about —40° C. Mendelow and Hamilton (1950) used a pressure of $2\text{–}5 \times 10^{-4}$ mm. Hg and a liquid N_2 water trap (—194° C.) placed within about 5 cm. of 1 mm. tissue blocks. This tissue was kept about —40° C. by means of a slush of diethyl oxalate (m.p., −40.6° C.). Under these circumstances drying was complete in about 7 hours. Danielli states that with solid CO_2 (—78.6° C.) as the drying agent, and a somewhat higher but unspecified pressure, drying time was 2–3 days.

There exist to the author's knowledge no observations comparing the distribution of antigenic material in frozen sections and freeze-dried sections. However, Schiller *et al.* (1953), Marshall (1951), and Mayersbach and Pearse (1956) have all employed freeze-dried sections in studies of the localization of foreign proteins, ACTH, or enzymes, with fluorescent antibody. Mayersbach and Pearse found the same distribution of ovalbumin in the kidney, using freeze-dried material, as Coons *et al.*, (1951) found using frozen sections.

Freeze-dried sections must be floated on some suitable fixative, or on some nonaqueous fluid.

III. PROCEDURES

A. Preparation of Antiserum

The crux of the problem in the use of fluorescent antibody is the preparation of the initial antibody solution. For this purpose the antigen preparation for injection must be as pure as possible, and usually an adjuvant must be employed in order to obtain a high titer. Either alum (see Proom, 1943) or water-in-oil emulsions containing killed mycobacteria (see Freund, 1947) should usually be employed with soluble proteins. There is also increasing evidence that the initial stimulus is of great importance in directing the subsequent course of immunization (see Davenport *et al.*, 1953; and especially Adler, 1956); the implication is that the first stimulus should be as pure as can be obtained, even if less pure material must be used later in the course of immunization.

Useful directions are given by Kabat and Mayer (1948).

The choice of animal in which the antiserum is to be prepared must depend upon the individual problem. Successful labeled antibody preparations have been obtained from the rabbit, monkey, chicken, cow, horse, dog, goat, and man.

B. Concentration of Antiserum

There are no data available by which to judge whether an antiserum is suitable for use as a cytochemical reagent. In general the higher the titer the more effective the preparation. In most cases so far explored, the titers have been close to the highest that could be attained for the antigen-antibody system under study. In two instances (ovalbumin and bovine albumin) where absolute amounts of antibody could have been determined (Coons *et al.*, 1951), no analyses were carried out. A discussion of the theoretical effects of antibody concentration on the sensitivity of cytochemistry with fluorescent antibody has been published elsewhere (Coons, 1956).

In order to avoid dilution of the final conjugated antibody solution with large amounts of inactive protein, it is usual to discard the serum albumin. This is accomplished in the author's laboratory by means of 50% saturation with ammonium sulfate. Others (Marshall, 1951, 1954a, b) have used alcohol fractionation for this purpose.

Dilute at least 30 ml. of high-titered antiserum with an equal volume of 0.15 *M* sodium chloride solution. Pour the diluted serum into a beaker cooled in an ice bath and fitted with a mechanical stirrer. When the solution has cooled to 4° C., add slowly (dropwise) a volume of cold ammonium sulfate solution, saturated at 4° C., equal to the volume of the diluted serum. Continue the stirring for about 30 minutes. (If desired, the precipitate may be stored at 4° C.) Pour the suspension into a cold centifuge tube and centrifuge at 4° C. until the supernatant is clear and the precipitate firmly packed. Pour off and discard the supernatant. Wash the precipitate once with cold ½ saturated ammonium sulfate, pour off as much as possible of the supernatant after centrifugation, and dissolve the precipitate by adding a volume of cold buffered saline equal to ⅓ that of the original serum. Hold back a little for use in washing the centrifuge tube and the funnel through which the solution is poured into a cellophane dialysis sac. Dialyze the solution against frequently changed buffered saline until free from ammonium ion. Add 1/10,000 merthiolate (sodium ethyl mercurithiosalicylate, Lilly), stopper securely and store at 4° C.

Before conjugation, the protein concentration must be determined. This can be carried out with an accuracy sufficient for the purpose on 0.1 ml. of the solution.

Carefully measure 0.1 ml. of the globulin solution with a 0.2 ml. graduated pipette calibrated "to contain." Wash out the pipette into 5.0 ml. 0.15 *M* sodium chloride and carry out nitrogen determinations on 0.5 ml. aliquots of this 1/51 dilution by the method of Koch and McMeekin (1924). Any other standard method for protein estimation is of course satisfactory.

The protein concentration should be at least 2%.

C. Conjugation with Fluorescein Isocyanate

The conditions for conjugating aromatic isocyanates with proteins, first described by Hopkins and Wormall (1933), were studied by Creech and Jones (1941a, b). In Table III of their paper (1941a) these conditions are summarized: In coupling horse serum albumin with various polynuclear aromatic hydrocarbons through the carbamido linkage, the protein concentration of the reaction mixture ranged from 9–25 mg./ml., the organic solvent (dioxane) concentration from 15–45%, and the pH from 8–10. The ratio of hydrocarbon to protein ranged from 0.065 to 0.35.

Fluorescein isocyanate differs from the compounds studied by Creech and Jones in its instability, and in its relative insolubility in dioxane. Otherwise the conditions for the reaction have been kept the same (Coons *et al.*, 1942; Coons and Kaplan, 1950).

Protein concentration	1%	(w/v)
1,4-dioxane (purified)	15%	(v/v)
Acetone	7.5%	(v/v)
pH	9.0	
Fluorescein isocyanate (as amine)/protein ratio		0.05
Saline (0.15 *M* NaCl)		q.s.

The concentrations specified are those after the addition of the isocyanate dissolved in acetone.

An example follows:

In this case the serum was purified therapeutic diphtheria antitoxin (horse) prepared by the Massachusetts Division of Biologic Laboratories, and contained about 10% protein. (Reagents to be added in order listed.)

Reaction Mixture	*Initial ml.*	*Final ml.*
Saline	28.45	28.45
$NaHCO_3$–Na_2CO_3 buffer 0.5 *M* (pH 9.0)	7.5	7.5
1,4-dioxane (purified)	7.5	7.5
Acetone (reagent)	1.25	3.75

These were put into a 250 ml. beaker fitted with mechanical stirring at a brisk rate and immersed in an ice bath. When the temperature of the mixture was 4° C. or below,

Serum (500 mg. protein)	2.8 ml.	

was added, followed by the dropwise addition of 2.5 ml. acetone containing 10 mg. fluorescein isocyanate (as amine)/ml.

	47.5 ml.	50 ml.

Continue to stir the mixture for 18 hours in the cold, then pour it into a cellophane dialysis sac, and dialyze it against buffered saline changed twice the first 24 hours and daily thereafter until the saline retains only the faintest green fluorescence (in a liter graduate no fluorescence in daylight when looking along the diameter, and faint green fluorescence when viewed from the top along the axis). This requires several days unless a stirring device is employed.

Remove from the sac, centrifuge to remove any precipitate, add 1/10,000 methiolate to prevent the growth of fungi, and store at 4° C. in the dark.

D. Purification of Conjugates

Before use, such conjugates must be shaken with tissue powder to remove green-fluorescing substances which react with normal tissue components, and which are present in normal (nonimmune) serum labeled in the same way. The nature of these substances is not known.

1. *Acetone-Dried Tissue Powder*

"Place 25 to 50 gm. of fresh or frozen tissue in a Waring blendor with an equal volume of 0.15 *M* NaCl solution. Homogenize the tissue with short repeated activations of the propeller, avoiding heating. Pour the homogenate into a beaker and add 4 volumes of acetone with stirring. After allowing the mixture to stand for a few minutes, decant and discard the supernatant. Pack the precipitate by centrifugation, and wash it in the centrifuge with several changes of saline solution until the supernatant is apparently free of hemoglobin. Suspend the washed precipitate in an amount of saline about equal to the volume of the precipitate, and to this suspension add 4 volumes of acetone with stirring. After allowing the suspension to settle for a few minutes, decant and discard the supernatant, add 4 more volumes of acetone, harvest with suction on a Buchner funnel, wash the precipitate with acetone, and allow it to dry on the funnel. Finally, dry the powder overnight at 37° C., and store it stoppered in a refrigerator at 4° C. The object of the procedure is to obtain dry, saline-insoluble material.

"For use, such a powder is stirred into an aliquot of a fluorescein-antibody solution (usually 5 ml.) in the amount of 100 mg./ml. It is our practice to do this in a 25 ml. lusteroid tube. After standing for about an hour at room temperature with occasional stirring, the supernatant is harvested by centrifugation in the cold at 18,000 r.p.m. in an angle head. The absorption is then repeated with an appropriate amount of powder (in this case about 300 mg.). The yield of conjugate is about 30%. The use of the high speed centrifuge packs the powder more tightly and increases the yield, but the procedure may be carried out at 3000 r.p.m. with larger mechanical losses.

"Aliquots of conjugate absorbed in this manner do not always keep well, and hence it is better to absorb small amounts as needed. At the end of the absorption procedure, 1/10,000 merthiolate is added, even though there is already some present in the stock conjugate solution." (Quoted from Coons *et al.*, 1955, with permission of the editors.)

2. *Choice of Tissue for Absorption of Conjugates*

No rules can be stated for this. Since the purpose is to exhaust the preparation of substances reacting with elements of the tissue to be examined, the ultimate choice is the tissue itself or at least tissue from the species to be examined. This choice is available when antigen foreign to the tissue is the object of study; it is not, or may not be, available for normal components of tissue.

As a matter of empirical practice, powder made from mouse liver is the first choice in this laboratory. Such a powder succeeds in abolishing nonspecific reactions with tissue sections of mouse and guinea pig tissue, and of rabbit tissue with the exception of polymorphonuclear leucocytes and eosinophiles. Abolition of these latter reactions can be achieved by absorption with powder prepared from the red bone marrow of rabbits (Sheldon, 1953; Coons *et al.*, 1955). Mouse liver powder is also successful in the purification of fluorescent antibody solutions for use on human cells grown in tissue culture (see Weller and Coons, 1954; Lebrun, 1956).

E. Fixation of Sections

Frozen or freeze-dried sections must be fixed before exposure to the aqueous solutions in which antigen-antibody reactions must be carried out. Antigenic substances differ to such an extent that the best fixative for each must be discovered by experiments on single sections.

1. *Proteins*

Ethanol (95% v/v) is a useful fixative for proteins, although some proteins are not well fixed by it, i.e., bovine albumin (Coons *et al.*, 1951) and human serum proteins (Gitlin *et al.*, 1953). Absolute methanol was employed by the latter. Antibody is well-fixed by ethanol (Coons *et al.*, 1955).

The slides are placed in Coplin jars containing ethanol preheated to 37° C. and held at 37° for 15 minutes. The slides are then removed and placed standing on edge in a 37° C. incubator for 30 minutes to remove the alcohol. Before the application of immunological reagents they should usually be rinsed in buffered saline.

Acetone is a poor fixative for proteins, but it is useful for the treatment of objects the antigenicity of which is destroyed by ethanol, e.g., viruses. Acetone extracts a fugitive, whitish fluorescing material.

Marshall (1954b) has used a fixative consisting of 50% 1,4-dioxane and 5% formalin in water for floating freeze-dried sections of pancreas for the demonstration of chymotrypsinogen and procarboxypeptidase.

2. *Polysaccharides*

Foreign polysaccharides are badly fixed in frozen sections, and so far have been more successfully studied by fixation of tissue blocks in picric acid–alcohol–formalin (Rossman, 1940; Kaplan *et al.*, 1950; Hill *et al.*, 1950), followed by paraffin embedding. Even here they are easily extractable and must be handled with care. Schmidt (1952) working with a polysaccharide of low molecular weight (Group A streptococcus) found that the best fixative was acetone over Ca_2SO_4, followed by paraffin embedding. It was necessary to float the sections on 80% ethanol to prevent extraction.

3. *Lipids*

The only lipid-containing substance of which there is yet experience (Forssman antigen) is alcohol-soluble, and is completely extractable from tissue sections by ethanol. It has been studied in frozen sections fixed in 10% formalin (Tanaka and Leduc, 1956).

F. Techniques of Staining with Labeled Antibody

The actual manipulations in "staining" the sections with labeled antibody solutions are simple. Care should be taken to time each step in order to achieve uniformly comparable results. Drying must be avoided. Small quantities of reagents can be used, and since a purified labeled antibody solution has become valuable from the effort expended on it, wastage should be minimized by good planning as well as by frugal use.

1. *Direct Staining of Antigen*

Rinse the slide (dry after fixation) briefly in buffered saline, and wipe it dry except for the area bearing the reaction. Do not allow the section to dry. With a capillary pipette put a small drop of absorbed labeled antibody solution over the section, cover the slide with a small glass dish with moist cotton in the top (a Petri dish cover is convenient and just fits two slides) to reduce evaporation, and let stand at room temperature for 20–30 minutes. Remove from the jar, wipe dry except for the area of the section, and mount in a small drop of buffered glycerol under a coverslip. Glycerol has a faint grayish blue fluorescence and also extracts fluorescent material from the sections. The layer of glycerol should therefore be kept thin. Diffusion of fluorescein into the glycerol indicates incomplete washing.

2. *Controls of Specificity*

(a) Specific inhibition with unlabeled antibody. Two slides should be processed in parallel, preferably consecutive sections. Over the section on Slide A put a small drop of unlabeled homologous antiserum, either neat, or in a low dilution. Over Section B put a similar concentration of unlabeled heterologous or normal serum from the same species. Cover and allow the reaction to proceed for 20 minutes. Wash both slides in buffered saline for 5 minutes, and over each section put a drop of labeled antibody solution. Cover and allow a reaction to take place for about 10 minutes. Wash both slides for 10 minutes in buffered saline, mount in glycerol and examine under the fluorescence microscope. If inhibition has been successful, the specific fluorescence present in Slide A should be markedly fainter or even absent, while Slide B should be stained with reasonable intensity. Successful inhibition with new antigen-antibody systems or with new antiserum requires experimentation, because specific replacement of deposited unlabeled antibody by supernatant labeled antibody readily takes place. Modification of reagent concentration and of time of reaction are therefore necessary until successful inhibition is clearly demonstrable.

Successful inhibition does not of course necessarily establish the specificity of a reaction. In effect, it establishes that there is some inhibitory substance present in the unlabeled antibody solution absent from the normal or heterologous antibody solution. Except under unusual circumstances, this inhibitory substance is most probably antibody, but it is not necessarily antibody against the antigen under investigation. Particularly in dealing with normally occurring antigenic material such as enzymes or hormones, or other structural components of normal tissue, the initial purification of the antigen may be so difficult that mixtures of antibodies appear in the serum produced for cytochemical purposes. Obviously, antibodies present in the unlabeled serum are also present in the labeled serum and inhibition of the latter by the former does not identify the antigen-antibody system involved.

A modification of this method of specific inhibition has been introduced by Goldman (1955). It consists of mixing unlabeled antibody with an aliquot of labeled antibody, and, to another aliquot of labeled antibody, an equal amount of normal or heterologous serum. In such mixtures, he reports that, with the proper adjustment of concentrations, the unlabeled antibody successfully competes with labeled antibody, while the normal or heterologous serum does not interfere with the reaction. He states that he could achieve good inhibition with systems in which convincing inhibition by the alteration method could not be attained.

(b) Specific removal of labeled antibody from an aliquot of labeled antisera. The addition of the proper amount of antigen to an aliquot of labeled antibody will remove by precipitation, or neutralize without precipitation, the antibody activity of the conjugate.

(c) Where foreign substances are under study, purified labeled antibody solution should not react with normal tissue.

(d) Heterologous labeled antibody solutions, or preparations made from normal serum of the same species, purified in the same way, should not stain the test material. This of course is not a strict control of specificity.

3. *The Use of Layers*

When the unlabeled specific antiserum is layered over a tissue section containing antigen, the antibody molecules react and are fixed in place. The excess is washed away, a deposit of gamma globulin remains behind. This gamma globulin is of course antigenic, and can in its turn be detected as an antigen *in situ*. Since antigen molecules have several reactive sites capable of reacting with specific antibody, and since these histochemical reactions are carried out in antibody excess, the use of such layers has the effect of multiplying the amount of antigen present. This considerably enhances the sensitivity of the method. It also increases the number of possible nonspecific reactions which can take place, and multiplies the problem of specific controls.

The basic limitation in the use of such layers is the restriction that the specific antibody must be derived from a species the gamma globulin of which does not cross-react immunologically with the gamma globulin of the species under study.

Antiserum against gamma globulin is best made by the injection of specific immune precipitates (Adler, 1956). However, this method is not always available and satisfactory antiglobulin sera reactive with specific immune globulin have been produced by the injection of crude gamma globulin fractions against rabbit globulin, human globulin, hen globulin, and bovine globulin. At present, goats are being used for the production of antirabbit gammaglobulin and antihuman gammaglobulin.

a. *Procedure.*

Specific unlabeled antiserum against the antigen sought, diluted 1:10, is layered over the tissue section and allowed to react for 20 minutes. It is then rinsed off and washed in buffered saline for 10 minutes. After being wiped dry except for the area of the section, a drop of fluorescein-labeled antiglobulin serum prepared against the species furnishing the specific serum is layered over the section, allowed to react for 20 minutes, and then washed and mounted as described above. As controls, a tissue section without any intervening serum should be processed in the same way with the fluorescent antiglobulin solution as a check on the nonspecific staining produced by this reagent. In addition, a third slide exposed to normal serum or heterologous serum drawn from the species producing the specific serum and diluted to the same concentration should be included to test for nonspecific reactions.

Should nonspecific reactions with the unlabeled specific antiserum be present, the serum must be absorbed with tissue powder as described above. This absorption is best carried out with undiluted serum, the dilution being made after absorption (Watson, 1955). These unlabeled absorbed sera do not keep well when diluted, and therefore only enough for a day's supply should be diluted for use.

The use of such antiglobulin fluorescent conjugates presents several conveniences in addition to the increased sensitivity. It makes possible the performance of a number of studies using the same labeled antiserum. It allows the use of convalescent serum or naturally occurring antibody which cannot be experimentally produced because no antigen in sufficient amount is available for injection. The titer of such sera does not need to be so high as that used in the direct staining of antigen because of the increase in sensitivity.

4. *Detection of Antibody*

By means of layers, antibody as well as antigen may be localized with fluorescent antibody. In this case, the middle layer is composed of antigen, the antibody to be detected is in the cell on the bottom, and the labeled antibody on the top.

a. *Procedure.*

Fix slides in 95% ethanol (v/v) at 37° C. for 15 minutes. Dry the slides upright in an incubator for 30 minutes, rinse in buffered saline, and place in a Coplin jar containing antigen specific for the antibodies sought in a concentration which must be determined by experiment.

With the antibodies so far studied, a concentration of approximately 1/1000 (w/v) is optimal. Wash at room temperature in buffered saline for 10 minutes, wipe the slide dry except for the area of the section, and apply purified specific labeled antibody. After 20–30 minutes, wash the slide in buffered saline for 10 minutes, mount in buffered glycerol, and examine under the fluorescence microscope for antibody-containing cells.

b. *Controls.*

The fluorescent antibody used in this procedure should not react with normal tissue nor with tissues making a response to a different antigen. In the case of rabbit leucocytes which stain nonspecifically as described above, the conjugate must be purified by absorption with tissue powder made from the red bone marrow of rabbits (femoral). The sections from the tissues under study exposed to labeled antibody without an intervening layer of specific antigen serve as controls for antigen persisting after injection. Any fluorescence detectable in such a slide should lead to further tests to see whether this fluorescence can be specifically inhibited as described in the methods for the detection of antigen. If this fluorescence cannot be inhibited, it must be ascribed to nonspecific staining.

The presence of specific antibody is revealed by the difference between a slide exposed directly to fluorescent antibody solution and one to which an intervening layer of antigen has been applied.

IV. CRITIQUE OF THE METHOD

The cytochemical problems which arise in the use of fluorescent antibodies are no different in principle from those of any cytochemical method. Movement of the substance under study in currents set up during the thawing of frozen sections, or by diffusion during manipulation must be controlled by the use of different fixatives and by comparison with freeze-dried material. A useful discussion of these problems has been given by Danielli (1953). As an example, it was reported by Coons *et al.* (1951) that hen's ovalbumin was present in high concentration in the nuclei of the renal tubular epithelium (and elsewhere) after its injection intravenously into the mouse. They discussed the possibility that this finding might be a diffusion artifact, or due to movement during the thawing of the frozen section. An experiment imitating diffusion was carried out by exposing the section of a normal kidney to ovalbumin during thawing; the nuclei were not selective loci of adsorption. Recently, the same nuclear localization of ovalbumin has been reported by Mayersbach and Pearse (1956) on freeze-dried sections. This excludes movement during thawing as the cause of the localization, which from the intrinsic evidence in the preparations was not likely, and appears to establish the validity of the finding.

The identity of the stained material is potentially more precisely proven by means of immunological reactions than by any other method including the use of enzymes, since immunological specificity depends on a larger area of the antigen molecule than do chemical or enzymatic reactions. However, cross-reactions and contaminating unsuspected antigens must be carefully guarded against by whatever methods are appropriate to the case at hand. The reader is referred, as an example, to the recent study by Vaughan and Kabat (1954) of an unidentified antigenic contaminant present in about 1% concentration in egg white, and to which antibody response was made on injection of crystalline ovalbumin. Such a contaminant would be of no importance in a study of the fate of ovalbumin injected into a mammal because of its low concentration, but might, for example, become very significant in a study of protein synthesis in the oviduct, where cells containing it in high concentration might be thought to contain ovalbumin instead.

Although immunologists and students of infectious disease are primarily concerned with the study of foreign antigens which have gained entrance into a cell from the environment, the cytologist's interest is focused on the location, identity, and changing concentration of enzymes, substrates, and synthetic products which are the intrinsic components and consequences of cellular activity. The study of molecules of this sort is theoretically within the reach of the methods described here provided the molecules are antigenic. However, the immunochemical problems raised by this approach are difficult. Enzymes and hormones, for example, despite the broad species spectrum of their biological activity, are antigenically species specific. This means that the necessary antiserum must be prepared by the injection of material derived from the same source which it is intended to study, and that any antibody to impurities present will have an opportunity to react with the cells from which they originally came. Such a situation may be quite difficult to untangle.

Methods for the quantitation of fluorescence due to labeled antibody are in their infancy. Mellors *et al.* (1955) have used microphotodensitometry to measure the amount of fluorescence, comparing one part of a section with another. An improvement in this method might be to scatter a standard fluorescent particle over the sections in order to provide an intrinsic standard. Even so, such measurements are of course purely relative, valid in making comparisons from cell to cell with the same labeled antibody solution applied under standard conditions. The calibration of a fluorescent conjugate for absolute measurements of the amount of antigen present appears to be a major technical challenge; it might perhaps be approached (after the solution of the problems of the reliable measurement of fluorescence) by measurements of an antigen labeled with a radioisotope.

V. ASSESSMENT OF RESULTS TO DATE

Antibodies labeled with fluorescein have been surprisingly successful in their applicability to a variety of antigenic substances: polysaccharides, proteins, and lipopolysaccharides. An increasing number of viruses have been examined inside cells both in infected animals and in tissue culture. In addition to those already listed elsewhere (Coons, 1956), vaccinia (Noyes and Watson, 1955) Egypt 101 (Noyes, 1955), measles (Cohen *et al.*, 1955), psittacosis (Buckley *et al.*, 1955), and poliomyelitis (Buckley, 1956) viruses have been demonstrated in tissue culture. The course of development of herpes simplex infection and of the inclusion bodies associated with it has been studied in tissue culture (Lebrun, 1956).

A limited number of studies of naturally occurring antigens have been carried out. Marshall (1951) localized adrenocorticotropic hormone as granular material in cells which he identified as basophiles in the anterior lobe of the swine pituitary. In his account, he discussed the problems which arise in the study of naturally occurring antigen not yet available in a pure state. Later (1954a), he used similar methods to localize chymotrypsinogen and procarboxypeptidase in the acinar cells of the beef pancreas. He found both substances in the zymogen granules at the apex of the cells; the nuclei, mitochondria, and basal part of the cytoplasm did not react with the labeled antibody. Similar experiments, of a preliminary nature, with ribonuclease and deoxyribonuclease did not give definitive results presumably because the enzyme preparations contained mixtures of pancreatic antigens.

Experiments with labeled nephrotoxic sera prepared in rabbits by the injection of rat kidney indicated that the antigen responsible for the nephritis lay in the basement membrane of the glomerulus, and that it or they were also common to basement membranes elsewhere, and to reticulum in many organs (Hill and Cruickshank, 1953; Cruickshank and Hill, 1953). They showed that their antiglomerular serum, which reacted only with basement membranes and reticulum (and with sacrolemma and neurolemma) would produce nephrotoxic nephritis on injection into rats.

In the case of ACTH and nephrotoxic serum, bioassay was of distinct aid in establishing the specificity of the reactions involved, and in general independent evidence of the presence of any naturally occurring substance will greatly strengthen the cytological evidence.

Recently, Tanaka and Leduc (1956) have described the cytological distribution of a lipopolysaccharide antigen of unknown function which occurs in several species, Forssman antigen. It was found in the collecting tubules of the kidney in the guinea pig, cat, dog, mouse, and chicken, and in the vascular endothelium and adventitia of all organs. Here, specificity could be established because of its heterophile distribution, and because of its peculiar physical properties. Thus, reactions could be abolished by absorption of an aliquot of fluorescent antibody with horse kidney, another organ known to contain the material, by treating the sections with ethanol, which extracted it rapidly and completely, and by demonstrating its stability on boiling.

ACKNOWLEDGMENTS

The methods described in this chapter have been developed largely in the author's laboratory with the active collaboration of a number of colleagues, especially Hugh J. Creech, Ernst Berliner, Melvin H. Kaplan, Barbara K. Watson, Ch'ien Liu, Alan G. S. Hill, and Robert G. White. They also owe much to the skillful and painstaking technical assistance of Mrs. Richard M. Clancy (Jeanne M. Connolly).

Support for the work from the Life Insurance Medical Research Fund, the Helen Hay Whitney Foundation, the America Heart Association, and the office of the Surgeon General, Department of the United States Army is gratefully acknowledged.

VI. APPENDICES

Appendix 1: Preparation of Fluorescein Isocyanate

The preparation of fluorescein isocyanate is reprinted from the Journal of Experimental Medicine, by permission of the editors. For literature, see original publication.

"The course of the synthesis was as follows: 4-nitrophthalic acid was heated with two equivalents of resorcinol, with the production of nitrofluorescein. The crude product was refluxed with acetic anhydride, and the resulting nitrofluorescein diacetate subjected to fractional crystallization. Two isomeric diacetates were separated. Each was saponified and the pure nitrofluorescein isomer recovered. Catalytic hydrogenation produced the corresponding aminofluorescein, which was converted as needed to the isocyanate by treatment with phosgene. No attempt was made to isolate the isocyanates.

"(a) *Preparation of Nitrofluorescein.*—100 gm. of 4-nitrophthalic acid (Eastman Kodak Company) and 100 gm. of resorcinol (Eastman Kodak Company) were intimately mixed in a beaker and heated on an oil bath at 195–200° C. until the mass was dry (12 hours). When cool, the melt was chipped from the beaker, ground in a mortar, and boiled with 1600 ml. 0.6 N HCl for 1 hour. After washing by decantation with three 300 ml. amounts of hot HCl solution, it was collected hot on a large Buchner funnel. The resulting brown paste was washed on the funnel with 5 liters of water, and dried in the oven at 110° C. Yield 176 gm. (98 per cent). *Crude nitrofluorescein.*

"(b) *Preparation of Nitrofluorescein Diacetate and Separation of Isomers.*—100 gm. of this crude nitrofluorescien was refluxed with 400 gm. of acetic anhydride for 2 hours, and set aside to cool. Crystallization was induced by seeding from a small preliminary run. The whitish yellow crystals of diacetate were collected the next day on a Buchner funnel, and washed with two 10 ml. portions of acetic anhydride, and 20 ml. ethanol (excluded from the filtrate). Yield 21 gm. (17 per cent) (fraction 1).

"The filtrate was concentrated by boiling to a volume of about 300 ml. and seeded as before. Yield 11.8 gm. (9.5 per cent) (fraction 2).

"Fractions 1 and 2 were combined, dissolved with heat in about 110 ml. acetic anhydride, filtered hot, and set aside to crystallize. Yield 26 gm. (20 per cent). M.p. 213–219° C. After repeated recrystallizations from benzene and ethanol, m.p. 221–222.5° C. (All melting points are uncorrected.) Calculated for $C_{24}H_{15}O_9N$: C 62.47; H 3.28; N 3.04 Found: C 62.62; H 3.06; N 3.06 (Microanalyses by Miss Shirley Katz.) *Nitrofluorescein diacetate* I.

"The filtrate from fraction 2 was concentrated to a thick paste *in vacuo* on the boiling water bath. On cooling, the dark brown mass became a viscous gum. To it was added 90 ml. benzene, the solution warmed to about 60° C., and stirred. The gum slowly mixed with the benzene, and crystallization began at once. After several hours, the yellow-white asbestos-like needles were collected and washed quickly on the funnel with 30 ml. of cold benzene. Yield 33 gm. (26 per cent).

"These needles were dissolved in 200 ml. benzene, filtered hot, and allowed to stand overnight. The resulting long white feathery needles were collected, washed quickly with about 40 ml. benzene, and dried in air. Yield 15 gm. (12 per cent). M.p. 189–190° C. After repeated crystallization from benzene and ethanol, m.p. 215–216° C. Calculated for $C_{24}H_{15}O_9N$: C 62.47; H 3.28; N 3.04. Found: C 62.61; H 3.19; N 3.03. *Nitrofluorescein diacetate* II.

"(c) *Recovery of Nitrofluorescein Isomers.*—5 gm. of diacetate I was added to 100 ml. hot filtered saturated alcoholic sodium hydroxide, warmed gently, and shaken for a few minutes. An immediate red color resulted. The solution was filtered and poured into 4 volumes of water, acidified with stirring with 2 ml. concentrated HCl, and allowed to stand for several hours. The nitrofluorescein isomer precipitated as a yellow powder which was separated with suction, washed with 500 ml. water, and dried. Yield 4.05 gm. (99 per cent). For analysis, a portion was crystallized from isopropanol. The orange crystals gradually darken on heating but do not melt up to 350° C. Calculated for $C_{20}H_{11}O_7N$: C 63.66; H. 2.93; N 3.71. Found: C 63.47; H 3.04; N 3.59. *Nitrofluorescein* I.

"To obtain the other nitrofluorescein isomer, 7 gm. of once recrystallized diacetate II was added to 100 ml. hot filtered saturated alcoholic NaOH with stirring. In this case, crystallization of the sodium salt began promptly, and after 2 hours 5 gm. of orange-red crystals were obtained (75 per cent). They were very soluble in water with a red color and faint green fluorescence. To obtain the nitrofluorescein itself, 3 gm. of these crystals was dissolved in 300 ml. water, and acidified with stirring by the addition of 2 ml. concentrated HCl. A prompt yellow precipitate formed which turned red on standing in its mother liquor overnight. Collected, washed, and dried. Yield: 2.1 gm. (74 per cent). A small portion was crystallized from isopropanol. It failed to melt up to 350° C. Calculated for $C_{20}H_{11}O_7N$: C 63.66 ;H 2.93; N 3.71. Found: C 63.41; H 3.16; N 3.43. *Nitrofluorescein* II.

"(d) *Reduction of Nitro Compounds.*—2 gm. of nitrofluorescein I was suspended in 100 ml. absolute ethanol and shaken with about 1.5 gm. Raney nickel in an atmosphere of hydrogen at room temperature and pressure. The reaction began promptly and at the end of 90 minutes the theoretical amount of H_2 had been taken up. The nickel was removed by centrifugation, washed with 15 ml. of absolute alcohol, and the washings added to the main lot. The alcohol-amine solution was diluted with an equal amount (115 ml.) of water and allowed to stand. Colorless needles slowly formed which were separated by filtration after standing overnight. Yield: 580 mg. of colorless matted needles which turned red slowly in air and instantly in the presence of water. Kept in the desiccator they slowly darkened to yellow and then to brownish-red. M.p. 215–220° C. (decomposed). (Addition of 50 ml. water to the mother liquor yielded another 510 mg.) Yield 60 per cent. *Amino-fluorescein* I. Micro analyses on two different samples of this compound gave low values for C, H, and N. Therefore, the hydrochloride was prepared for analysis. 400 mg. of this amine was dissolved with heat in 15 ml. 2 *N* HCl, filtered hot, and the dark red crystals collected and dried in the desiccator over H_2SO_4 and solid NaOH. Yield 330 mg. (81 per cent). This compound does not melt. Calculated for $C_{20}H_{14}O_5N \cdot HCl$: C 62.58; H 3.68; N 3.65; Cl 9.24. Found: C 62.71; H 3.66; N 3.61; Cl 9.26. *Amino-fluorescein HCl* I.

"2 gm. of nitrofluorescein II was reduced and crystallized as above except that 4 volumes of water was added to the alcohol-amine solution. Canary yellow crystals which did not darken in air and which were stable on storage resulted. Yield: 1.12 gm. (67 per cent). M.p. 315–316° C. (decomposed) (put in bath at 285°). *Amino-fluorescein* II. Micro analyses on two different samples of this compound gave low values for C and N, a high value for H. For analysis, therefore, 60 mg. of amine II was dissolved in 1.5 ml. 2 *N* HCl with heat, filtered, redissolved by warming and the addition of a few drops of 2 N HCl, and allowed to stand. The red crystals were collected, dried in the desiccator with NaOH, and weighed. Yield: 50 mg. (82 per cent). This compound does not melt. Calculated for $C_{20}H_{14}O_5N \cdot HCl$: C 62.58; H. 3.68; N 3.65; Cl 9.24. Found: C 62.66; H 3.85; N 3.67; Cl 9.56: *Amino-fluorescein HCl II.*

"In common with fluorescein itself (20), each isomer of both nitro- and amino-fluorescein exists in a red and yellow form; the red form of amino-fluorescein II has been observed only on heating to about 150° C.

"We have not attempted to determine which of the two possible positions in the molecule is occupied by the N atom in either of the two isomeric series described above.

"(e) *Preparation of Fluorescein Isocyanate.*—The required amount of fluorescein amine (10 to 60 mg.) was added to 5 ml. of dry acetone (dried over $CaSO_4$), and added dropwise from a dropping funnel to 15 ml. acetone saturated with phosgene, and through which phosgene was constantly bubbled. (This procedure was carried out in a hood with good forced draft. The reaction vessel formed part of a closed system under slight negative pressure. The phosgene was led from a tank through concentrated H_2SO_4, thence to the reaction flask, and thence through a trap to a solution of 20 per cent NaOH where the excess was destroyed.) As each drop of amine solution entered the reaction flask a yellow precipitate formed which rapidly dissolved. The solution in the flask became slightly warm. The reaction was allowed to continue for 30 minutes, by which time the flask had cooled again. The reaction flask was removed from the phosgene-train, three small anthracite chips added as an anti-bumping device, and the solution taken to dryness *in vacuo* over a water bath at 45° C. (10 to 15 minutes). This step served to remove the excess phosgene and acetone. The greenish brown gum was immediately dissolved in 2 volumes (1 to 2 ml.) of acetone and 1 volume of dioxane [see below], and this solution of fluorescein isocyanate added dropwise to the stirred chilled protein solution described below. Care must be taken to exclude water from the isocyanate solution until the moment of use, as it decomposes rapidly at room temperature in the presence of water." (Coons and Kaplan, 1950.)

More recently, it was reported by Marshall (1951) that solutions of the isocyanate retained their activity for as long as a year when protected from heat, light, and moisture.

We have found that 500 mg. of fluorescein amine can be treated with phosgene as described above (100 ml. acetone). At the end of the vacuum distillation, take up the residual gum in 50 ml. acetone and distribute the solution in 5 ml. ampoules in 2.5 ml. amounts. Seal the ampoules and store in the dark.

During the next 24 hours an acetone-insoluble film is deposited on the wall of the ampoule, which increases somewhat on longer storage. However, such solutions retain their activity for 2–3 months when kept in the dark at room temperature, and no doubt would keep longer in the cold.

Such solutions are very convenient to use.

Appendix 2: Other Technical Notes

a. "Buffered saline." 0.8% (w/v) NaCl with 0.01 *M* phosphate pH 7.0.

b. 1,4-dioxane (*p*-dioxane, diethylene dioxide). Purify by refluxing for 30 hours with sodium metal, distill from sodium, and store over sodium.

c. Buffered glycerol. Reagent glycerol 9 parts, buffered saline (see above) 1 part. Should be made fresh weekly, and should have a pH of 7.0.

d. Slides. Standard glass slides for microscopy are satisfactory. Their thickness should conform to the focal length of the dark field condenser. Since experiments with immunological reagents are frequently complicated, it is convenient to use slides with a frosted glass area for labeling with carbon pencil. Wax glass-marking pencils are undesirable because the dyes used in them are often fluorescent and sometimes contaminate the tissue section.

e. Coverslips. Standard glass coverslips are satisfactory.

f. Cleaning glasswork. Glassware, including sides and coverslips should be cleaned with chromic acid cleaning solution.

g. Condensing lens. The specifications for the condensing lens follow:

Stock: General Electric Co. fused quartz (nonoptical grade), refractive index 1.458.

Focal length	2.400 inches
Diameter	2.418 inches
Axial thickness	0.930 inches

R1 + 12.50 diopters (1.6693 inches)

R2 — 7.62 diopters (2.7366 inches)

This lens was calculated and ground by the A. D. Jones Optical Works, Cambridge, Mass.

h. Immersion oil. Mineral oil of low fluorescence is required. Shillaber's immersion oil, low viscosity, very low fluorescence is satisfactory, although it fluoresces appreciably.

Appendix 3: Additional Technical Points

1. *Purification of Conjugates*

a. Anion exchange resins remove an orange-colored substance from conjugates which is responsible for much of the nonspecific reaction between fluorescein-labeled antibody solutions and leucocytes, connective tissue, and other tissue elements. Absorption with exotic materials such as bone marrow powder is no longer necessary. However, it is usually still necessary to absorb conjugates with mouse liver powder once or occasionally twice, in addition to treatment with resin.

Dowex-2 chloride (Dow Chemical Company), 20–50 mesh, 4% cross-linked, is satisfactory for this purpose, when used in a column 1 cm. in diameter and about 20 cm. long. No doubt other ion-exchange resins of similar properties would serve as well. By employing buffered saline to start the column, and to displace the conjugate still present in the column at the end of the operation, and mechanically arranging a sharp interface between saline and conjugate near the top of the column, this operation can be carried out with practically no loss of material.

Dineen and Ada (1957) have recently reported that free fluorescein derivatives can be extracted from conjugates at pH 7.0 with 2 volumes of ethyl acetate. They also state that fluorescein is slowly liberated from conjugates on storage either at 4° C. or at −10° C., and recommend an additional extraction before use.

b. Goldman and Carver (1957) have recently reported that fluorescein isocyanate solutions may be dried on thick filter paper and stored until use in a desiccator. The subsequent addition of a strip of appropriate size of such an impregnated paper to a stirred, buffered antibody solution results in conjugation without use of organic solvents.

c. Goldwasser and Shepard (1957) have recently reported the use of fluorescein antibody against guinea pig complement. This conjugate detected guinea pig complement incorporated into antigen-antibody complexes formed by smears of leptospirae and rabbit antiserum when the reaction on the slide was carried out in the presence of normal guinea pig serum. The use of this reagent on tissue sections has not so far been reported, but it is potentially very convenient, because it could be used with any antigen-antibody system which will fix guinea pig complement. Further, it could presumably be used in conjunction with a labeled antiglobulin serum to increase sensitivity.

REFERENCES

Adler, F. L. (1956). *J. Immunol.* **76,** 217.

Bell, L. G. E. (1952). *Intern. Rev. Cytol.* **1,** 35.

Buckley, S. M. (1956). *Arch. ges. Virusforsch.* **6,** 388.

Buckley, S. M., Whitney, E., and Rapp, F. (1955). *Proc. Soc. Exptl. Biol. Med.* **90,** 226.

Cohen, S. M., Gordon, I., Rapp, F., Macaulay, J. C., and Buckley, S. M. (1955). *Proc. Soc. Exptl. Biol. Med.* **90,** 118.

Coons, A. H. (1956). *Intern. Rev. Cytol.* **5,** 1.

Coons, A. H., and Kaplan, M. H. (1950). *J. Exptl. Med.* **91,** 1.

Coons, A. H., Creech, H. J., Jones, R. N., and Berliner, E. (1942). *J. Immunol.* **45,** 159.

Coons, A. H., Leduc, E. H., and Kaplan, M. H. (1951). *J. Exptl. Med.* **93,** 173.

Coons, A. H., Leduc, E. H., and Connolly, J. M. (1955). *J. Exptl. Med.* **102,** 49.

Creech, H. J., and Jones, R. N. (1941a). *J. Am. Chem. Soc.* **63,** 1661.

Creech, H. J., and Jones, R. N. (1941b). *J Am. Chem. Soc.* **63,** 1670.

Cruickshank, B., and Hill, A. G. S. (1953). *J. Pathol. Bacteriol.* **66,** 283.

Danielli, J. F. (1953). "Cytochemistry: A Critical Approach." Wiley, New York.

Davenport, F. M., Hennessy, A. V., and Francis, T., Jr. (1953). *J. Exptl. Med.* **98,** 641.

Dineen, J. K., and Ada, G. L. (1957). *Nature* **180,** 1284.

Freund, J. (1947). *Ann. Rev. Microbiol.* **1,** 291.

Gitlin, D., Landing, B. H., and Whipple, A. (1953). *J. Exptl. Med.* **97,** 163.

Glick, D., and Malmstrom, B. G. (1952). *Exptl. Cell Research* **3,** 125.

Goldman, M. (1957). *J. Exptl. Med.* **105,** 557.

Goldman, M., and Carver, R. K. (1957). *Science* **126,** 839.

Goldwasser, R. A., and Shepard, C. C. (1957). *J. Immunol.* (In press).

Hill, A. G. S., and Cruickshank, B. (1953). *Brit. J. Exptl. Pathol.* **34,** 27.

Hill, A. G. S., Deane, H. W., and Coons, A. H. (1950). *J. Exptl. Med.* **92,** 35.

Hopkins, S. J., and Wormall, A. (1933). *Biochem. J.* **27,** 740.

Kabat, E. A., and Mayer, M. M. (1948). "Experimental Immunochemistry." C. C Thomas, Springfield, Illinois.

Koch, F. C., and McMeekin, T. L. (1924). *J. Am. Chem. Soc.* **46,** 2066.

Kaplan, M. H., Coons, A. H., and Deane, H. W. (1950). *J. Exptl. Med.* **91,** 15.

Lebrun, J. (1956). *Virology* **2,** 496.

Linderstrøm-Lang, K., and Mogensen, K. R. (1938). *Compt. rend. trav. lab. Carlsberg, Sér. chim.* **23,** 27.

Malmstrom, B. G. (1951). *Exptl. Cell Research* **2,** 688.

Marshall, J. M. (1951). *J. Exptl. Med.* **94,** 21.

Marshall, J. M. (1954a). *Exptl. Cell Research* **6,** 240.

Marshall, J. M. (1954b). Cytochemical localization of certain proteins by fluorescent antibody techniques. Thesis, Univ. Michigan (Doctoral Dissertation Series, Publication No. 9606, University Microfilms, Ann Arbor, Michigan).

Mayersbach, H., and Pearse, A. G. E. (1956). *Brit. J. Exptl. Pathol.* **37,** 81.

Mellors, R. C., Siegel, M., and Pressman, D. (1955). *Lab. Invest.* **4,** 69.

Mendelow, H., and Hamilton, J. B. (1950). *Anat. Rec.* **107,** 443.

Noyes, W. F. (1955). *J. Exptl. Med.* **102,** 243.

Noyes, W. F., and Watson, B. K. (1955). *J. Exptl. Med.* **102,** 237.

Proom, H. (1943). *J. Pathol. Bacteriol.* **55,** 419.

Rossman, I. (1940). *Am. J. Anat.* **66,** 277.

Schiller, A. A., Schayer, R. W., and Hess, E. L. (1953). *J. Gen. Physiol.* **36,** 489.
Schmidt, W. C. (1952). *J. Exptl. Med.* **95,** 105.
Sheldon, W. H. (1953). *Proc. Soc. Exptl. Biol. Med.* **84,** 165.
Tanaka, N., and Leduc, E. H. (1956). *J. Immunol.* **77,** 198.
Vaughan, J. H., and Kabat, E. A. (1954). *J. Immunol.* **73,** 205.
Watson, B. K. (1955). Personal communication.
Weller, T. H., and Coons, A. H. (1954). *Proc. Soc. Exptl. Biol. Med.* **86,** 789.

THE CALCIUM PHOSPHATE PRECIPITATION METHOD FOR ALKALINE PHOSPHATASE

J. F. Danielli

King's College, London, England

I. INTRODUCTION

The alkaline phosphatases are enzymes which catalyze the hydrolysis of mono-esters of orthophosphoric acid, and have optima in the region of pH 9 or higher. Thus the diagnostic reaction for these enzymes is

$$ROPO_3H_2 + H_2O \rightleftharpoons ROH + H_3PO_4 \quad (1)$$

However, Axelrod (1948) and Meyerhof and Green (1950) found that these enzymes can also catalyze the transfer of phosphoric acid from one hydroxylic compound to another, thus:

$$ROPO_3H_2 + R'OH \rightleftharpoons R'OPO_3H_2 + ROH \quad (2)$$

The functions of this group of enzymes in cells and tissues are not well understood. In regions where calcification occurs it is probable that extracellular phosphatase is concerned in hydrolyzing phosphate esters, thus leading to the precipitation of a calcium phosphate. But it is by no means certain that hydrolysis is the only, or even the primary, function of intracellular alkaline phosphatase. The intracellular enzyme may act as a phosphokinase, or even have some activity which has not as yet been surmised. It is not inconceivable that the phosphatase activity, in some instances, is a fortuitous by-product of some other important property of the macromolecule.

Histochemical studies of phosphatase were first made by Robison (1923), who showed that if thick sections of hypertrophic cartilage were incubated with a phosphate ester in the presence of calcium ion a precipitate of a calcium phosphate would form in the sections, i.e.,

$$x\,Ca^{++} + y\,PO_4^{\equiv} \rightarrow Ca_x(PO_4)_y \qquad (3)$$

The composition of the calcium salt which is first precipitated is quite unknown, and the subscripts x and y are inserted in Eq. (3) merely to emphasize that this is so. Robison found this technique of much value in studies of the calcification of cartilage, but, so far as is known, did not attempt to use the method except on a gross histological scale. Development of a qualitative cytochemical method was effected by Gomori (1939) and Takamatsu (1939). These authors showed that, (1) the phosphatase was to some degree resistant to the procedures of histology, so that thin sections could be prepared from wax-embedded material; (2) the calcium phosphate which is deposited by appropriate incubation can be rendered visible by treating the sections first with cobalt ion and then with sulfide ion; and (3) the cobalt sulfide so deposited is found in cytological positions which are characteristic for any type of cell. The additional reactions introduced by Gomori and Takamatsu are

$$Ca_x(PO_4)_y + x\,Co^{++} \rightarrow Co_x(PO_4)_y \qquad (4)$$

and

$$Co_x(PO_4)_y + S^{=} \rightarrow CoS \qquad (5)$$

The stoichiometry of these reactions is unknown, but it is certain that the composition of the precipitates is more complex than is indicated by the above equations, e.g., Doyle (1953) has shown that the exchange of $S^{=}$ for $PO_4^{\equiv}$ indicated by Eqs. (4) and (5) is far from complete.

This technique has been critically examined from a wide variety of points of view—more critically than any other cytochemical technique.

It has been shown that, although there are numerous sources of error to be avoided, nevertheless if properly critical use is made of the method, and of various accessory techniques, it is capable of demonstrating the localization of alkaline phosphatases with remarkable precision. The problem of avoidance of errors will be discussed in Section V.

Development of a quantitative technique became possible as a result of the development of interference microscopes for routine measurements (see chapter by Davies in this volume, p. 55). By a combination of chemical analytical studies and interference microscopy it has been shown that, (1) under appropriate conditions of incubation the phosphate liberated in a tissue section is precipitated to the extent of at least 97.5% in the vicinity of the enzyme; and (2) no subsequent diffusion of calcium phosphate occurs; (3) no significant loss of enzyme activity occurs as a result of incubation. Consequently quantitative estimation of phosphatase activity is possible in areas of the order of $1\mu^2$, and it is possible to characterize the enzyme in a site of this magnitude by measurement of the dependence of activity on pH and substrate concentration, (Barter *et al.*, 1955).

The investigations so far outlined resolved most of the problems involved in the quantitative cytochemistry of alkaline phosphatases. However, there is necessarily some error in the determination by chemical methods of the amount of phosphate deposited in a section, and as we shall see later it is theoretically impossible for *all* of the phosphate liberated by enzyme activity to be trapped immediately. Consequently there is a fraction of the phosphate, of the order of 2.5% or less, which is not accounted for. Consequently, when as a result of incubation a site of low apparent activity is found adjacent to a site of high activity, there must be doubt whether the site of low activity has an intrinsic activity, or whether the apparent activity is an artifact. This possibility can now be eliminated by using a fine beam of ultraviolet light to destroy activity in the site of low activity; if after action of ultraviolet light the activity in the site disappears, the activity was intrinsic, whereas if the activity persists it must be an artifact. Thus it seems probable that all the difficulties limiting this technique can now be evaluated.

II. THEORY OF THE METHOD

The theory of quantitative precipitation has long been a part of analytical chemistry, and all of the considerations which have been developed in standard analytical practice are relevant here. However, the problems involved in a cytochemical technique are more complex than those usually met by the analytical chemist. This is partly because the cytochemist, unlike the analytical chemist, is obliged to precipitate the products of enzyme action in a highly heterogeneous environment, and partly because in addition to the analytical chemist's requirement of quantitive precipitation, the cytochemist adds a topological consideration—the precipitate must be formed in the immediate vicinity of the enzyme under investigation. In addition to the theory of precipitation, we must also consider certain aspects of the theory of fixation and of interferometry.

A. Precipitation Zone Theory

At equilibrium, the concentrations of calcium and of phosphate present in solution will be defined by the solubility product of the least soluble of the calcium phosphates present in the precipitated phase. We do not know the exact composition of the calcium phosphate which is precipitated in tissue sections. But let us suppose that it is $Ca_3(PO_4)_2$. If the precipitated phase had this composition, then the solubility product would be

$$[Ca^{++}]^3[PO_4^{\equiv}]^2 = K_S \tag{6}$$

The solubility product K_S is constant only if the environmental conditions are constant, e.g., temperature, pressure, pH, ionic strength.

If the concentration of calcium is set at an arbitrary figure, as is customary in this cytochemical method, then calcium phosphate may be precipitated if the phosphate concentration exceeds the value S defined by the value of K_S and of $[Ca^{++}]$, i.e.,

$$S = [K_S/[Ca^{++}]^3]^{1/2} \tag{7}$$

There will be no precipitation if the phosphate concentration does not exceed this value; this is quite certain. Also, as has been emphasized by Johansen and Linderstrøm-Lang (1951, 1952), although precipitation *may* occur if the solubility product is exceeded, it does not *necessarily* occur immediately. Immediate precipitation will occur only if there are adequate numbers of crystallization nuclei or other precipitation centers in the system. If precipitation does not occur immediately, phosphate ion or calcium phosphate will be free to diffuse, and will in fact do so until a suitable precipitation center occurs. Thus we see immediately two potential sources of major errors in techniques involving precipitation: (a) the solubility product may not be exceeded in the section, so that no precipitation occurs, and (b) the phosphate will, in the absence of sufficient precipitation centers, diffuse away from its enzymic origin.

Now consider Fig. 1. The upper part of this shows diagrammatically the situation in the vicinity of a small concentration of enzyme E, in the steady state. When a tissue section is plunged into a substrate bath for incubation there will initially be a brief period of varying conditions, during which the substrate, etc., is penetrating into the section; this will be followed by steady state conditions. The duration of the transient state will be determined by the thickness of the section and the diffusion constants of the substrate, etc., in the section. Normally the duration will be less than ten seconds and may be ignored. In the steady state, as is depicted in the upper part of Fig. 1, phosphate will

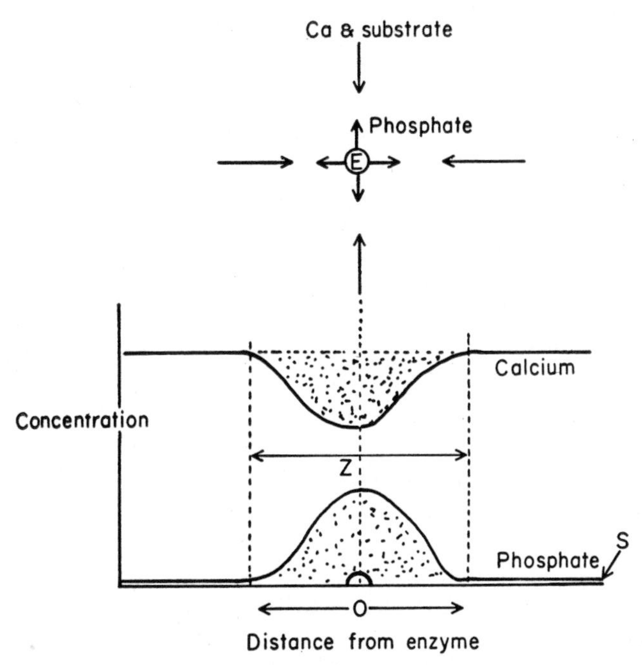

FIG. 1. Diagram to illustrate the kinetic situation in the region of a concentration of phosphatase. E = site of enzyme, Z = width of precipitation zone, S = solubility limit for phosphate = $[K_S[Ca^{++}]^3]^{1/2}$.

be continuously liberated by the enzyme and tend to diffuse away from the enzymic center. If at any point the concentration of phosphate exceeds the solubility limit S it will be precipitated, (subject to presence of adequate precipitation nuclei) and the local concentration of calcium will be depleted. Consequently calcium and substrate will be continuously diffusing toward E, and phosphate will be continuously diffusing away from E. The lower part of Fig. 1 shows the variation in concentration of calcium and phosphate with distance from E. It emphasizes that the actual size of the zone in which precipitation occurs is defined by the distance Z over which the $[PO_4^{\equiv}]$ exceeds the solubility product S, and is only secondarily related to the size of the enzymic site *E*. The actual size of the zone of precipitation will be determined by the availability of precipitation centers, the rate of precipitation at a precipitation center, the rate of liberation of phosphate ion, and the rates of diffusion of phosphate ion and calcium ion.

With this picture in mind, let us now consider in turn the problems mentioned above (a) arising from the solubility product, and (b) arising from presence of precipitation centers.

Since phosphate will escape from the section unless the solubility product is exceeded, it at first sight would appear that the incubation medium should be saturated with phosphate by addition of orthophosphate just before incubation. However, it is now known that the precipitate first formed in sections is a relatively unstable calcium phosphate, which in sites of high activity such as the intestinal brush border may readily recrystallize to a more stable form, possibly hydroxyapatite. On the other hand the precipitate formed on adding phosphate to the incubation medium is likely to be hydroxyapatite. The initial precipitate in sections may be a complex involving organic constituents of the cells. Its solubility product is unknown. In practice the gain in localization achieved by adding phosphate is negligible, and a more important consideration seems to be to keep the concentration of calcium high so that the solubility limit for phosphate (S) is as low as possible. However, it is clear that, unless the concentration of phosphate in the incubatory medium reaches the limiting value S, some phosphate inevitably escapes into the medium.

Let us now consider the statistical aspects of precipitation, and problems which arise from uneven distribution of precipitation nuclei.

Precipitation will never be *completely* quantitative: there will always, even if the environment is saturated with respect to the precipitate, be a small leakage of ions away from the precipitation zone. But, within certain limits, the leakage will constitute only a small, possibly negligible, fraction of the total phosphate liberated. Consider Fig. 2, which shows a diagram of percentage precipitation of phosphate ions as a function of rate of liberation of ions. It is assumed that the substrate and calcium concentrations, and the buffer concentration, are kept constant, that the solution is saturated with respect to the precipitated form of calcium phosphate, and that the variation of rate of phosphate ion liberation is due to variation in amount of enzyme. If the rate of liberation of the phosphate ion is very low, as over the region AB, very little will be precipitated in the critical zone. For example, if one phosphate ion were liberated per second, the probability that it would be precipitated within the critical zone, rather than elsewhere, would be negligible. The reasons for this are as follows. If no precipitate has yet formed, the formation of a precipitate depends upon the probability of a minimum number (the *precipitation* minimum) of calcium and phosphate ions coming together in a very small volume of space. In a saturated solution the addition of one phosphate ion at a point will not significantly modify the probability of the precipitation minimum being reached within the critical zone, and consequently there is a large probability that the excess ion will have diffused away from the critical

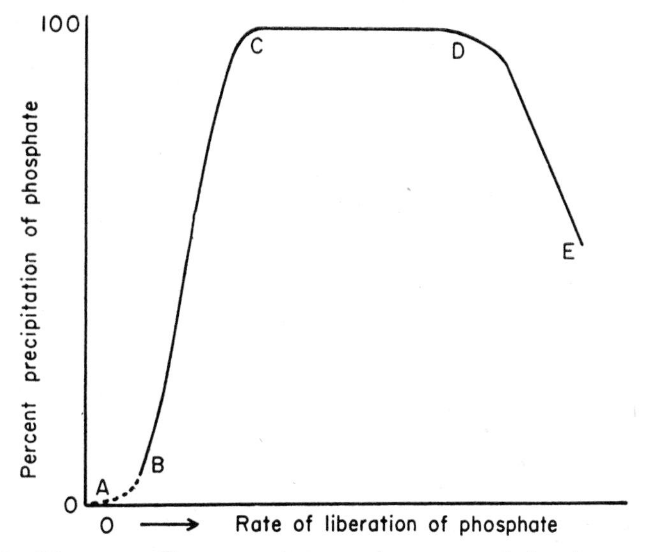

FIG. 2. Diagram to illustrate variation in the amount of phosphate trapped with increase in rate of liberation of phosphate from substrate.

zone before it is precipitated. If on the other hand a precipitate has already formed in the precipitation zone, the rate of further precipitation will depend upon the relative probabilities of addition and loss of calcium and phosphate ions at the surface of the precipitate. These probabilities also will not be greatly affected by addition of one phosphate ion at one point. Consequently in this case also it is not possible to predict the position at which the additional phosphate will be precipitated.

As the rate of release of phosphate ions per unit volume is increased, the statistical significance of the additional ions rapidly becomes greater, and consequently so does the probability of precipitation in the precipitation zone, giving rise to the region *BC*. The curve *BC* rises rapidly towards 100% precipitation, but the asymptote which it approaches will always be less than 100%, since there is bound to be a finite probability that phosphate ions will escape from the critical zone, and be precipitated elsewhere. Consequently *BC* levels off to the plateau *CD*, which will correspond to a percentage precipitation of less than 100%. As the rate of release of phosphate continues to increase, a region will be reached in which the percentage precipitation in the critical zone will again fall. This will occur for two reasons, giving rise to the curve *DE*. The first reason is that for maximal precipitation to occur, the calcium concentration must be kept constant. But as the rate of liberation of phosphate ion increases, the rate of removal of calcium as calcium phosphate becomes significant, and replacement can occur only at the rate at which Ca^{++} can diffuse into the critical zone from the surroundings. Hence a point must be reached at which precipitation in the critical zone will decline. Secondly, there will be a release of hydrogen ions within the precipitation zone as a result of the hydrolysis. When a point is reached at which the change in pH in the precipitation zone is significant, the solubility of calcium phosphate will increase, and so correspondingly will the probability of loss from the precipitation zone.

In other methods for the localization of phosphatase which involve precipitation of a reaction product the same factors will be involved, but variants of the curve *ABCDE* will be involved in each case. For this reason, in the localization of low concentrations of enzymes, different results must be expected from different methods (Loveless and Danielli, 1949). Precipitation will only occur within the critical zone if the increment in precipitant concentration produced by enzyme activity is statistically significant compared with the concentration of the precipitant when present in saturated solution. Hence low concentrations of enzyme cannot be localized. The lower limit of enzyme concentration which can be localized is, in part, determined by the statistical factor we have just mentioned: the less the concentration of the precipitant in the saturated solution, the lower will be the level of enzyme solubility which can produce a statistically significant alteration in concentration in the critical zone. Consequently, if other factors are equal, the smaller the solubility of the enzyme reaction products, the lower the concentration of enzyme which can be localized. It is partly for this reason that regions of low concentration of phosphatase can often be detected by use of one substrate, and not by another substrate.

Johansen and Linderstrøm-Lang (1951, 1952) have pointed out that the number of p. nuclei (precipitation or crystallization nuclei) per unit volume is a limiting factor in the determination of the precision of a cytochemical method which depends upon precipitation. When a solution becomes supersaturated, precipitation does not necessarily occur, and will only occur if appropriate precipitation or crystallization nuclei are present. The number of such p. nuclei varies enormously from precipitant to precipitant. If there were only one such p. nucleus per cell, of random distribution within the cell, it would be impossible for the localization of the precipitate to correspond to that of the enzyme. The possibility of obtaining a correct localization depends upon there being a sufficient number of nuclei in the immediate vicinity of the enzyme to be localized.

Johansen and Linderstrøm-Lang (1952) have estimated the number of p. nuclei for calcium phosphate present in homogeneous solution in a volume equivalent to a single kidney proximal tubule cell, as of the order of 10^5. This is a sufficient number to permit accurate localization of alkaline phosphatase at the level of resolution possible with the light microscope, but difficulties might well arise at the electron microscope level. However, one cannot simply transfer this estimate for homogeneous aqueous solution to the complex situation in a fixed cell. The number and nature of p. nuclei are likely to be affected by organic molecules, which of course abound in great variety in a fixed cell. The number of p. nuclei may be either increased above, or decreased below, the number present in an equivalent volume of homogeneous aqueous solution. Consequently it is necessary to have information about the actual distribution of p. nuclei in the material under investigation. Information of this nature may be obtained by two methods.

In the first method (Danielli, 1946) phosphate is slowly liberated throughout a solution containing an excess of calcium ions, so that a slow precipitation of calcium phosphate occurs. This is achieved by oxidizing the glycerophosphate of the normal incubation medium with hydrogen peroxide, when an unstable derivative is formed which decomposes spontaneously. The precipitation which occurs within the section reveals the distribution of p. nuclei. By this method, as will be shown later, p. nuclei are found to be widely distributed in duodenal epithelium and kidney proximal tubule cells. This distribution is not uniform, and the cell nuclei usually contain the largest amount of calcium phosphate when studied in this way, and brush borders relatively little. As the phosphatase activity of the brush borders is usually high and that of the cell nuclei small, it is evident that in these two cell types the quantitative distribution of p. nuclei is not such as to distort the precipitation picture.

A second method is to dissect sections, as was done by Martin and Jacoby (1949), or cut very thin sections in carefully selected planes, so that parts of cells which are normally contiguous are now separated.

For example, it is just possible that precipitation of calcium phosphate in a brush border is due, not to intrinsic enzyme, but to phosphatase activity in an adjoining region of the cytoplasm. But since intense activity persists in a thin section containing only brush border, distortion by uneven distribution of p. nuclei must be small.

B. Theory of Fixation

Here there will be a discussion only of those points which are essential in the phosphatase technique. The ideal objective in fixation is to obtain a solid object which resembles the living tissue in having every macromolecule in its original position and with its original activity unimpaired. This cannot be achieved. In cytochemistry the endeavor is limited to obtaining a solid object in which (a) the multimolecular structures are retained, (b) the displacement of the species of macromolecules which are under investigation is less than the resolving power of the microscope, and (c) the activities of these macromolecular species are unchanged. Probably the most difficult artifacts to avoid are diffusion artifacts. Many chemical fixatives appear to maintain structure surprisingly well.

In the case of alkaline phosphatase the methods of fixation which may be considered are chemical fixation, freeze-substitution, and freeze-drying. The action of a large number of chemical fixatives has been studied (Danielli, 1953). Many give what at first sight appears to be sharp localization of phosphatase, and qualitatively the distribution appears to be the same with all fixatives which preserve cell morphology. However, many chemical fixatives, e.g., those containing formaldehyde, reduce phosphatase activity. Activity is also lost during wax embedding. Furthermore, Ruyter and Neumann (1949) have shown that although alkaline phosphatase is virtually insoluble in water and in absolute ethanol, it is soluble in ethanol–water mixtures. This means that there is a serious risk of diffusion of phosphatase during fixation in alcohol, during any dehydration or rehydration procedure, and probably during exposure to other mixtures of water and organic solvents. This contention has been borne out by comparison of alcohol-fixed and frozen-dried material (Danielli, 1953, 1954). It still does not seem to be sufficiently realized how rapid diffusion is over distances of cellular dimensions. Bell (1956b) has recently published calculations which emphasize again how great the danger of diffusion may be. Thus consider a body of the size of a mitochondrion suspended in a medium of the viscosity of cytoplasm, and containing a constituent X to which the mitochondrial membrane is impermeable. Upon a sudden breakdown of the membrane permeability, as is likely to follow the incursion of a fixative, the concentration of X in the mitochondrion will rise rapidly, at a rate inversely proportional to the particle size of X. Even for a substance of molecular weight 10^6, the concentration will rise almost to the value in the cytoplasm in one second. It is thus clearly apparent that even a slight solubility of phosphatase in the solvent employed at any stage may lead to serious diffusion artifacts. The evidence seems to indicate very strongly that chemical fixatives should be avoided.

In the technique of freeze-substitution the specimen is first frozen at liquid air temperatures, and then placed in, e.g., ethanol, at a low temperature, usually between —40 and —80°C., at which temperature water is dissolved out of the specimen. The preliminary step of rapid freezing seems to be relatively sound, but some hesitation must be felt about the second step. Close to the ice–solvent interface in the specimen there must be a mixing zone of relatively high water content in which phosphatase will be soluble. At present we know little about this mixing zone, and consequently the technique must be regarded with some suspicion. Although clearly superior to chemical fixation, freeze-substitution, so far as can at present be discerned, is inferior to freeze-drying.

We should also note here that any modification of fixing procedure carries with it the necessity of checking, by analytical methods, whether any loss of phosphatase activity occurs. It has been shown that fixation of some mammalian tissues in ethanol, and freeze-drying, cause no significant loss of activity. The brief infiltration with wax which is necessary with frozen-dried material also carries no hazard. But the relatively prolonged infiltration necessary with chemically fixed material causes up to 75% loss of activity (Danielli, 1946, 1953; Stafford and Atkinson, 1948). It is equally true that analytical checking is necessary on changing to new biological material; thus some plant phosphatases are readily destroyed by alcohol.

C. Theory of Quantitative Determination of Activity

This depends upon the use of an interferometer microscope, to determine the mass of the calcium phosphate precipitate which is formed as a result of enzyme activity (Barter *et al.*, 1955). A detailed discussion of the use of mass determination by this method is given in the chapter by Davies on page 55 of this volume. Consequently only certain additional points will be mentioned here.

Davies and Wilkins (1951, 1952) showed that the mass of a homogeneous object in water is related to its thickness t and its refractive index by the equation

$$m = \frac{(\mu_0 - \mu_w)t}{\chi} = \frac{\phi_w}{\chi} \tag{8}$$

where μ_o and μ_w are the refractive indices of the object and of water, ϕ_w is the optical path difference and χ is a constant characteristic of the object. Thus, provided χ is known, the mass m is clearly derivable from measurement of ϕ_w, which is readily measurable even in areas as small as $1\mu^2$. However, we have already pointed out that the composition of the calcium phosphate precipitated in this technique is unknown and may change during the experiment. Barter *et al.* (1955) pointed out that, since the value of χ varies with the composition of a calcium phosphate (e.g., with the ratio of Ca to P), there is some uncertainty in the value of χ which is appropriate. They concluded that the value lay within the limits of $0.11 \pm 20\%$. This clearly leaves a good deal to be desired in a quantitative determination of enzyme activity.

D. Summary of Conclusions from Discussion

Summing up the preceding discussion, and adding certain other obvious points, we see that there are five main criteria which must be tested:

(1) *Enzyme must not be destroyed during the procedure, or, if it is destroyed, the extent of destruction and the variation in destruction at certain sites must be measured.*

This point can be met by chemical assay of enzyme, in which the activity of fresh tissue is compared with that of prepared tissue. In the investigations made so far on this point it has been shown that there is no significant destruction of enzyme in major sites, but in minor sites of activity, e.g., nuclei of intestinal epithelium, no conclusive evidence has yet been obtained.

(2) *There must be assurance that the enzyme in the tissue sections is in its physiologically normal position.*

So far as fixation is concerned this appears to be met by freeze-drying, and possibly by freeze-substitution, but the evidence in favor of the latter method is not yet adequate. Chemical fixation is to be avoided.

There is, however, always some risk of diffusion of enzyme during, e.g., the incubation period. This can be detected by incubating a section containing enzyme after it has been superimposed upon an inert section. Appearance of calcium phosphate in the inert section is in indication of diffusion of phosphatase.

(3) *There must be assurance that the substrate is able to penetrate to all sites of enzyme activity freely.*

Some evidence bearing on this point may be obtained by using substrates with differing physical properties, e.g., lipophilic and hydrophilic substrates, such as glycerophosphate and naphthol phosphate. But the interpretation of differences between the results obtained with different substrates is complicated by the problem of variation in enzyme specificity. Consequently the best method for settling this point is that of Pratt (1953). In this technique alternate sections are incubated using the present cytochemical technique, for an appropriate period, whereas the remaining sections are homogenized and then incubated under corresponding conditions. The phosphate liberated is then estimated in both instances. In his observations Pratt found that, within experimental error, the same amount of phosphate is liberated in normal sections and in homogenized sections, indicating that access of substrate was adequate in the cytochemical technique.

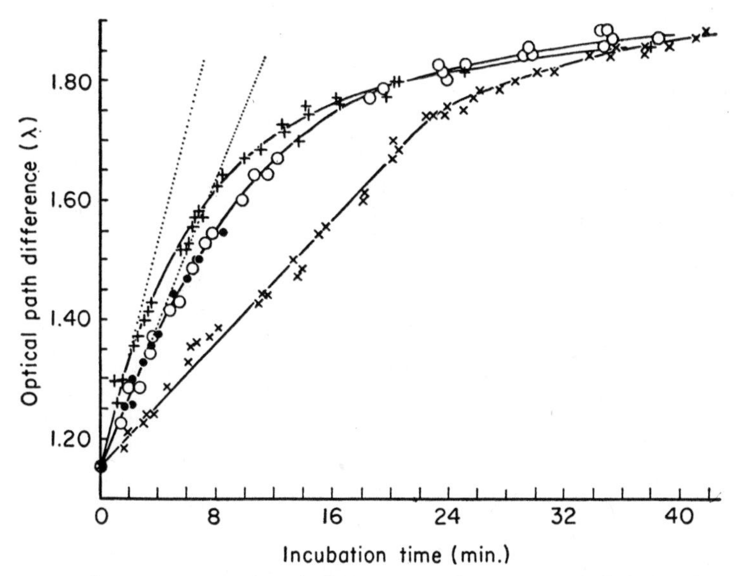

FIG. 3. Changes in optical path difference with time. Rat kidney brush border. pH values: ○, 9.3; ×, 8.8; +, 9.7; ●, repeated at 9.3.

Once again, this method of Pratt does not provide an answer to the problem of studying sites of low activity in the presence of sites of high activity.

(4) *Evidence must be provided to show that the precipitate of calcium phosphate is formed at the site of the enzyme activity, rather than at sites of special affinity for calcium phosphate.*

As indicated above, this depends upon having an adequate distribution of precipitation nuclei, to prevent supersaturation, and on keeping the solubility limit for phosphate as low as is compatible with high enzyme activity.

The tests available are (a) study of the distribution of precipitation nuclei by nonenzymic precipitation of phosphate; (b) use of different substrates; (c) use of superimposed sections; (d) use of narrow ultraviolet beams to destroy enzyme locally. The combined use of these methods suffices to establish rigorously the extent to which the calcium phosphate is precipitated at the actual site of enzyme activity.

(5) *For quantitative work it is necessary to show that the amount of calcium phosphate precipitated is proportional to the amount of enzyme present.*

So far, strict *experimental* evidence has not been adduced to show that this condition is in fact met at all the sites in a tissue section. On theoretical grounds it seems certain that the criterion will be met if the enzyme is combined with the substrate to the same extent at all sites. This is most readily achieved if the enzyme is in fact completely saturated with substrate. In the case of the enzyme of rat kidney and duodenal brush border phosphatase this requires a glycerophosphate concentration between 0.12 and 0.5 molar.

Figure 3 shows that under these conditions the initial change of optical path difference with time is linear. The slope of this line is an appropriate measure of the amount of enzyme present.

III. INSTRUMENTATION

Comparatively little special instrumentation is required. For freeze-drying any of the instruments mentioned by Bell (1956b) is adequate. For those who are not in a position to construct their own dryer, the instrument manufactured by Edwards & Son appears to be the best value for money.

For inactivation of enzyme in highly localized sites any ultraviolet lamp suitable for microscopy is effective. The area irradiated is controlled by varying the aperture of the lamp diaphragm. The specimen can be mounted on any good microscope. It is simplest to mount the microscope horizontally so as to avoid use of a mirror to illuminate the specimen. If this is not done a "first surface" mirror is desirable. The complete transmission provided by a reflecting condenser gives maximum rate of inactivation, but the transmission of longer wavelength ultraviolet light by glass condensers is sufficient to cause inactivation, although the exposure times required may be lengthy. The specimen should be irradiated in water, immediately preceding incubation.

Details of the interference method used for quantitative work are described in the chapter by Davies on p. 140.

IV. GENERAL TECHNIQUE

A. Fixation, etc.

Specimens not more than 0.5 mm. in thickness are plunged into a mixture of propane and isopentane maintained at liquid nitrogen temperatures. The mixture must be well stirred. After freezing the specimens are transferred to a tray (precooled in liquid nitrogen) and quickly transferred either to the drying chamber of a freeze-dryer, or failing this to ethanol at —60 to —80° C., for freeze-substitution.

Frozen-dried tissue is warmed to room temperature in the dryer (to avoid condensation) and vacuum embedded in paraffin. In the case of freeze-substitution the alcohol should be kept cold for at least 12 hours, then allowed to return to room temperature. After a further hour in alcohol the specimen is cleared in xylene and embedded in paraffin. The embedding temperature should not exceed 60° C. Lower embedding temperatures are sometimes necessary.

B. Sections

Wax-embedded material should be stood in a desiccator in the cold store. Sections are cut on an ordinary microtome and flattened on 95% alcohol or on acetonitrile. The sections are dried onto slides at room temperature or at 37° C., preferably in a vacuum desiccator, and may be stored for a few days in a desiccator in the cold store.

C. Procedure before Incubation

Removal of wax from sections, three changes of xylol	3 minutes
Three changes of absolute alcohol	1 hour
Dry in air	—
Distilled water	5–60 minutes

It is essential to keep frozen-dried sections in absolute alcohol for 1 hour to complete fixation. Freeze-substituted sections need only 3 minutes. The sections are dried in air to avoid passage through alcohol–water mixtures, in which phosphatase is soluble. Passage through distilled water removes traces of phosphate, etc.

All water should be either from a glass or quartz still, or obtained by passage over ion-exchange resin, to ensure absence of heavy metal contaminant.

D. Incubation for Qualitative Observations

Incubate for the desired period in the following mixture:

2% Sodium veronal	20 ml.
2% Sodium β-glycerophosphate	20 ml.

2% Calcium nitrate $[Ca(NO_3)_2 \cdot 6H_2O]$	20 ml.
2% Magnesium chloride	2 ml.
Water	38 ml.
	100 ml.

The pH should be adjusted to pH 9.3 using 1*N* HCl and 1*N* NaOH, with bromothymol blue as indicator. For quantitative work the glycerophosphate should be recrystallized. Veronal and glycerophosphate solutions should be made up anew each week, and kept in a refrigerator.

For a preliminary study the greatest amount of information is provided by incubating a series of slides for a geometrical series of times, e.g., 5, 20, 80, 320 minutes.

E. Procedure after Incubation

Wash in 2% calcium nitrate	5 minutes
In 1% cobalt nitrate	2 minutes
In distilled water, two changes	2 minutes
In 1% (v/v) ammonium sulfide	1 minute

Before use 0.5 ml. of 2% sodium veronal should be added to each 100 ml. of the calcium nitrate and cobalt nitrate solutions and of the distilled water. This is necessary to avoid any slightly acid condition, in which phosphates are soluble in water.

Slides may be left in the calcium nitrate or cobalt nitrate solution for hours. Exposure to distilled water should not exceed the stated two minutes.

It must be realized that the times given in this section have been worked out for sections between 1 and 8 μ in thickness. For sections of other thicknesses the times must be adjusted (Danielli, 1953, p. 43).

The procedure given above has one serious disadvantage. In sites of high activity the initial precipitate may recrystallize forming a compound which does not readily exchange Ca^{++} for Co^{++}. Thus sites of high activity may not appear black, and would therefore be judged to lack activity. This may be avoided by exposing the sections to silver nitrate instead of cobalt solution, followed by blackening in ultraviolet light (Ruyter, 1952).

Material in which the initially formed calcium phosphate has been converted into black cobalt sulfide or silver may be dehydrated in alcohol, cleared in xylol, and mounted in balsam or other mountants using the normal histochemical procedures. However, in the cobalt method the exposure to solvents should be as brief as is consistent with good dehydration, or some decoloration may occur. The alcohols and xylols used for dewaxing sections should be maintained as a separate series, and not used for mounting, so as to minimize the risk of contaminating sections with enzyme inhibitor before incubation.

F. Incubation for Quantitative Observations

The procedure used differs from that given above only in the following respects.

(a) The veronal buffer in the incubation medium is replaced by an equal volume of 0.05 *M* Clark and Lubbs borate buffer, pH 9.3. This is necessary to ensure adequate buffering with the small volumes used.

(b) The sections are mounted on coverslips. These are mounted section downwards on glass slides, but separated from the slide by two spacers about 0.2 mm. thick. Since the optical properties of the system must be invarient during observations, the spacers must be fixed to the slide (e.g., with balsam) and as close together as is consistent with maintaining a satisfactorily large observation field. The arrangement is kept rigid by adding a brass plate (with a suitable observation hole) above the coverslip, which is then held in position by a clamp.

(c) Illumination in the interference microscope is by means of monochromatic light, e.g., λ 546 mμ isolated by filters from a mercury arc. Heat filters and an ultraviolet-excluding filter must also be used, the latter to exclude ultraviolet light which would otherwise inactivate enzyme. See chapter by Davies on p. 55 of this volume for other details of interference microscopy.

(d) At the beginning of a series of measurements the space between coverslip and slide (originally containing distilled water) is replaced by incubation medium lacking substrate. The optical path difference ϕ is then observed at the site chosen for measurement of enzyme activity. Then the solution is replaced by normal incubation medium (using borate buffer) and the variation of ϕ with time observed. If more than one activity measurement is made on a section it is necessary to remove calcium phosphate from the section. This is done by irrigating with phosphate buffer at pH 7.

The replacement of fluid between coverslip and slide can be carried out quickly and efficiently by filling a small reservoir space beyond one end of the coverslip, and then rapidly removing all excess fluid from the other end with coarse filter paper.

A series of measurements should always commence with a measurement of activity at pH 9.3 and end with a second measurement at this pH. This both ensures a common standard of reference and checks that no activity has been destroyed.

V. SUPPLEMENTARY TECHNIQUES

A. Alternative Substrates

For checking certain points, e.g., ability of substrate to reach the enzyme, it is often desirable to use substrates having physical properties different to those of glycerophosphate. Of value in this connection are techniques in which a phenol phosphate is incubated in presence of a diazonium hydroxide. As the phenol is released from its phosphate ester it is attacked by the diazonium ion, forming an insoluble dye (Menten *et al.*, 1944; Danielli, 1946; Manheimer and Seligman, 1948). The substrates most used in this connection have been α- and β-naphthol phosphates.

An alternative procedure has been to use the phosphate of a dye, e.g., *p*-nitrophenylazo-α-naphthyl phosphate. This substance is soluble in water, but when the phosphate moiety is split off an insoluble dye is precipitated (Loveless and Danielli, 1949).

At present none of these methods has been really rigorously studied.

B. Superimposed Sections

This method is particularly useful in the study of the capacity of phosphatase to diffuse during incubation, etc. Sections are mounted on slides, and then exposed to distilled water at 95° C. for 30 seconds. This should destroy their intrinsic enzyme content, but it is better to check this by incubating a few sections overnight. If the incubated sections lack activity, a second active section is then superimposed upon the first, and the two sections incubated in this relationship. If any significant diffusion occurs, whether of enzyme or of the products of enzymic reaction, an apparent activity will be found in the underlying, nominally inert, section (Danielli, 1946; Martin and Jacoby, 1949; Leduc and Dempsey, 1951).

The weakness of this technique is that the physical properties of the inert section must differ from those of the normal section as a result of the inactivation procedure.

C. Detection of Precipitation Centers

To 100 ml. of normal incubation medium is added 1 ml. concentrated hydrogen peroxide solution. The pH is adjusted to 9.3, using bromothymol blue as indicator. Sections inactivated as described under (B) are placed in this medium. Phosphate is slowly liberated throughout the solution over 2 or more hours. Slides are removed at intervals and deposits of calcium phosphate demonstrated by the cobalt or silver methods (Danielli, 1946).

D. Detection of Centers Having High Affinity for Phosphatase

Sections are inactivated as in (B) and then placed at 37° C. in a suspension containing, e.g., fresh kidney homogenate of the same species. The suspension has 0.4% sodium veronal and 0.4% calcium nitrate added immediately after homogenizing. After 1 hour in this suspension the sections are washed briefly in distilled water and incubated as usual (Barter, 1954). Barter found that, with kidney tissues, enzyme is taken up at the same cellular sites which displayed enzyme prior to heating. Earlier attempts to study sites of affinity for enzyme were made using purified alkaline phosphatase. These largely failed to display sites of affinity, probably because the purification procedure had separated the enzyme from its natural macromolecular associate molecules.

E. Inactivation by Ultraviolet Light

The necessary details are given under "Instrumentation" on page 436. But it should be added that, before "negative" results at a site after irradiation can be taken to demonstrate inactivation of enzyme at that site, it is also necessary to demonstrate that the site contains precipitation nuclei for calcium phosphate after irradiation, using technique (C). Otherwise it could well be that irradiation was simply destroying precipitation nuclei.

F. Characterization of Enzyme

An enzyme may be characterized by determining the variation in activity with variation (a) in nature of substrate, (b) in concentration of substrate, (c) pH, and (d) nature of activator. Work of this type is much better done using the quantitative interferometric method of determination than by relying upon visual estimates of amount of blackening of sections using the cobalt or silver methods. However, to date almost all published results have been based upon visual assessment of blackening. The results are often striking when substrates such as glycerophosphate, hexose phosphate, aneurin phosphate, estrogen phosphate, and cortisone phosphate are compared. However, at present it has usually been assumed that in passing from one substrate to another it is not necessary to check diffusibility of enzyme, etc. Unfortunately most of the control experiments necessary must be repeated with each change of substrate (and also of tissue) if the results obtained are fully to be relied upon.

VI. NATURE OF RESULTS OBTAINED

Qualitative studies have been very widely reviewed elsewhere—see, for example the volumes of the International Review of Cytology. In general, alkaline phosphatase most commonly appears at sites of formation of connective tissue or of calcification, in cell nuclei, and in the surface layers of many types of cell having a secretory function. The amount of phosphatase present in a cellular site varies with stage in development, with cell type, and with substrate used. It is a striking fact that, although much attention has been given to this enzyme, by biochemists, physiologists, and cytochemists, it is only in the case of the extracellular phosphatase concerned in calcification that we have any real understanding of the function of this enzyme. And it must equally be admitted that Robison, thirty years ago, knew almost as much about its function in calcification as we do today. It is just possible that the development of the quantitative technique for studying alkaline phosphatase may lead on to a proper understanding of its intracellular functions. If this is to be so, it appears to the present author that studies are necessary to cover the following points.

(1) The differences between phosphatases revealed by use of different substrates. It is necessary that the enzymes revealed at different sites should be quantitatively characterized by determination of Michaelis constants and of sensitivity to different inhibitors. This can be done using the quantitative technique. Then, by using cell fractionation techniques, it should be possible to discover whether the enzymes demonstrated in the different sites by different substrates are in fact different, or whether their substrate specificities are produced by a secondary phenomenon.

(2) An approach must be found to the elucidation of the actual function of the phosphatases. There are many potential lines of approach, e.g. (a) isolation of the molecular complexes containing phosphatase, followed by biophysical study; (b) observation of changes of cell behavior after treatment with specific phosphatase poisons; (c) observation of changes in cell function in presence and absence of phosphatase in those cases where phosphatase content is under humoral control. We should note that we are at present inadequately provided with specific phosphatase poisons; such could probably be derived either by designing substrate analogs, suitable end group or prosthetic group reagents, or by antibody formation.

(3) The mechanisms, both genetical and humoral, whereby the absolute amount of phosphatase in a given site is controlled, must be investigated.

All three approaches are quite feasible, but all require a good deal of preliminary spadework.

REFERENCES

Axelrod, P. (1948). *J. Biol. Chem.* **172,** 1.
Barter, R. (1954). *Nature* **173,** 1233.

Barter, R., Danielli, J. F., and Davies, H. G. (1955). *Proc. Roy. Soc.* **B144,** 412.
Bell, L. G. E. (1956a). *Nature.*
Bell, L. G. E. (1956b). *In* "Physical Techniques in Biological Research" (G. Oster and A. W. Pollister, eds.), Vol. III, pp. 1–25. Academic Press, New York.
Bhattacharjee, D., and Sharma, A. K. (1951). *Sci. and Culture (Calcutta)* **17,** 268.
Chevremont, M., and Firket, H. (1953). *Intern. Rev. Cytol.* **2,** 261.
Cleland, K. W. (1950). *Proc. Linnean Soc. N. S. Wales* **75,** 25, 74.
Danielli, J. F. (1946). *J. Exptl. Biol.* **22,** 110.
Danielli, J. F. (1953). "Cytochemistry: A Critical Approach." Wiley, New York.
Danielli, J. F. (1954). *Proc. Roy. Soc.* **B142,** 146.
Danielli, J. F., and Catcheside, D. G. (1945). *Nature* **156,** 294.
Dounce, A. L. (1950). *In* "The Enzymes" (J. B. Sumner and K. Myrbäck, eds.), Vol. I, Part 1, Chapter 5. Academic Press, New York.
Doyle, W. L. (1953). *Intern. Rev. Cytol.* **2,** 249.
Doyle, W. L., Omoto, J., and Doyle, M. E. (1951). *Exptl. Cell Research* **2,** 20.
Firket, H. (1952). *Bull. microscop. appl.* **2,** 57.
Gomori, G. (1939). *Proc. Soc. Exptl. Biol. Med.* **42,** 23.
Johansen, G., and Linderstrøm-Lang, K. (1951). *Acta Chem. Scand.* **5,** 965.
Johansen, G., and Linderstrøm-Lang, K. (1952). *Acta Med. Scand.* **266,** 601.
Krugelis, E. J. (1945). *Genetics* **30,** 12.
Leduc, E. H., and Dempsey, E. W. (1951). *J. Anat.* **85,** 305.
Lorch, I. J. (1947). *Quart. J. Microscop. Sci.* **88,** 159.
Loveless, A., and Danielli, J. F. (1949). *Quart. J. Microscop. Sci.* **90,** 57.
Manheimer, I. H., and Seligman, A. M. (1948). *J. Natl. Cancer Inst.* **9,** 181.
Martin, B. F., and Jacoby, F. (1949). *J. Anat.* **83,** 351.
Menten, M. L., Junge, J., and Green, M. H. (1944). *Proc. Soc. Exptl. Biol. Med.* **57,** 82.
Meyerhoff, A., and Green, G. (1950). *J. Biol. Chem.* **183,** 377.
Moe, H. (1952). *Anat. Record* **112,** 217.
Novikoff, A. (1951). *Science* **113,** 320.
Pratt, O. E. (1953). *Biochem. J.* **55,** 140.
Robison, R. (1923). *Biochem, J.* **17,** 286.
Ross, M. H., and Ely, J. O. (1951). *Exptl. Cell. Research* **2,** 339.
Ruyter, J. H. C. (1952). *Acta Anat.* **16,** 209.
Ruyter, J. H. C., and Neumann, H. (1949). *Biochim. et Biophys. Acta* **3,** 125.
Simpson, W. L. (1941). *Anat. Record* **80,** 173.
Stafford, O. R., and Atkinson, W. B. (1948). *Science* **107,** 279.
Takamatsu, H. (1939). *Trans. Soc. Pathol. Japon* **29,** 492.
Verzar, F., Sailer, F., and Richterch, R. (1952). *Helv. Physiol. et Pharmacol. Acta* **10,** 231.

AUTHOR INDEX

Italic numbers indicate pages on which complete references may be found.

A

B

C

M

SUBJECT INDEX

B

D

E

G

K

L

N

T